The Brahmaputra River in Assam

This holistic book covers the richest area in Northeast India in terms of both explored and foreseen reserves of fossil fuels and other natural resources. Using a multi-disciplinary approach, GIS, and geospatial data gathered from different case studies included, this book helps readers develop a thorough understanding of a highly dynamic big river, the Brahmaputra, and use it as a comprehensive resource for further understanding the science of rivers. It discusses the causal factors of decadal-scale fluvial dynamics, the nature of fluvial dynamics, lateral variability of the older flood plains and neotectonics in the shallow subsurface, and the overall trend of basin evolution at different depths.

The Brahmaputra River in Assam

Geomorphology, Hazards, and Natural Resources

Siddhartha Kumar Lahiri

CRC Press
Taylor & Francis Group
Boca Raton London New York

CRC Press is an imprint of the
Taylor & Francis Group, an **informa** business

First edition published 2023
by CRC Press
6000 Broken Sound Parkway NW, Suite 300, Boca Raton, FL 33487-2742

and by CRC Press
4 Park Square, Milton Park, Abingdon, Oxon, OX14 4RN

CRC Press is an imprint of Taylor & Francis Group, LLC

© 2023 Siddhartha Kumar Lahiri

ISBN: 9781032298528 (hbk)
ISBN: 9781032298535 (pbk)
ISBN: 9781003302353 (ebk)

DOI: 10.1201/9781003302353

Typeset in Palatino
by codeMantra

Dedicated to the people of the Brahmaputra

valley of Assam for their resilience

and courage while facing flood and

erosion disasters every year

Contents

List of Figures

List of Tables

Preface

The Brahmaputra River is considered one of the top ten large anabranching mega-rivers of the world. The Brahmaputra is also the seventh-largest tropical river in the world in terms of mean annual discharge (20,000 m³/s in Bangladesh) and passes through the three most populous countries: China, India, and Bangladesh. The present study concentrates on a 240-km × 100-km corridor of the upper Assam valley for documenting the fluvial dynamics of the Brahmaputra River system during the period 1915–2015 using the archived satellite imagery and maps. It also uses geophysical data of the shallow subsurface in smaller windows to map the buried channels to get some handle on fluvial dynamics during the late Quaternary period. Furthermore, the basin fills down to the basement have also been examined using seismic data to understand the basin evolution since early Tertiary. One of the major findings of the present study is a first-order tectono-geomorphic zonation of the valley, in terms of basement-borne uplifts, slopes, and depressions and the relationships of these features with co-seismic and inter-seismic forcings, which explains variable residence time of the Brahmaputra River orchestrating uneven fluvial sediment stack across the valley. Secondly, a new model was evolved to explain the genesis of Majuli-type relict river islands, which are observed to have a stronger connection with the basement configuration and the tectonic setting in contrary to the previous explanation based merely on the geomorphic process. Thirdly, new utilities of the lateral seismic wave velocity discrimination were explored in selective oil-bearing windows to understand neotectonic forcings on palaeo-fluvial dynamics. Fourthly, a multi-parametric micro-zonation scheme was evolved based on the superposition of seismic wave velocity and thickness variability of different shallow subsurface layers. Fifthly, the evolution of the poly-history basin was traced to explain the inversion observed in the stratigraphic sequence of the Brahmaputra basin by a post-Miocene strong episodic wrench causing basin tilt and subsequent valley fill characteristics, which not only explain the nature of accommodation space generation in the foreland basin margins regulating the residence time of the major sediment-carrying river system of the valley but also help understand why the oil fields are mostly located along with the Naga-Patkai thrust belt in the south bank of the valley. A brief study was conducted on the variability of the bed–bank relations in different stretches of the Brahmaputra River, which is supposed to help address flood disaster risk reduction programs on a better rationale. Bringing the observations of some of the pioneer workers a synoptic appraisal was added to highlight why the upper reach of the Brahmaputra valley should be called the golden corridor, not metaphorically, but due to substantial evidence from the surface, shallow subsurface, and basin-scale deep subsurface.

Acknowledgements

The urge to understand the Brahmaputra system came from a delegation from a severely erosion-affected area called Rohmoria near the Brahmaputra River in the Dibrugarh district of Assam. Working as a faculty in the Department of Applied Geology, Dibrugarh University, teaching subjects related to oil exploration in the subsurface, the author did not have the faintest idea what a disastrous event was going on barely 20 km away from his workplace on the surface. The village folks asked for certain feasible solutions to arrest erosion. In an agrarian setup, flood is tolerable; losing land forever is simply horrific, and it is realized suddenly that the basis of the whole existence has collapsed and become part of the river. And it does not happen to a few persons or families; the entire village, thousands of people whose families have lived in the area for hundreds of years, gets uprooted. I acknowledge and salute first all those village people of my native land, those needy faces, who felt, perhaps too naively, that this man could be of some use to find solutions for them.

The late Mukeshwar Yadav, my old friend who distributed *Peoples' Democracy*, was so excited about the project said, 'Jaiye aur deri maat kijiye' (Go, don't delay any longer!). I acknowledge and remember with great respect our Yadavji. Alas, he could not see my work!

Prof. Rajiv Sinha's method of indirect guidance is acknowledged with deep gratitude. Prof. J.N. Sarma helped the author a lot during the field visits in Majuli and recognized some of the palaeochannels in the field. A field visit with Prof. Robert James Wasson helped form a better understanding of the bed–bank relationship.

The author is thankful to IIT Kanpur for providing the institutional support to conduct this study. The author is grateful for the services provided by the USGS website for the DEM data from the SRTM source, and the India Office Library and Records, London, UK, for providing the topographic map of the study area prepared during the 1912–1926 seasons, to Oil India Limited, Duliajan, for sharing some of the rarely available data. The author is more specifically thankful to late Dr Rahul Dasgupta and Mr Kanailal Mandal for their encouragement. The author thanks Dr Chandrashekhar Bhuiyan, Dr Sagnik Dey, and Dr Noni Gopal Roy for the coffee house chats.

The favourable attitude shown by the faculty members of the Department of Applied Geology, Dibrugarh University, is acknowledged. In particular, the author immensely benefited from discussions on many geological issues with Dr Devojit Bezbaruah.

The author acknowledges the *Natun Padatik* (new foot soldiers) circle, a monthly Assamese magazine, and thanks the Editor, Prof. Hiren Gohain, and Nabajyoty Borgohain, the Managing Editor of the aforementioned magazine for continuous encouragement.

The author remembers the faces of his late parents, particularly his mother who sacrificed for the sake of the higher education of her children. Curiosity expressed by his sisters Suparna Lahiri Baruah and Sangeeta Lahiri Srivastava, and brothers Sanjay Lahiri, Sagar Lahiri, and Surya Lahiri helped to keep the author motivated.

Juli, the wife of the author, managed everything on her own during periods of his long absence. She also helped understand the socio-economic issues related to bank erosion.

During the recessionary time, despite the difficulties compounded due to the pandemic, the encouragement particularly from Ms Irma Shagla Britton, Senior Editor, CRC Press/Taylor & Francis Group, helped to boost up the spirit. Thanks to the anonymous reviewers for their critical comments, which helped rectify several language issues.

Despite repeated verifications, the author is solely responsible if certain mistakes remain.

Author

Siddhartha Kumar Lahiri is an associate professor and coordinator at the Department of Applied Geology in Dibrugarh University, Assam, India. He joined the ONGC as a Surface Geophysicist in 1985 to explore basin-scale oil prospects in the Brahmaputra and Barak valleys of Assam. In 1993, he joined Dibrugarh University, India, as a faculty member, and he was subsequently instrumental in developing the Applied Geophysics Program at the university. He earned his PhD from IIT Kanpur on basin evolution, morphotectonics, and fluvial processes in the Brahmaputra River system. Currently, he is engaged with connecting surface processes with different scales of subsurface geophysics.

1

Introduction

Before landing at the Dibrugarh Airport, while looking below, a few big pythons seem to majestically guard a huge tract of lush greenery. The pythons are the interwoven channels of the Brahmaputra River, and the widespread green stretches having an exceptionally high degree of geometric consistency maintaining almost uniform hues are, of course, the tea gardens. Many famous tea gardens of Assam, still bearing the old colonial heritage, were developed by the side of the Brahmaputra due to cheaper navigability and the typical growth-promoting climate – tropical and temperate – necessary for various plants belonging to the tea family. The commonly available tea species is *Camellia sinensis*. Kudos to blending, if Darjeeling tea is famous for its flavour, Assam tea has remained equally famous for its strength and colour for the last 100 and more years. Along with green tea bushes at the top, there are tall oil rigs on the surface. High-quality coal reserves found in the shallow subsurface with the presence of iron sulfide having catalytic property is known worldwide (Bergius process for which Friedrich Bergius was awarded the Nobel prize in 1931) as an excellent candidate for the production of liquid hydrocarbons for use of synthetic fuel by hydrogenation. Unfortunately, at present, they are wasted for making coke out of it due to lopsided policies. The estimated hydro-electricity potential around the Brahmaputra valley is more than 60,000 MW, whereas at the current level of industrialization, the peak hour electricity need of all the seven states of northeast India is less than 4,000 MW.

Putting the area in a larger backdrop, the Upper Assam Brahmaputra valley is situated in a classic geological setting representing the tri-junction of Eurasian, Burmese, and Indian plates (Nandy, 1983) with high mountains in its boundaries. On average, a repository of 5+ km-thick sediments, partly marine and partly fluvial, is stacked in the valley under veritable tectonic controls over the crystalline basement complex that is associated sometimes with the Gondwana of the main Indian craton. The fluvial component of the sediment build-up was orchestrated by an array of mountain-fed rivers (Sinha and Friend, 1994) to enrich ultimately the great valley divider, the Brahmaputra. Intermontane basins located in the tectonically active areas witness rapid changes in the landscape even on the historical time scale. Fluvial processes are influenced by tectonic forcings. The processes related to basin filling mechanisms are almost continuous on a longer geological scale. Orographic precipitation and ice melts are the two principal contributors to flow in the channels. The nature of sediments, the grading of particle sizes, and the dissolved radicals in the surface runoffs and the groundwater flows are principally controlled by the rock/soil types intercepted and the slope of the ground surface. However, sedimentation processes are also highly influenced

DOI: 10.1201/9781003302353-1

episodically by structural changes of different orders. For example, a surface that is more or less horizontal, or having a mild slope and acting thereby as the appropriate place for aggradation, with a slight change in tilt can become a place not all suitable for deposition. Any surface subjected to rapid erosion or prolonged non-deposition fails to conform in age to the adjacent places going through continuous deposition and thus 'unconformities' form. Regional unconformities in the subsurface are assumed to represent the time-lapse during which the basins were modified (Kingston et al., 1983). The sediment stack between two unconformities represents continuous facies and is termed as one 'sequence'. The presence of one unconformity between two sequences acts as a discriminator to understand two different stages of basin evolution. If basins and basin margin areas are treated as inseparable 'pairs' of a system, then it is observed that 'modifiers' responsible for reshaping basin margins also influence the basin forming processes in various ways. Some of the strain marks resulting from the actions of the basin modifiers are directly visible either in the mountain belts or in the typical geomorphology of the fluvial systems. Highly active basins give us rare opportunities to understand basin evolution on different scales. Tectonic geomorphology of mountains, based on paleoseismology and other pieces of evidence, has been studied by Bull (2007) and others (England and Molnar, 1990; Raymo and Ruddiman, 1992; Anderson and Densmore, 1997; Keller and Pinter, 2002; Dadson et al., 2003, 2004). Techniques have been developed by other authors to study neotectonics from the behaviour of the alluvial rivers (Schumm, 1986, 1993; Schumm et al., 2000). Different geomorphic indices based mostly on geometrical variations have been developed to express temporal changes and comparative studies in quantitative terms (Friend and Sinha, 1993). Efforts are under way to understand the controls responsible for different types of structural modifications in the subsurface by analysing surface geomorphology. The knowledge about ongoing processes in the subsurface can help us understand causes behind surface processes, and attainment of better predictability about natural disasters is gaining stronger ground among earth scientists and environmentalists. Thus, an increasing need is being felt by earth scientists to integrate surface, shallow subsurface, and deep subsurface processes within the common ambit of earth system science.

Geophysical data, especially the seismic sections, provide highly reliable information about subsurface lithology in terms of lateral continuities and variations of strata on the regional scale. Using these data sets, models of basin evolution can be prepared during different stages, where each stage is confined between two basin-modifying events marked by unconformities. Short-term geomorphological changes on the surface can be monitored by comparing properly georeferenced topographic maps and satellite imageries procured at different intervals of time. Rivers with high stream powers and flowing through deep gorges are not always 'dynamic' in the planar view in response even to highly significant tectonic changes. However, smaller channels with low stream power respond even to very minute slope variations.

Issues such as separating the climate–tectonics coupling, sediment–structure combination, coupling of climatic and anthropogenic factors, and preparing models have not yet reached a very convincing stage. However, there are possibilities to integrate subsurface stratigraphy, structural elements, and the short-term changes in geomorphology to understand neo-tectonics and basin evolution as a continuous process.

The Brahmaputra valley in Assam shares its borders with Bhutan, Bangladesh, and Burma and lies very close to China (Figure 1.1).

Very recently, Majuli Island, a famous world heritage site and one of the largest colonized islands in the world and the largest in Asia with habitation of 0.16 million people, has come to light because of its faster rate of erosion. However, the river-borne erosion is not going on only around Majuli Island. The perennial erosion is perhaps the biggest 'natural disaster' that affects most adversely the peasant population living by the side of the Brahmaputra River and some of its major tributaries of this narrow valley. Recently, the construction of big river dams for hydroelectricity power projects is being carried out in the mountain ranges bordering the Brahmaputra valley, which are supposed to drastically change the existing fluvial processes and the sediment architecture in the coming decades.

A long history, spanning almost 150 years, of oil exploration activities in the Brahmaputra valley (Pascoe, 1914; Sale and Evans, 1940; Evans and Mathur, 1964; Murthy et al., 1969; Karunakaran and Ranga Rao, 1979; Murthy 1983; Ranga Rao, 1983; Mallick, 1992; Mallick et al., 1995, 1997; Kent and Dasgupta, 2004) has yielded an enormous volume of subsurface geophysical data. The prospect of integrating the surface observations with the subsurface data was conceived in this volume with the idea to connect basin

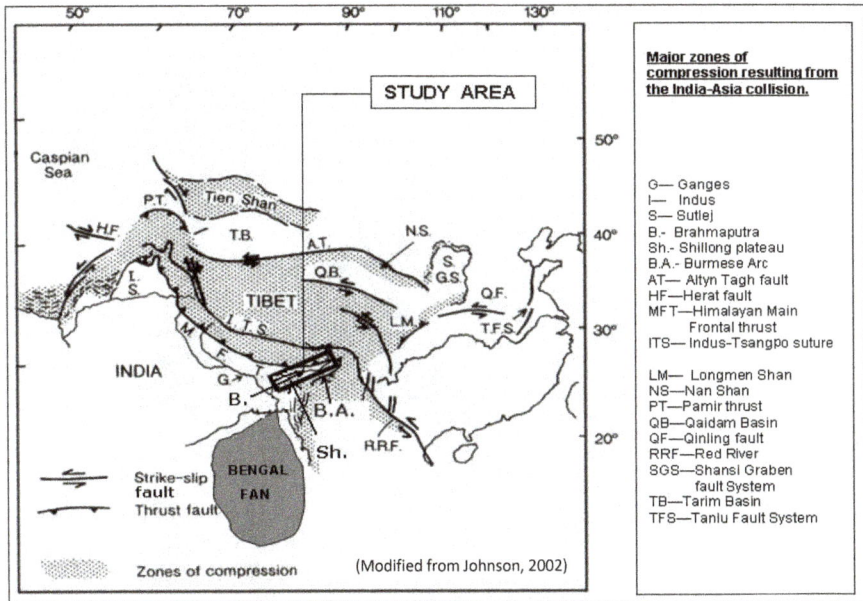

FIGURE 1.1
Brahmaputra valley is shown within a broader perspective of major structural elements.

evolution and fluvial dynamics in part of the foothills of the Himalayas where a comparatively faster rate of tectonic and sedimentological changes is expected. This involved a multi-disciplinary approach using fluvial geomorphology, remote sensing, image processing, GIS, shallow subsurface geophysics, reflection seismology, seismic sequence stratigraphy, neotectonics, and, of course, basin analysis.

The Brahmaputra River is a conduit through which a huge mass transfer of sediments takes place from the source that broadly represents the active zone of continent-to-continent collision between the Indian and the Eurasian plate and a sink – the Bay of Bengal (Figure 1.1), where the sedimentary architecture of the great Bengal Fan has been taking place since the late Eocene, that is, not less than the last 35 million years. The catchment responsible for supplying vast quantities of sediment includes the erosion of actively uplifting mountains of the Himalayas, slope erosion of the Himalayan foothills, and movement of alluvial deposits from the Assam valley (Thorne et al., 1993). In the entire Indian subcontinent, the Brahmaputra has been carrying the highest load of sediments – about 648 million tons per year (Venkatachary et al., 2001). This yearly transport of sediments amounts approximately to a sheet of 25 km in length and breadth and 10 m in thickness. The present-day upper Brahmaputra valley (Figure 1.2), an NE-SW trending intermountain alluvial

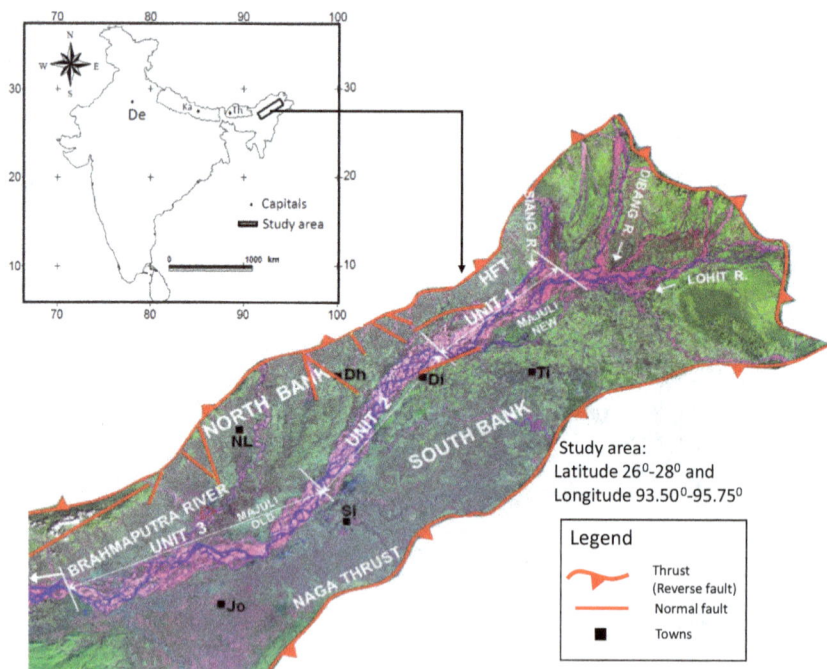

FIGURE 1.2
Study area comprises an NE-SW trending intermountain valley sandwiched between the eastern Himalayas and the Naga-Patkai Hills.

relief, was earlier part of the Assam–Arakan basin, and to be more precise, it constituted mainly the shelf part of the basin. Interestingly, basins undergoing active tectonic adjustments are not considered suitable for hydrocarbon prospects. However, the Brahmaputra valley, despite being a seat of large-scale seismic activities, is also an excellent preserver of hydrocarbons.

The focus of this book is to develop a synergistic approach and discover connections between basin evolution and fluvial dynamics. The motivation behind the present study was the urge to consolidate the interdisciplinary mode of pursuing enquiries to bridge three layers of knowledge gaps (Figure 1.3) in the backdrop of 'processes' about *surface, shallow subsurface (≤100 m),* and *deep subsurface (>100 m and <10 km)* of the upper Brahmaputra valley.

The work presented in this book provides important insights for understanding the morphotectonic evolution of a highly promising 'upper reach' of the Brahmaputra valley of Assam, India.

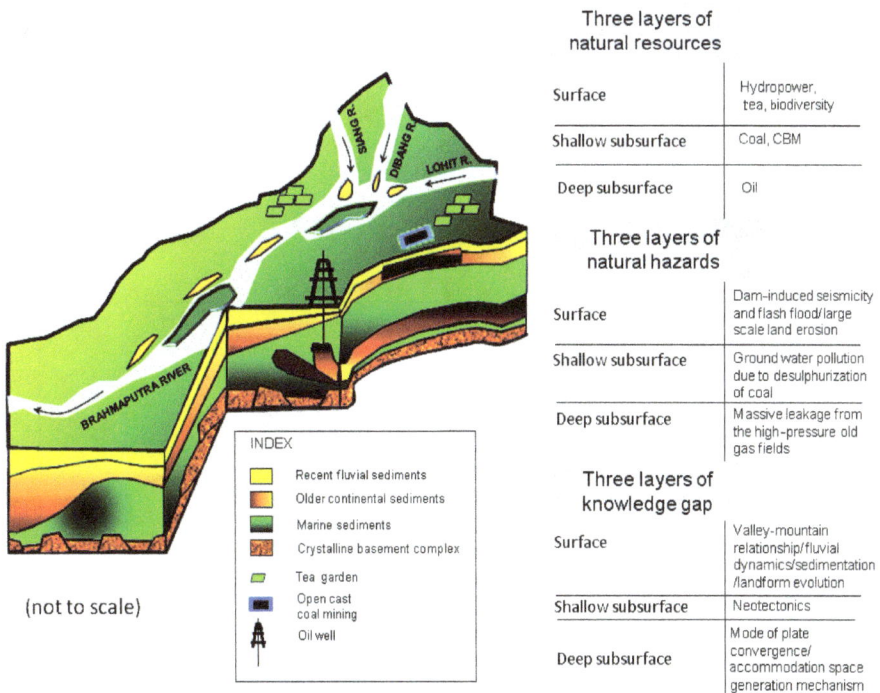

FIGURE 1.3
Figure shows three layers of natural resources, three layers of natural hazards, and three layers of knowledge gaps in the upper Brahmaputra valley, Assam.

The Upper Reach of the Brahmaputra Valley

For the sake of convenience, the Brahmaputra River and the valleys through which it passes can be divided into at least three reaches. The upper reach of the Brahmaputra (Figure 1.4a) contains a northeast–southwest flow that originates from the confluence of three rivers – the Siang, the Dibang, and the Lohit, wherefrom the river gets its name the Brahmaputra.

On the downstream side, this reach extends up to the Mikir Hills. In the middle reach, the westbound, near-parallel continuation of the Brahmaputra with the Eastern Himalayas, extends from the Mikir Hills up to the western tip of the Shillong massif. The Brahmaputra, in its lower reach, takes an L-turn and flows in the southward direction and enters Bangladesh. So, the lower reach extends from the upper tip of the western boundary of the Shillong massif, where the river takes suddenly a 90° turn and subsequently merges with the Ganges. Afterwards, the river is known as the Meghna. Having a bias towards geophysical data, we have focused our attention on the upper reach of the Brahmaputra valley, confined within the latitudes 26°–28° and longitudes 93.50°–95.75°, where most of the old and new oil fields, as well as highly prospective oil fields, are located.

We have divided the reach into three units (Figure 1.4b). The most upstream unit 1 is characterized by the former confluence of three rivers – the Lohit,

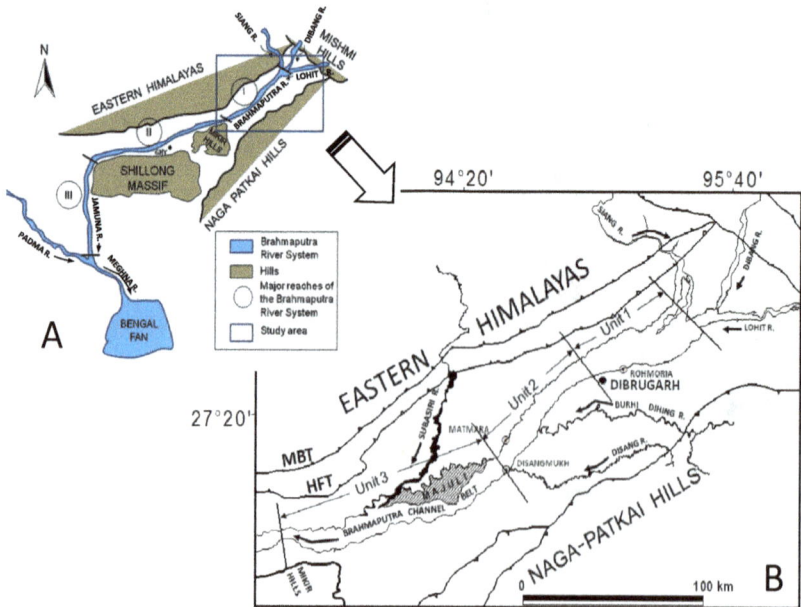

FIGURE 1.4
Three major reaches of the Brahmaputra River in Upper Assam.

the Dibang and the Siang, and a recently evolved island referred to by the locals as the 'New Majuli'. The mid-stream unit 2 continues up to the upper tip of Majuli Island and shows much less temporal changes than the other two. The most downstream unit 3 includes the famous Majuli Island.

Precipitation and Hydrology

The Brahmaputra basin comes under a powerful monsoon rainfall regime under a humid climatic zone. The average annual rainfall is about 230 mm, but the distribution of precipitation over the catchment is highly uneven. For example, a catchment called Jia dhol in the northeastern fringe of the valley receives 410 mm of rainfall annually, whereas the Kopili catchment in the south-central part receives only 175 mm.

Rainfall in the Himalayan sector amounts to 500 mm per year, with the lower ranges receiving more than the higher areas. For example, the annual rainfall at Dibrugarh in the eastern part of the valley is 285 mm, whereas at Pasighat – located in the foothill region, it is 507 mm, and at Tuting – located further up in the Himalayas, it is 274 mm. Monsoon rains from June to September account for 60%–70% of the annual rainfall (Figure 1.5). These rains, which contribute a large portion of the runoff in the Brahmaputra and its tributaries, are primarily controlled by the position of a belt of depression

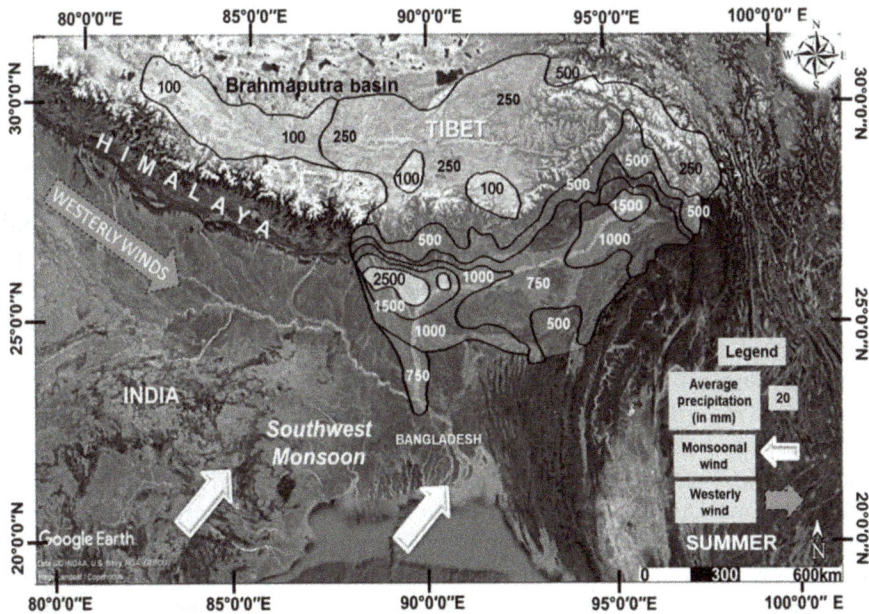

FIGURE 1.5
Major monsoonal precipitation in the Brahmaputra valley and the adjacent areas (after Wasson et al., 2022).

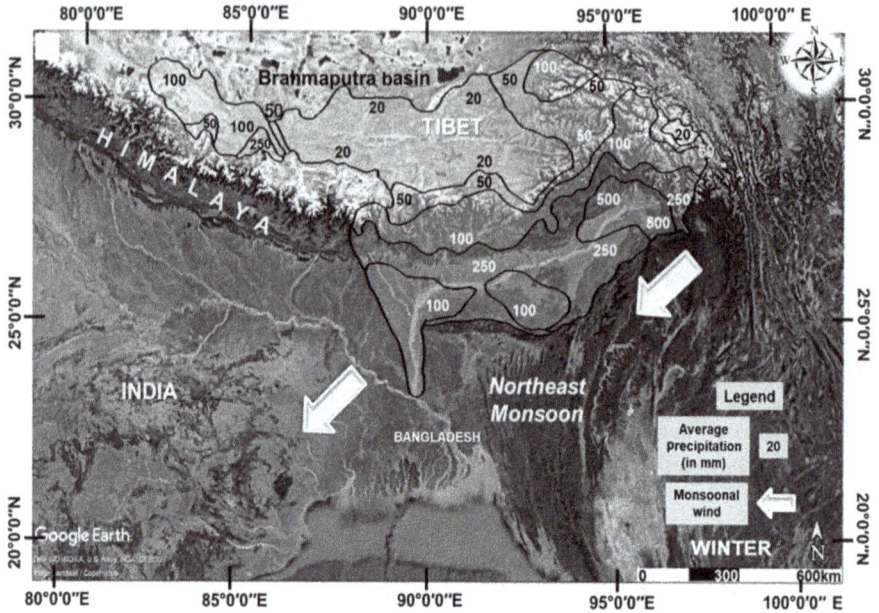

FIGURE 1.6
Minor monsoonal precipitation in the Brahmaputra valley and the adjacent areas (after Wasson et al., 2022).

called the monsoon trough, extending from the northeast of India to the head of the Bay of Bengal. In the course of its north–south oscillations in summer, when this axis moves to the foothills of the Himalayas, heavy precipitation is caused in Assam and adjoining highlands. The pre-monsoon season covering March, April, and May produces 20%–25% of annual rainfall caused primarily by depressions moving from the west and by local convectional storms (Figure 1.6). Precipitation in the Meghalaya plateau to the south of the valley is of the order of 1,000 mm/year (Purkait, 2004). For selective areas of the Meghalaya plateau, the precipitation can exceed 2,000 mm per year (Shrestha et al., 2015; Wasson et al., 2022).

Meteorological data also distinctly show that some portion of the Brahmaputra valley in Assam and the adjoining mountain regions receive heavy to very heavy rainfall in a spell of 2–3 days or sometimes rainfall even continues for 7–8 days during monsoon months under synoptic meteorological conditions. Under the influence of such situations, 1-hour rainfall of 70 mm is common in areas favourable for heavy rainfall. The highest recorded 1-, 3-, and 24-hour rainfalls within the basin area are 97.5, 200, and 805 mm, respectively. Such a spell of rainfall gives rise to flooding in the concerned tributaries or the main river along with the tributaries in the Brahmaputra valley of Assam and Bangladesh (Datta and Singh, 2004).

The hydrographs of the annual flow and sediment discharge of the Brahmaputra River reflect a repetitive pattern of rise and fall of flow and sediment load in the river corresponding to the seasonal variation of the monsoonal precipitation and the freeze–thaw cycle of the Himalayan flow. The average discharge at the mouth of the Brahmaputra ranks fourth ($19.83 \times 10^3 \, \text{m}^3/\text{s}$) as compared with other large rivers of the world, for example, the Amazon ($99.15 \times 10^3 \, \text{m}^3/\text{s}$), Congo ($39.66 \times 10^3 \, \text{m}^3/\text{s}$), Yangtze ($21.80 \times 10^3/\text{s}$), and Hwang Ho ($19.83 \times 10^3 \, \text{m}^3/\text{s}$) (Goswami, 1998). In terms of sediment yield, the Brahmaputra ranks the second highest (1,128 tons/km^2/year at Bahadurabad, Bangladesh, and 804 tons/km^2/year at Pandu, Assam) in the world, after the Yellow River (1,403 tons/km^2/year) (Goswami, 1998).

Natural Hazards

The typical geological and tectonic framework coupled with structural complexities has rendered the Brahmaputra basin geomorphology a most complicated one. A variety of landforms under varied climatic conditions have formed over the geologic and tectonic base of the region. The periglacial, glacial–fluvial, and fluvial processes are dominantly operative in the basin at varying altitudes. Higher elevations of the Himalayas experience periglacial solifluction. The low hill ranges with hot and humid climate and heavy rainfall concentrated to a few months of the year experience solifluction, sheet erosion, and landslides. The incidence of landslides is high in the Himalayan foothills, where heavy rainfall, high seismicity, and toe cutting of hill slopes by the streams are most frequent. Heavy rains often loosen soil and the soft rocks of the young Himalayan ranges. Rainwater percolates through the joints, fractures, foliations, and pores of rocks and soils and finally makes them loose and heavy, which causes heavy slope failure. Fluvial processes are, on the other hand, significantly dominant on the valley bottoms and plains, where alluvial deposition takes place due to erosion of the higher surface by rivers and flooding in the valleys. Also, the erosional and depositional processes conspicuously intensified by copious rainfall and frequent seismic movements play a dominant role in creating various fluvial-geomorphic environments in the basin (Bora, 2004).

Essential Objectives

As mentioned earlier, geoscientific investigations in the upper reach of the Brahmaputra valley have been carried out more or less steadily for the last 150 years. Understanding a valley as complex as the Brahmaputra on a regional scale needs multi-disciplinary data sets. A variety of data used for this study include satellite remote sensing images, shallow refraction, and the uphole data sets shared for the first time by Oil India Limited for academic research purposes. The major objectives of this study are as follows:

1. Understanding the decadal-scale fluvial dynamics and changing interfluves using older topographic maps and remote sensing data in a GIS environment.
2. The nature of fluvial dynamics in the shallow subsurface using shallow refraction and uphole data sets to understand late Quaternary basin evolution.
3. Multi-parametric micro-zonation of the selective windows based on the 'degree of consolidation of the top 2–3 layers of the topsoil and understanding of the influence of neotectonics in the valley areas peripheral to the thrust belts with the help of the refraction and uphole data sets.
4. Understanding the spatial inhomogeneity in the Paleogene and Neogene sediment sequestration from the seismic data.
5. Understanding fluvial dynamics in the backdrop of the multi-scale basin evolution process from the integration of all the above data sets.
6. Variability in the bed–bank relationship in different reaches within the study area of the Brahmaputra River and the valley and assessment of vulnerability due to flood and erosion disasters.

Accordingly, in the first phase, surface changes in the landscape from 1915 to 2005 were monitored using satellite images and maps. Detail measurements of various morphometric parameters and their temporal variability were recorded for three different periods, 1915, 1975, and 2005. Temporal changes in channel belts, channels and braid bars, bank-line shift, and other morphological parameters such as sinuosity, braid channel ratio, etc. were measured for different smaller reaches to document fluvial dynamics and to understand the causal factors. The pattern of the recent spurt in the erosion of Majuli, located in the Brahmaputra River system of the Upper Assam basin, one of the largest riverine islands in the world and the largest in Asia, was studied in detail. For the authenticity of measurements, short field visits were conducted.

In the second phase, shallow subsurface changes were monitored in the most prolific oil-bearing provinces of the south bank of the Brahmaputra River with the help of shallow refraction and uphole seismic survey data. Strategic windows were selected close to two proven subsurface faults running almost parallel to the Naga-Patkai Thrust belt. Older fluvial dynamics of the late Quaternary period were identified from the lateral variation of the seismic wave velocities. The purpose is to extrapolate recent fluvial dynamics (decadal-scale) to historical-scale (1,000 years or more) understanding of fluvial dynamics to identify 'forcings' and thereby form a better understanding of the morphotectonics. Multi-parametric micro-zonation was carried out to understand neotectonics and the influence of the Naga-Patkai Thrust on the valley landscape.

In the third phase, deep subsurface changes in the sedimentary basin were documented with the help of recent seismic data. Sediment sequestration patterns, the accommodation space generation mechanism, and basin-forming and basin-modifying tectonics were studied to understand the subsurface control on the surface processes from the broader understanding of basin evolution.

In the fourth phase, integration of data was performed to suggest models explaining the probable causes of certain patterns observed in the fluvial dynamics and landform evolution in the context of 'forcings' connected to geomorphology, sediment dispersal, fault activation, and broader tectonics in the upper reach of the Brahmaputra valley. To fulfil these objectives, as per the demand of the situation, other sources of the available database were also used liberally.

Finally, changing river bed and river bank relations in different reaches were studied to quantify vulnerability due to flood disasters for the period 1915–2015.

2

A Quick Review: Geoscientific Understanding of the Brahmaputra Valley

There were oil seepages at different places of Upper Assam bordering the present-day Arunachal Pradesh and the surface identification of coal seams drew the attention of pioneering geologists and others as early as 1860 and subsequently. Of course, the first preliminary reconnaissance surveys were conducted much earlier in 1825 by Wilcox (1832) in the Lohit Frontier Division of Arunachal Pradesh, and from there, the news of oil seepages in the remote forests near the head of the Brahmaputra valley spread, which raised geographical curiosity. The world's first oil well was drilled by Drake in 1859 in Pennsylvania, and just after 7 years, that is, in 1866, the first oil well was hand-dug to a depth of 102 ft at Naharpung (in the Jaipur area) located at the top of the Brahmaputra valley. In 1889, Digboi Well No. 1 was spudded by Assam Railway and Trading Co., which was completed as a producer in November 1890. To map the oil-bearing structural trap at Digboi and search out other probable structures in the vicinity of the area, geophysical methods were used in the aftermath of the First World War. Dr Pekar of the Eötvös Geophysical Institute, Budapest, was assigned to carry out a gravity survey with a torsion balance by Burmah Oil Company Limited during 1925–1928. These surveys showed that the torsion balance could readily detect buried thrust faults. From 1936 to 1938, electrical resistivity surveys were conducted to understand the deeper parts of an incompletely exposed anticline by removing the masking effect of alluvium. In October 1938, for the first time in Asia, a seismic reflection survey was conducted by Petty Geophysical Engineering Company. This was a collaborative work involving three companies, namely, Burmah Oil Co, British Petroleum (then Anglo-Iranian Oil Co.), and Shell. The Second World War halted the fast-advancing pace of oil exploration in Assam. However, based on the interpretation of the seismic reflection data during 1938–1939, a site for drilling was selected at Nahorkatiya in 1951 on the south bank of the Burhi Dihing River, a tributary of the Brahmaputra. A well was drilled for a depth of 11,715 ft (3,751 m), and it proved to be a prolific oil-producing well (Singh et al., 1996). The oil strike in Nahorkatiya in 1953 was the first landmark event of exploration geophysics in India. All doubts relating effectiveness of geophysical methods were silenced by this single discovery.

The search for new trade routes from Assam to China prompted the British to search for the origin of the Brahmaputra River, which was identified later on as the most important and interesting geomorphological exploration in this part of the world. The connection between the Tsangpo and Brahmaputra was suspected long back by Major James Rennell (1765). He was a celebrated adventurer and cartographer, appointed as the first Surveyor General of India at the age of 24 years in 1767 by the then Major General Robert Clive, the governor of colonized Bengal and Bihar. Rennell, who is called the father of Indian geography, proposed the identity of these two rivers in the form of a theory La Touche, 1910. But, despite conducting several expeditions from time to time, only after mapping the syntaxial bend between the Namcha Barwa and Gyala Peri in 1924 by two explorers, Kingdon-Ward and Earl Cawdor, who were soldier-botanists, the final piece of the jigsaw puzzle could be solved.

Although the volume of study undertaken in the Brahmaputra valley is much less than that in other areas of the world having the comparable geological setting, the history of systematic studies in this region, broadly along with three independent directions – namely, stratigraphy, geomorphology, and seismology – is quite old.

Stratigraphic studies (Medlicott, 1865; Mallet, 1882, 1875, 1876; La Touche, 1883, 1885, 1886; Simpson, 1896, 1906a,b; McLaren, 1904; Hayden, 1910; Pascoe, 1912, 1914; Evans, 1932, 1935, 1964; Bhandari et al., 1973; Acharya and Sastry, 1976; Karunakaran and Ranga Rao, 1979; Das Gupta and Biswas, 2000) were mainly governed by the identification of coal units and prospective reservoir rocks (principally sands) capped by shale from the vantage angle of hydrocarbon exploration. Understanding of geomorphology (Wilcox, 1830; Ghosh, 1935; Coleman, 1969; Goswami, 1985, 1998; Bristow, 1987, 1993; Thorne et al., 1993; Goodbred and Kuehl, 1998, 2000; Goodbred et al., 2003; Uddin and Lundberg, 1999; Sarma, 2005; Sarma and Phukan, 2006; Latrubesse, 2008; Lahiri and Sinha, 2012, 2014) was important to identify soil types and mapping of different types of water bodies to explore a rational basis of revenue collection and development of a cheap transportation system to facilitate trade and business. On the other hand, seismology with a special emphasis on monitoring earthquakes (Godwin, 1869; Bond, 1899; Dunn et al., 1939; Poddar, 1953; Tandon, 1954; Verma and Mukhopadhyay, 1976, Verma et al., 1993; Das Gupta and Nandy, 1982; Sukhija et al., 1999; Ambraseys, 2000; Bilham and England, 2001) was important because from the past experiences, it was known to the colonial masters that sometimes a single tremor could be sufficient enough to dump the whole myth of civilization into a heap of rubbles. Only at a much later stage did scientists start recognizing the close interrelationship among the aforesaid subjects.

Oldham (1899), in his *Report on the Great* (Assam) *Earthquake of 12 June 1897*, had mentioned that this event reduced to rubble all masonry buildings within a region of Northeastern India roughly the size of England and was felt over an area exceeding that of the great 1755 Lisbon earthquake. It was believed that this was the consequence of a north-dipping Himalayan thrust fault propagating south of Bhutan. Although Bond (1899) reported 8 m of uplift and 4 m of displacement of parts of the Shillong Plateau after comparing meticulously with the original points of the 1862 trigonometrical survey across the Shillong Plateau, the results and claims were dismissed on the ground that they failed to meet the triangle closure standards of the Survey of India. It took another 100 years for Bilham and England (2001) to show that contrary to earlier explanations that the earthquake was due to the activation of the Dauki fault in the southern boundary of the Shillong massif, a more probable cause was rupture of a buried reverse fault in the northern boundary that was approximately 110 km in length and dipping steeply away from the Himalayas. The earthquake caused the northern edge of the Shillong Plateau to rise violently by at least 11 m. Their explanation is given as follows:

"The Shillong plateau is not being built as part of a system of thin-skinned thrusting but is bounded by a high-angle reverse fault to its north, and probably also to its south. The Shillong plateau thus resembles the pop-up structures that border thrust belts elsewhere."

(Bilham and England, 2001)

We want to emphasize that about one and a half century of exploration history in the Assam–Arakan basin has helped accumulate a storehouse of data that should have helped understand the subsurface processes in a much better manner. But, in most probability, highly uneven data density, quick 'Drill-barrel-sell' tendency had encouraged industrial geoscientists to go for highly focused localized observations that resulted in either too conservative, local situation-specific data interpretation or abnormally high degree of extrapolation of data and sweeping generalizations that may not be of much practical use to develop a widespread regional perspective. Very few works are available relating the impacts of certain new concepts that amount to paradigm shifts on the existing framework of the knowledge base. For example, we do not know the implications of the new explanation put forward by Bilham and England (2001) about the rise of the Shillong Plateau as a pop-up structure and its consequences on the overall strain pattern of the region. In the subsequent discussion, we will try to review the present status of understanding the Brahmaputra valley and its tectonic relationship with some of the important geological elements in the surrounding area.

General Geology and Geomorphology

The Brahmaputra River basin is bounded by the Eastern Himalayas on the north and northwest, the Naga-Patkai ranges on the east and southeast, and the Shillong Plateau and the Mikir Hills in the southern part. The Brahmaputra valley can be subdivided into the north Brahmaputra valley adjacent to the Himalayan foothills, and the south Brahmaputra valley adjacent to Naga foothills. The south Brahmaputra valley extends south-westwards across Dhansiri valley after a break in the region of the Barail ranges into Cachar and Tripura ranges. This part of the region, that is, the valley southwest of the Barail ranges, is sometimes called the proto-Brahmaputra delta. The north Brahmaputra valley and south Brahmaputra valley up to the Barail ranges are sometimes called the Upper Assam shelf (Singh et al., 1996).

The Himalayan zone comprises three topographic units that rise progressively to the north. The lowermost ranges called the sub-Himalayas, with an average elevation of 1,000 m, mainly consist of Tertiary sandstones and are conspicuous by the presence of many raised, relatively young terraces (Gansser, 1964). The Middle Himalayas, having an average elevation of 4,000 m and underlain by lower Gondwana (Palaeozoic) deposits comprising shales, slates, and phyllites, are overlain by a thick horizon of basaltic rocks. The greater Himalayas, with an average elevation of 6,000 m, primarily consist of granites and gneisses (Goswami, 1985). Wadia (1968) observed that the Himalayan Mountains with their syntaxial N-E bend originated out of the 'Tethyan Geosyncline'. The concept of 'Geosyncline' has become obsolete. So, it is better to say the northern fringe of the Brahmaputra valley having the Himalayan Mountain with their syntaxial N-E bend represents tectonically deformed and metamorphosed lithofacies that earlier used to be the forearc and was composed of pre-orogenic pelagic sediments overlain by synorogenic flysch facies (Selley, 2000). The sub-Himalayas and the lower Himalayas are characterized by piedmont zones, low discontinuous ridges, low linear ridges, high rugged hills, and upland valley depressions. The Naga-Patkai ranges, with an average elevation of 1,000 m, are composed of Tertiary sediments and characterized by the presence of a large number of active faults. This zone consists of piedmont plains, anticlinal ridges and synclinal valleys with terraced alluvial fills, undifferentiated sharp ridges and narrow valleys, upland valley depressions, and plateau remnants. The Shillong Plateau and the Mikir Hills attaining an elevation ranging from 600 and 1,800 m are primarily made up of gneisses and schists. It is characterized by plateau remnants, inselbergs, deeply dissected uplands with faulted monoclines of Tertiary cover, denuded hills, basement-controlled structural ridges covered with Tertiary rocks, and upland valley depressions (Bora, 2004).

Ranging in the average elevation from 50 to 120 m (above the mean sea level), the landscape of the Brahmaputra valley is about 800 km long, and the maximum width is about 130 km. There are several channel cut-offs or bills, oxbows, and marshy tracts, which occupy part of the flood plain deposits. Ten million years back, two important geological elements, the Shillong Plateau and Mikir Hills, were not uplifted, and the valley was then much wider, at least more than 200 km in the lower part. The catchment area of the entire Brahmaputra River is about 580,000 km², out of which a 195,000-km² area lies in India. The average discharge of the Brahmaputra at its mouth is 19,830 m³/s. The maximum discharge of the Brahmaputra as measured at Pandu in 1962 (on 23.08.1962) was of the order of 72,794 m³/s, while the minimum discharge was 1,757 m³/s in 1968 (on 20.02.1968) (Goswami, 1998). The Brahmaputra flows for about 670 km through the state of Assam, and within Assam, the Brahmaputra constitutes 103 tributaries – 65 on the right (north) bank and 38 on the left (south) bank (Sarma, 2005). The catchment of the north bank tributaries lies in the Himalayan range, and both in size and height, they are larger and receive higher rainfall than the tributaries flowing from the Assam range of hills. The north bank tributaries have a very steep slope and shallow braided channels for a considerable distance from the foothills in some cases right up to the outfall. They are subjected to flash floods and carry heavy silt charges. On the other hand, the south bank tributaries have comparatively flatter slopes; deep meandering channels almost from the foothills, bed, and banks of fine alluvial soil; and comparatively low silt charge. The important north bank tributaries are Subansiri, Puthimari, Pagladiya, and Manas. The south bank tributaries are Noa Dihing, Burhi Dihing, Dikhau, Dhansiri, Kopili, etc. The major points of differences between the north and south bank rivers are as follows:

i. The northern tributaries have a very steep channel gradient, but the southern tributaries have comparatively flatter gradients.

ii. The northern tributaries have shallow braided channels, while the southern ones have their meandering channels over the plains.

iii. In the case of northern tributaries, the long section of the river course is in the hilly terrain and a short distance is over the plains, whereas the southern tributaries have their longer courses over the plains.

iv. The northern tributaries have generally coarse sandy beds with occasional gravel beds up to some distance from the foothills. But the southern tributaries have beds and banks composed of fine alluvial soils.

v. The northern tributaries carry an enormous sediment load as compared with the southern tributaries. On average, the sediment yield of the north bank tributaries is three times higher than that of the south bank tributaries.

vi. The northern tributaries generally have flash floods, but the southern ones have generally fewer flash floods.

vii. The northern tributaries are characterized by frequent shifting of their channels during floods, but the southern tributaries change their courses less frequently (Bora, 2004).

The drainage pattern of the Brahmaputra network within India, in general, is dendritic in the northeast and northwest, sub-parallel in the south, and reticulate or anastomosing in the main course. The drainage pattern in the valley is of antecedent type (Purkait, 2004).

Stratigraphy

Outcrop-based stratigraphic succession studies of rock units, extrapolated to the subsurface, based on their dip direction and strikes are usually supposed to present comparable and correlatable vertical columns of rock units, which are subsequently called generalized stratigraphy of a particular area. The uniqueness of the Brahmaputra valley is that it could preserve all the stages of Tertiary successions (Table 2.1). But the degree of stratigraphic variation

TABLE 2.1

General stratigraphy

Erathem	System	Series	Stages (Standardized as per I.C.S.N. & N.A.C.S.N.)	Super Sequence	NW Himalaya	Nepal Himalaya	Arunachal Himalaya	Assam			
								Upper Assam Plain	Upper	Assam	Shelf — Dhansiri Valley
CENOZOIC	QUATERNARY	HOLOCENE	VERSILIAN	XV	Newer Alluvium (T₀, T₁, T₂)		Newer Alluvium; Channel Alv.(T₀); Terrace Alv.(T₁); Alluvial Fans	Newer Alluvium; Channel Alv.(T₀); Terrace Alv.(T₁); Alluvial Fans		Newer Alluvium	
		PLEISTOCENE — Upper	MILAZZIAN	XIV	Older Alluvium (T₃, T₄, T₅) Shympur Formation; U. Karewa		Older Alluvium; High level Terraces(T₃ & T₄) & Hapoli Fm.	Older Alluvium; High level Terraces (T₂ & T₃)		Older Alluvium	
		Middle — SICILIAN									
		EMILIAN			U.Siwalik Group; L.Karewa	Upper Siwalik	Kimin Formation(U. Siwalik)	Dihing Group	Dhekiajuli Beds/Formations (Kimin Fm.)	Dihing Group	Dhekiajuli Beds/Formations Demulgaon Fm.
		Lower — CALABRIAN									
		PLIOCENE — Upper	PIACENZIAN		M.Siwalik Group	Middle Siwalik	Subansiri Fm. Dafla Fm. (M. Siwalik)	Dupitila Group	Namsang Beds/Formations (Subansiri Fm.)	Dupitila Group	Namsang Beds/Formations
		Lower	ZANCLIAN								
	NEOGENE	MIOCENE — Upper	MESSINIAN	XIII	L. Siwalik Group	Lower Siwalik	Kimi Fm. (L. Siwalik)	Tipam Group	Nazira Fm.	Tipam Group (Undifferentiated)	
			TORTONIAN								
		Middle	BERRAVALLIAN						Girujan Fm. Lakwa Fm.	Surma Group(?)	Khoraghat SSt. Member
			LANGHIAN						Geleki Fm.		Bokabil Fm.
		Lower	BURDIGALIAN AQUITANIAN								
	TERTIARY	OLIGOCENE — Upper	CHATTIAN		Murree;Dharamsala; Kasauli–Dagshai	Dumri/ Suntar		Barail Group(?)	Demulgaon Fm.	Barail Group(?)	
		Lower	RUPELIAN	XII							
		EOCENE — Upper	PRIABONIAN						Disangmukh Fm.		Disangmukh Fm.
		Middle	BARTONIAN			Bhaniskati / Swat			U. Member M. Member	Kop- ili Fm.	Amguri M. Charali M.
			LUTETIAN				Dalbuing Formation	Kopili Fm.			
		Lower	YPRESIAN	XI	Subathu; Kanji				L. Member	Jaintia Group	Sylhet Fm.
		PALEOCENE — Upper	THANETIAN				Goku Formation		Prang M.		
								Sylhet Fm.	Narpuh M.		Tura Fm.
									Lakadong M.		
		Lower	DANIAN		Manikot(Tal) shell Limestone;			Tura/ Theria Fm.			
MESOZOIC	CRETACEOUS		Upper — MAAS-TRICHTIAN	X	Sangcha Malla Fm. Balachadhura Vol.	Amile		Khasi Group	Langpar Fm. Mahadeo Fm.		
			Lower	IX → VII				Dergaon Group	Moabund Fm. Mikir Trap Bamangaon Fm.		
								Sylhet Trap		
PALAEOZOIC	PERMIAN		Upper — TATARIAN(?)	VI		Yamne Formation	Abor Volcanic Fm.	Gondwana			
			Lower			Lower Gondwana Group	Bhareli Fm. Bichom Fm. Miri Fm.				
PRE CAMBRIAN BASEMENT											

from one well to another is so wide that different authors have proposed different stratigraphic successions.

The most frequently referred stratigraphic succession of the upper Brahmaputra Valley is by Mathur and Evans (1964). Other works dedicated to such research are by Bhandari et al. (1973); Handique et al. (1989);

Arakan	Basin			Tripura		Lithology	
Naga-Patkai Range (Schuppen belt)	South Shillong			Cachar			
	Garo & W. Khasi Hills	SE Khasi & Jaintia Hills					
Newer Alluvium; Channel Alv.(T₃); Terrace Alv.(T₁); Alluvial Fans	Newer Alluvium	Newer Alluvium		Newer Alluvium		Unoxidised sediments of active channel boulders, cobbles, pebbles with unoxidised sand.	
Older Alluvium; High level Terraces (T₂ & T₁)	Older Alluvium	Older Alluvium		Older Alluvium		Boulders, cobbles, pebbles with oxidised sand and sandy clay, Lacustrine deposits for Hapoli & Karewa Formation	
Dihing Group — Dhekiajuli Beds/Fm.	Dihing Group			Dihing Group		Coarse bouldery conglomerates and sands. Carbonised wood.	
Dupitila Group — Namsang Beds/Formation	Dupitila Group	Dupitila Group		Dupitila Group	U.	Coarse to fine loose sands with minor clays.	
					L.	Bluish to dark grey clays with intercalated silts and sands.	
Tipam Group — Girujan Fm. / Jaipur Fm.		Tipam Gr.(undiff.)		Tipam Gr.	Gobindpur Fm. / Jaipur Fm.	Massively bedded white to grey sandstone with intercalated clay stone(Nazira Fm., Lakwa Fm., Geleki Fm. Jaipur Fm.)/ Mottled and variegated brick red to ochreous claystones with silt and sand intercalations(Girujan Fm., Gobindpur Fm.)	
Surma Group — Bokabil Fm. / Bhuban Fm. U./L.	Surma Gr.(Undiff.)	Sur-ma Gr. — Bokabil Fm. / Bhuban Fm. U./M./L.		Surma Group	Bokabil Fm. / Bhuban Fm. U./M./L.	Interbedded sst. shale , Light to dark grey shale with few silt/sand intercalation (Khoraghat M.)/Shales with minor sandstone(Bokabil Fm.)/Interbedded sst. and shale.(Bhuban U.)/Shale with intercalations of siltstone & sandstone.(Bhuban M.)/ Interbedded sst. and shale.(Bhuban L.)	
Barail Group — Tikak Parbat Fm. / Baragolai Fm. / Naogaon Fm.	Barail Group (Undiff.)	Barail Gr. — Renji Fm. / Jenam Fm. / Laisong Fm.		Barail Group	Renji Fm. / Jenam Fm. / Laisong Fm.	Massive sst. With few shale intercalations(Renji/Demulgaon Fm.) / Interbedded shale and sst.(Jenam/Disangmukh Fm.) / Massive sst. (Laisong Fm.)	
Disang Group — Upper Disang	Kopili Fm. (Jaintia Group)	Kopili Fm. (Jaintia Group)	Prang Lst. M. / Narpuh Sst. M. / Umlatdoh Lst. M.	Disang Group (Undifferentiated)		Light to dark grey greenish shale with intercalated sst. And Marls.(Kopili Amguri M.)/ Shale with lst. And sst. Intercalations (Kopili Charail M.)/ Limestone with	Shales with minor silt and sands. Occasionally carbonaceous shales(Disang Upper).
	Sylhet Fm.	Sylhet Fm.	Lunsunong Sst. M. / Lakadong Lst. M.			sand shale intercalations (Sylhet Fm.)/Sst., kaolinitic, Pyritic	
	Tura Fm.	Therria Fm.	U. Sst. M. / L. Lst. M.			with occasional coals(Th.Fm. U.sst.m.)/Lst. With shale/sst.interbeds(Th.Fm.L.lst.m.)	
Disang Group — Lower Disang	Khasi Gr. / Maha-deo Fm.	Khasi Group	Langpar Fm. / Mahadeo Fm.			Shale with lst. Intercalations. Glauconitic shale.	Shales with minor sst. And shale. Lst. Towards base, slightly metamorphosed. (Disang Lower)
	Sylhet Trap	Sylhet Trap				Glauconitic sandstone Forr. And coarse near base (Dergaon Gr. Moabund Fm.)/Basaltic flows with intertrappean (Mikir trap)/Sandstone with brown to gray shales (Bamangaon Fm.)	
						Basaltic and Andesitic flows	
	Gondwana	Gondwana				Sandstones with Kaolinitic beds	
						Granites, Metamorphics, Gneisses and intrusives	

Mallick et al. (1997); Das Gupta and Biswas (2000); Kent and Dasgupta (2004). A comparative study reveals that the influence of the geosynclinal theory were declining in the later works. Also, the lowest part of the Tertiary sediments belonging to Paleocene and Eocene remained undifferentiated in the early works (e.g., Mathur & Evans (1964) identified a thick column of over 3,000 m thickness as the 'Disang' group) but with increasing oil prospects in the Paleogene sediments, further subdivisions were identified. The debate regarding 'group', 'subgroup', 'member', and 'formations' still lingers. For example, an important rock type of Miocene age, the Tipam unit, was referred to by some authors as 'group' (Purkait, 2004) and others as 'formation' (Kent and Dasgupta, 2004). The general rock type description that emerges is as follows:

Archaean

Archaean formation comprises mainly biotite–hornblende, sillimanite, gneisses, and schists associated with feldspathic biotite granulites, pyroxene–hornblende granulites, calc granulites, and aplites intruded at places by massive porphyritic, coarse-grained granite, pegmatite, quartz veins, and basic intrusives. They are found in the eastern part of Nowgong district and as isolated inselbergs in the Brahmaputra valley.

Precambrian

The Shillong group of Precambrian rocks comprise quartzite, phyllites, and schists, unconformably overlying the Archaean sediments in the western flank of the Mikir Hills across the Kopili valley of Assam. In the central Mikir Hills, occurrences of the weathered trap of Sylhet suite (Jurassic) of Meghalaya are reported. Regionally, the foliation trends EW with frequent changes of the direction from NE-SW with steep to vertical dip.

Mesozoic

In the foothills of the Shillong Plateau, there is a basal conglomerate, followed by coarse, massive, ferruginous, and glauconitic marine sandstones, which, in turn, are overlain by and may also locally pass into thin shales and limestones. The sandstones make up the Mahadek stage and the higher bed Langpar stage. These occurrences are all of the shoreline deposits, but in the eastern part of Assam (Manipur), there is a thin limestone associated with the Disang shales, the lower part of which may be of the Cretaceous age.

Tertiary

Tertiary sediments are broadly divided into Paleogene and Neogene strata. Some authors (e.g., Kent and Dasgupta, 2004) term them 'sequence'. But this does not match with the rapidly catching definition of 'sequence' from sequence stratigraphy, where a sequence is invariably overlain and underlain by unconformities.

Paleogene Strata

The Paleogene strata can be divided into three major units. The basal unit is the Paleocene Langpar Formation, which is composed of marine quartzitic sandstone and minor claystone. The middle unit is the Jaintia Group, which consists of the lower Eocene Sylhet Formation composed of siliceous shales with limestone and sandstone interbeds, and the middle to the upper Eocene Kopili Formation, which consists of shale and fine-grained sandstone. The third major Paleogene unit is the upper Eocene to lower Oligocene Barail Group, which is composed of sandstones of the Tinali Formation and overlying mudstones and coals of the Moran Formation.

Neogene Strata

From bottom to top, the Neogene strata can be divided into five formations, namely, Surma, Tipam, Girujan, Namsang, and Dihing/Dhekiajuli.

The Surma Formation is sand/shale alteration that is not identifiable at all the places of the Brahmaputra valley. The lower Miocene Tipam Formation sandstone is mostly of fluvial origin, and the heavy mineral content of the unit (Bhandari et al., 1973) indicates derivation from the rising Himalayas.

The Girujan Formation consists of lacustrine shales and fluvial deposits and may reflect changes in drainage patterns related to the initial development of the thrust belt to the south of lacustrine flooding related to foreland basin subsidence.

The Namsang Formation lies unconformably over the Girujan Formation and consists of poorly consolidated fluvial sandstones with interbedded clay and lignite.

The Dihing or Dhekiajuli Formation of the Plio-Pleistocene age consists of unconsolidated fluvial and alluvial fan deposits, lying unconformably over the Namsang Formation. The subsurface equivalent of the outcropping Dihing rock type is called Dhekiajuli Formation.

Quaternary Deposits

These strata thicken northwards of the Brahmaputra valley, and when approached closer to the Mishmi Hills, the vertical extent exceeds 2,000 m. Based on different types of local hiatuses, earlier authors divided this thick column on different bases, such as 'terrace deposits', 'older or high-level Alluvium', 'unstratified drifts', and 'red bank soils'. However, currently, the deposits lying unconformably over the Dhekiajuli Formation are referred to as 'Brahmaputra alluvium' (Das Gupta and Biswas, 2000). From the borehole samples, the older deposits appear as hardened yellowish, brownish, or reddish clay mixed up with sands, shingles, gravel, and boulder deposits. From the borehole log data, the cyclic nature of sedimentation is indicated. The recent alluvium includes flood plain deposits and comprises clay, coarse sand, shingles, gravel, and boulder and attains a thickness of approximately 200–300 m.

Structural and Tectonic Setting

The ophiolite suite that outcrops in the Indus and Tsangpo valley (Indus suture zone), northeast of the Himalayas (Gansser, 1964, 1980), marks the pre-collision boundary, along which the Indian plate collided with the Eurasian plate. The present configuration of the Brahmaputra valley is the result of the uplift and subsidence of the Precambrian crystalline landmasses, the remnant of which is now represented by the Mikir Hills (Karbi–Anglong hills) of Assam and the Shillong Plateau of Meghalaya.

Four geotectonic provinces through which the Brahmaputra flows are as follows:

 i. The stable shield area of the Shillong Plateau and Mikir Hills;
 ii. The platform area peripheral to the shield, now covering the North Cachar Hills, Bangladesh plains, and the Brahmaputra valley;
 iii. The Eastern Himalayan thrust belt Ghosh, 1956; and
 iv. The Assam–Arakan Fold and thrust belt.

These four geotectonic provinces are bounded by major tectonic lineaments that are as follows:

 a. The frontal Himalayan thrust belt is trending from EW to NE-SW.
 b. The NW-SE trending the Mishmi thrust along the Lohit foothills.

c. The NE-SW trending the Burmese arc that has its continuation further down in the southward direction to become the Andaman and Nicobar Islands.

d. The 'Belt of Schuppen' is an NE-SW trending structural feature of imbricate thrusts over the northern part of the Naga-Patkai range.

e. EW trending the Dauki fault along the southern margin of the Shillong Plateau, extending up to Haflong, which is a system of faults that change from a low angle thrust to a high angle reverse, to vertical faults with increasing depth (Murthy et al., 1969).

Seismicity

The data on historical seismicity show that all the tectonic zones of Northeastern India have been active for the past several 100 years. Besides the two devastating earthquakes of 1897 and 1950, events with an intensity exceeding IX (on the Rossi–Forel intensity scale) have been observed for earthquakes at Dhubri (1930) near Darjeeling (1899) and near Cachar (1869). It is usually seen that the seismicity follows the trend of major tectonic zones in the area. The trend is N-S to NE-SW along the Tripura fold belt, the Arakan Yoma belt and the Burmese molasse basin between lat. 22° N to 27° N, and long. 93° E to 97° E. Near lat. 27° N, the trend changes sharply to NW-SE along the Assam syntaxial zone and further changes to NE-SW near 29° N, 95° E.

Seismicity follows the well-known Himalayan thrusts between longitudes 95° E and 88° E and continues further west along the Nepal Himalayas. The entire area of the Shillong Plateau appears to be very active as seen from the epicentres located near its margins and in the central region. Three earthquakes of large magnitude (exceeding 7.0) in the years 1897, 1923, and 1930 have been recorded from this plateau. The 1897 earthquake was the largest, with a magnitude of 8.7 (Gutenberg and Richter, 1954) and the intensity reaching a value of XII (Oldham, 1899). This earthquake was felt over an area of nearly 4.5 million km². The latest large earthquake in Assam took place on 28 April 2021. Having a magnitude of 6.4 on the Richter scale, the epicentre was located at latitude 26.690° N and longitude 92.360° E, about 80 km northeast of Guwahati and lasted for about 30 seconds. Since the focal depth of the earthquake was shallow [at ~17 km depth as reported by the National Centre for Seismology (NCS)], its impact was devastating. Six aftershocks ranging from 4.7 to 3.2 occurred within two and half hours after the first tremor. Many lower magnitude aftershocks followed. Liquefaction, tilting and major cracks in the concrete structures, rockfall from the mountains in the near vicinity, and at least two deaths due to heart attack were reported in the media (Source: https://indianexpress.com/article/explained/

explained-in-assam-earthquake-reminder-of-seismic-hazard-along-hft-faultline-7292770/).

Gupta and Singh (1982) have studied the seismicity and its relationship to faults and lineaments and have noted that all the major lineaments surrounding the Shillong Plateau, namely, the Brahmaputra lineament to the north, Dauki fault to the south, and Kopili fault to the east, are seismically active.

The entire Arakan Yoma fold belt is characterized by high seismicity with the depth of foci gradually increasing towards the east (Verma et al., 1976a, b; Verma and Mukhopadhyay, 1977; Verma et al., 1993). The deepest earthquakes in this belt occurred nearly 250 km to the east of the Arakan Yoma Mountains (Verma and Mukhopadhyay, 1976). It may be mentioned here that this seismic belt continues southwards towards the Andaman Islands into Sumatra, as shown by Verma et al. (1978). Along the Assam syntaxial zone, several thrusts (including Lohit and Mishmi thrusts) trending NW-SE appear to be quite active up to 100 km depth (Verma and Mukhopadhyay, 1977).

Tectonic Setting of the Brahmaputra Valley

The Brahmaputra valley – currently acting as the principal conduit for sediment supply from the continent–continent collision zone triangle having a frontal Himalayan thrust belt as one side, Naga-Patkai thrust belt as the second side, and Dauki fault as the third side – plays a major role in the Assam–Arakan basin fill history of evolution. This is now believed to have a poly-history. This poly-history was guided by the Himalayan orogeny, the raising of the Tibetan Plateau, cycles of glacial and interglacial periods, changes in the rates of precipitation, sediment dislocation, stress pattern changes in the interacting plates, developments of new faults for stress release, the appearance of new structural elements, and local influences exerted by these neo-structural elements.

We have got different plate collision models to explain the Himalayan orogeny; the central task for each of the models is to raise Tibet; an equally important task is to explain consequence of these models on sediment budget and the process of basin evolution. In this section, we consider some interesting aspects of the contemporary debates, open as well as subtle, that influence the geoscientist community until now so that it could be possible to identify the areas of more precise knowledge gaps.

Limitations of 'Geosynclinal Theory'

Mathur and Evans (1964) summarized the geological history of the northeastern region by stating that during the earliest Tertiary times, there was a geosyncline that included the Bay of Bengal, much of Burma, and the eastern part of Assam. They also theorized that along the Burmese arc, at various times, there was a partial separation of the geosyncline into two closely related parts. To the west and northwest of the geosyncline, there was an extensive shelf that, before the Oligocene–Miocene break in sedimentation, includes much of the Eastern Himalayas. After the post-Oligocene movements, the northern part of the shelf area became depressed to form a geosyncline that covered the site of the present Eastern Himalayas. In no time, these geosynclinal basins became very deep, and for the most part, subsidence was matched by sedimentation. This interpretation influenced the later workers tremendously (e.g., Wadia, 1968; Mallick and Raju, 1995; Purkait, 2004). The concept of geosynclines as understood by most geologists was dividing troughs into several tectonic-morphic zones. Four elements were found. A stable foreland on the craton passes laterally into a shallow trough termed the miogeosynclinal furrow, is separated by a positive axis, the miogeanticline, from the eugeosynclinal furrow, and is the main active and unstable trough. The eugeosyncline is separated from the open oceanic basin by a second ridge, the eugeanticline; this second positive feature is an island arc of rising volcanics. The eugeosynclinal basin (now fore-arc) is composed of pre-orogenic pelagic sediments overlain by synorogenic flysch facies. These are generally tectonically deformed and metamorphosed as the fore-arc basin migrates outwards over the trench. The main flysch trough is often hosting minor rifted troughs filled with post-orogenic continental clastics. The former miogeosyncline, with its fill of pre-orogenic shelf sediments and post-orogenic molasse, is now termed the 'back-arc basin'. Where this is developed on the continental crust and is encroached by an active thrust system, it is termed a foreland basin. Now, for any modern sedimentary theory, it must be in a position to explain a characteristic morphology, plate tectonic setting, time of development, and type of fill (Selley, 2000). The 'geosynclinal theory' was a very simplified means to describe observations relating to morphology, time of development, and type of fill, but it was not at all good as a correlation tool. Also, one needed to fix up different types of troughs at several odd places and raise those places at very high altitudes by simply attributing them to 'orogenic forces'.

Implications of 'Pop-Up' Theory

It seems that the upper reach of the Brahmaputra valley is subducting like a wedge, except for its far southwest end where the basement is exposed in the form of the Mikir Hills. A dominant opinion suggests a pop-up mechanism (Bilham and England, 2001) for the Mikir Hills along with the Shillong Massif. There are certain contentious issues related to the resultant landforms which are misfit with the conventional foreland areas of the active convergent tectonics. The 1897 Shillong earthquake (Oldham, 1899), so far one of the bigger intra-plate earthquakes (mentioned before) was studied mainly on the basis of outcomes of geodetic data and the absence of geological evidence (or, perhaps the lack of subsurface geophysical data) raised debates (Nandy, 2017). We have mentioned earlier that on the southern boundary of the Shillong massif, there is a north-dipping high-angle reverse fault, the Dauki. Some authors attributed the origin of the 1897 earthquake to this fault (Seeber et al., 1981; Gahalaut and Chander, 1992). Based on the analysis of geodetic data, Bilham and England (2001) attributed the origin of the 1897 earthquake to the south-dipping reverse fault in the northern boundary of the Shillong massif and proposed the pop-up mechanism to explain the co-seismic rise of the Shillong plateau. Since there was no geological evidence of this fault on the surface, it was further postulated as a blind fault and named after Oldham. The absence of any surface geological evidence of faulting along the northern boundary of the Shillong massif divided the opinions of geologists. Some opinions endorse pop-ups partly by recognizing the presence of transverse tectonics presently active along the northwest-southeast trending Kopili fault in the southern part of the Mikir Hills (Kayal et al., 2012). Based on subsurface evidence from geophysical data, the south-dipping fault was further extended northward by Rajendran et al. (2004, 2005) and instead of relocating the Oldham fault, a new name, the Brahmaputra fault, was coined in conformity with the almost linear flow of the Brahmaputra River along the claimed reach of fault activation. Later on, despite a more detailed explanation from England and Bilham (2015), the controversy and additional questions related to the genesis of the Shillong plateau remained (Mukhopadhyay, 2015). New data sets emphasize the role being played by the north-dipping reverse faults along the southern boundary of the Shillong massif (Morino et al., 2014; Mallick et al., 2020). The pop-up mechanism recognizes an important role for uneven sediment balances, thicker deposition of sediment stacks at two ends of a slab, and increasing stress at the middle zone having thinner stacks of sediments due to the up-arching trend of tectonic origin. Thus, unevenness in sediment budgeting acts rather as positive feedback to accentuate stress which is followed by sudden upliftment, the 'pop-up'. The influence of sedimentary processes on plate tectonics, in the light of the 'pop-up' theory as well as the counter-arguments, is an important area of investigation for the Brahmaputra valley and the adjoining area.

Different Collision Models and Their Influence on the Basin Evolution

Models proposed for the origin of crustal thickening in the Himalayas and SE Asia (Tibet, Tien Shan, etc.) can be classified into two main categories (Figure 2.1):

India–Asia Collision

Indian lithosphere confined to the south of Indo-Tsangpo suture

Indian lithosphere lie (mantle/crust or both) under part of all of Tibet

A. Indentation model

B. Mantle delamination and roll back model (Willett and Beaumount, 1994)

C. Under thrusting of India beneath all of Tibet (Powell, 1986; Powell and Conaghan, 1973; Beghoul et al., 1993; Matte et al., 1997)

D. Injection of Indian crust into Tibetan crust (Zhou and Morgan, 1995; Westerway, 1995)

A(i). Diffused homogeneous thickening, India the rigid indenter (Molnar and Tapponnier, 1975; Dewey et al., 1988)

A(ii). Indentation followed by convective removal of lithospheric mantle (England and Houseman, 1988)

roll back

KEY

☐ Crust | Indian
☐ Mantle | lithosphere
▨ Crust | Asian
▨ Mantle | lithosphere

INDO-TSANGPO SUTURE

INDIA   TIBET   ASIA

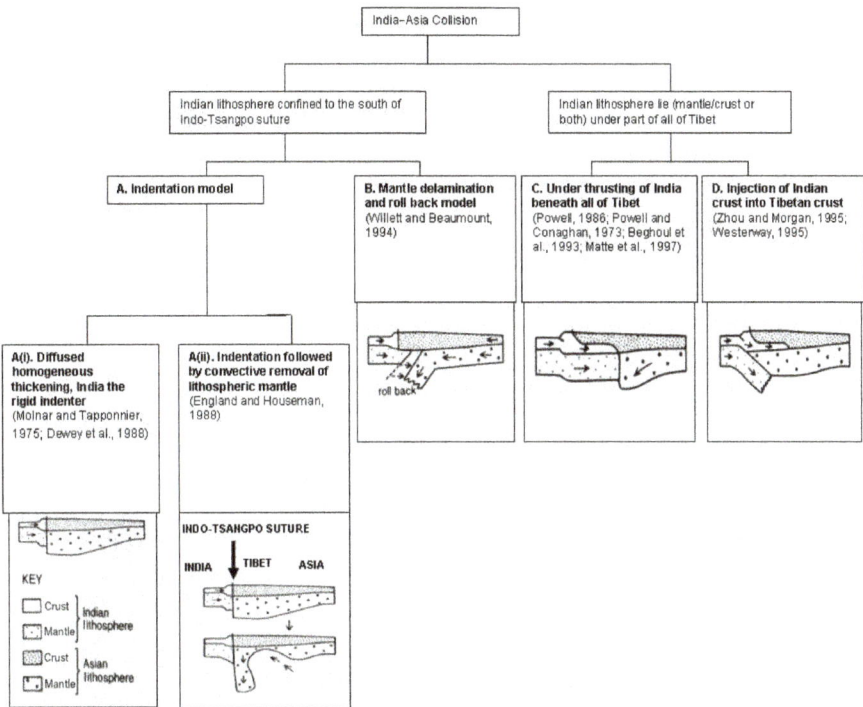

FIGURE 2.1
A generalized representation of India–Asia collision.

1. Those in which the Indian lithosphere is confined to the south of the Tsangpo suture within the Himalayas and

2. Those in which the Indian lithosphere – either the mantle or crust or both the mantle and crust – extends to the north of the suture to lie under part or all of Tibet. This is an oversimplification because few would now maintain that there has been no under-thrusting by the Indian lithosphere. Rather, the controversy centres on how much under-thrusting has occurred. Within each category, there is a great diversity of mechanisms.

Examples of the first category are given as follows:

a. The indentation model, with India acting as a rigid indenter, resulted in continental expulsion (Molnar and Tapponnier, 1975) or homogenous thickening in Tibet and areas to the north (Dewey et al., 1988; England and Houseman, 1988; Platt and England, 1993);

b. Mantle delamination and roll-back (Willett and Beaumount, 1994); and

c. Migratory subduction of the continental lithosphere (Meyer et al., 1998).

Examples of the second category are as follows:

i. Under-thrusting of India beneath all of Tibet (Powell, 1986; Powell and Conaghan, 1973; Beghoul et al., 1993; Matte et al., 1997) and

ii. Injection of the Indian crust into the Tibetan crust, either in the form of a solid 'piston' (Zhou and Morgan, 1995) or as a viscous fluid (Westerway, 1995).

Zhou and Morgan's hypothesis, to provide the 'weak' lower crust, states the need for injection in Tibet presumably to achieve major crustal thickening, a feature which they are attempting to explain.

However, the weakening could have been provided by the injection of magma by Andean-type granite intrusion at the base of the Tibetan crust, thereby increasing the temperature. An Yin and Harrison (2000) have given a recent comprehensive review of the geological evolution of the Himalayan-Tibetan orogen. Another review by Johnson (2002) considers the evidence for the amount and the rate of the convergence between India and Asia taking place after the initial collision of these continents at roughly 55–50 Ma.

Neotectonic Elements

Besides the nature of the seismicity mentioned earlier, which shows the active tectonic activities, there is some geomorphic and structural evidence to prove the same. These are described in the following text:

Geomorphic Evidence

i. Unpaired terraces occupying different elevations of the same age.

ii. The presence of younger sediments in upstream areas, followed by older sediments in some streams.

iii. The number of meander cut-offs, numerous bils and back-swamps indicate the transient nature of the streams. The subsidence of the Brahmaputra valley floor is indicated by these back-swamps.

iv. The rapid southerly migration of the Brahmaputra and the channel metamorphosis support the subsiding nature of the valley floor.

v. Localized reversal of surface water flood and gently upstream dips of sediments.

Structural Evidence

i. The straight course of most of the streams and the drainage patterns like rectangular, subangular, and dendritic, directly show NE-SW, NW-SE, ENE-WSW, and NS trending linear segments within the Quaternary alluvium. They indicate the presence of fracture/faults along with these directions, which are either reactivated during the Quaternary or created due to rupture in recent times. The Kalong and Kopili rivers are aligned along the major lineaments of the area (NW-SE lineament).

ii. The ENE-WSW and NE-SW trending fractures, present in the Archaean Group of crystallines, are seen extending into the alluvium, for example, the NE-SW trending fractures within the Karbi–Anglong crystalline extend into the alluvium where a segment of the Mara Diphlu River is aligned parallel to this fracture.

iii. The ENE-WSW and NE-SW trending fractures/faults within the Archaean Group of crystalline of Karbi–Anglong are found to be still active as they are associated with thermal springs.

Fluvial Dynamics and Morphotectonics

The pioneering work by Coleman (1969) on channel processes and sedimentation due to the Brahmaputra River was carried out with the principal objective to form an understanding of the nature of facies change to be expected in interpreting ancient rock sequences formed out of the deltaic sediments for a basin which was already producing a substantial amount of oil. Goswami (1985) aimed to find a realistic estimation of the suspended load budget from the denudation rate of the Eastern Himalayas and channel aggradation at different reaches of the Brahmaputra valley. A major problem in the alluvial sediments is uncertainty in the formation of the proper capping mechanism of a reservoir. With increasing depth, due to the low value of the reflection coefficient, it becomes very difficult to identify sandbar tops in the seismic survey records. Of course, sandbar thickness can be known by studying electrofacies from the well logs. But these thicknesses can be simply misleading if the nature of their lateral extent is not known. Bristow (1993) studied sedimentary structures exposed in bar tops in the Brahmaputra River, along with the vertical and lateral arrangement of sandbars formed in large braided rivers. The idea was also to correlate the lateral and vertical extents of sandbars. However, in a river basin that belongs to a highly active tectonic regime, the vertical and lateral correlations can again be very deceptive if the tectonic cyclicity is not incorporated with the aggradation–degradation alternation. Rivers are extremely sensitive to subtle changes in their grade caused by tectonic tilting. Holbrook and Schumm (1999) discussed the geomorphic and sedimentary response of rivers to tectonic deformation. They have discussed recognition of tectonic tilting effects on rivers and their resultant sediments, which can be a useful tool for identifying the often-cryptic warping associated with incipient and smaller scale epeirogenic deformation in both modern and ancient settings. Tectonic warping may result in either longitudinal (parallel to flood plain orientation) or lateral (normal to flood plain orientation) tilting of alluvial river profiles. Alluvial rivers may respond to deformation of the longitudinal profile by (i) deflection around zones of uplift and zones of subsidence, (ii) aggradation in back-tilted and degradation in fore-tilted reaches, (iii) compensation of slope alteration by shifts in the channel pattern, (iv) increase in the frequency of overbank flooding for fore-tilted and a decrease for back-tilted reaches, and (v) increased bedload grain size in fore-tilted reaches and decreased bedload grain size in back-tilted reaches. Lateral tilting causes down-tilt avulsion of streams where tilt rates are high and steady down-tilt migration (combing) where tilt rates are lower. Each of these effects may have profound impacts on lithofacies geometry and distribution, which may potentially be preserved in the rock record. They have further added that caution must be exercised when using these effects as criteria for past or current tectonic warping as these effects may be caused by non-tectonic factors. These non-tectonic causes must be eliminated before tectonic interpretations are made.

The early initiation of oil exploration activities, now about 150 years old, has resulted in interesting correlations between the meticulous outcrop studies and surface observations carried out by some pioneer geologists and information derived from deep drilling in different strategic locations of the Brahmaputra valley in Assam.

Despite having a rich archive of subsurface data of sedimentary units extended to the depth of crystalline basement top, the history of basin evolution of the Brahmaputra Basin has many contentious issues which cause differences in opinions among the geoscience workers representing different oil industries due to a common approach of the stratigraphic theory being followed by the majority. It is the highly conservative lithostratigraphic approach which essentially tries to implement Steno's laws (1669) of stratigraphy in a rigid manner. For example, when the sea level falls, coral reefs will form at increasing distances from the ancient shoreline. At a given point in time, we expect the simultaneous existence of three different types of facies: between the shoreline and the coral reef wall-sandstone facies, followed by limestone facies of the coral reef and finally in the deeper sea, shaly facies. With falling sea levels, the locations of this triad will keep on shifting towards the deeper sea. Rigid adherence to the original lateral continuity of rock units suggested by Steno will make the conservative interpreter draw fault lines for the discontinuous limestones or perhaps suggest a basin sag. For the Brahmaputra Valley, particularly in the basin margin areas, the simultaneous occurrence of tectonic dislocations and uneven accumulation of sediments, make the task of correlation difficult due to remarkable lateral variability in lithology.

As geophysical tools are now more mature to interpret thrust belt tectonics, geoscientists can eliminate many ambiguities. Yet, there are many knowledge gaps, particularly in raising multi-order basin evolution models integrating pieces of evidence of the surface, shallow subsurface, and deep subsurface. The present work is a modest attempt to generate new ideas to fill up the existing gaps by adopting a multi-disciplinary approach.

3

Tools and Methods: A Brief Appraisal

For addressing deeper issues in the subsurface as part of landform evolution in different stages of spatio-temporal developments in sedimentary basin analysis, the prime role of geophysical methods was treated as an accepted fact of life by the leading researchers since the 1960s of the 20th century (Koefoed, 1965; Claerbout, 1971; Sheriff 1973; Sangree and Widmier, 1979; Sheriff and Geldart, 1983, 1995; Telford et al., 1990; Cowan & Cowan, 1991; Green et al., 1998). For shallow subsurface stratigraphy too, with the increasing degree of complexity in correlating different strata and the need to monitor the subsurface as closely as possible, avoiding multiple cost escalation at the same time, scientists are endeavouring to apply finely calibrated geophysical tools with the known wells and then undertake as many virtual drillings as possible by making use of different types of geophysical tools (Park et al., 2007; Yadav et al., 2010). As the present work banks heavily on geophysical data, an attempt will be made in this chapter to provide a quick understanding of selective geophysical methods, the data of which will be used in the subsequent chapters along with other methods. It is known to every geologist that intelligent extrapolation of the outcrops can help reconstruct the spatial distribution of the geological bodies in the subsurface (Figure 3.1).

However, for the situations like layered earth models and blind faults, which do not have any surface manifestation, either we use drilling or apply geophysical tools on the surface in direct contact with the earth or use some airborne means above the ground surface.

Some of the geophysical tools are based on natural field-based methods and others on artificial fields. A lateral contrast (Figure 3.2) in common physical properties such as density, magnetic susceptibility, and electrical resistivity of the causative bodies with the surrounding host rocks may influence gravitational or magnetic fields or magneto-telluric current densities on the surface and thereby generating 'anomalies' that can be measured at different stations and compared. These are examples of natural field-based methods.

Unless there is lateral discontinuity, differences in physical properties of the layered earth can show at best a change in the scale factor of the 'zero-anomaly line'. A vertical contrast in physical properties for layers of different rocks stacked over each other may not have anomalies manifested 'naturally' on the surface, but these situations provide us boundaries that might act as reflectors or refractors for the artificial signal sources created on or near the surface, and the reflected or refracted responses can be monitored by properly calibrated geo-sensors of our choice. These are the artificial source-based

DOI: 10.1201/9781003302353-3

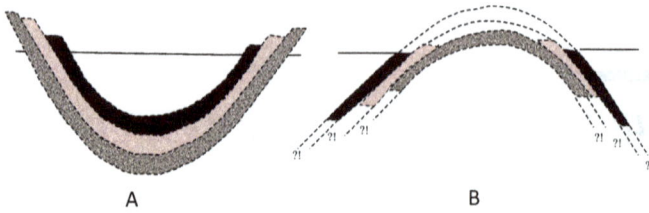

FIGURE 3.1
Problems of outcrop-based extrapolation in the subsurface.

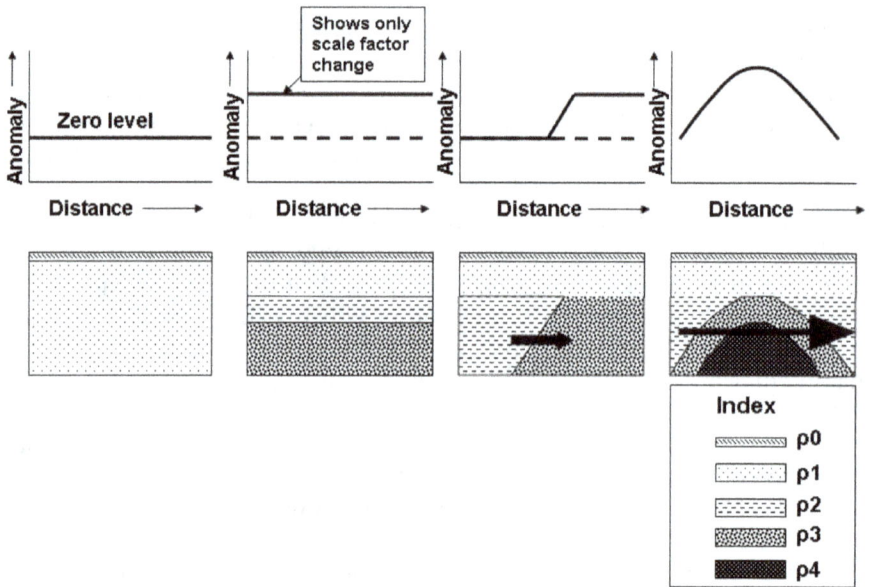

FIGURE 3.2
Idealized geophysical anomaly generated due to the variation in the subsurface geology.

methods, where the earth 'system' is forced to respond (the output) to certain 'input' fields. The assumption is that if the 'input' and 'output' are known, the 'system' characteristics can be predicted faithfully by making use of the communication theory.

In the last few years, the technology has improved remarkably; we are now in a better position to acquire high-resolution data, devise new techniques to interpret the existing pool of data, and also exercise highly improved coordination of different methods to realize the dynamics of the subsurface changes. It is in this context exploration geophysics has started playing a very meaningful role.

Common Geological Problems

In geophysical exploration, common geological problems which pose as a challenge to the explorer can be broadly classified into two categories, structural and stratigraphic. Some of the important *structural* situations are as follows:

i. The folded structures, anticlines and domes, salt domes, anticlines over igneous intrusives, and buried ridges (folds due to uplift and differential compaction of sedimentary layers);

ii. Faulted structures include faults in folded structures and homoclinal structures;

iii. Homoclinal structures cut by igneous dykes, etc.; and

iv. Faults, fissures, joints, etc.

Some of the important *stratigraphic* situations are as follows:

i. Features associated with carbonate reefs;

ii. Permeability barriers associated with erosional truncations such as pinch-outs;

iii. Sand bodies such as lenses or stream channels surrounded by impermeable materials; and

iv. Facies change from permeable to impermeable lithology, etc.

The deployment of methods is governed by the amount of data access. The area is called virgin if the exploration work has not at all been carried out earlier. Also, even wildcat wells were not drilled. The area concerned might have some data, such as airborne magnetic survey data and wildcat well information. A fairly good database for undertaking basin analysis means detailed geological mapping from outcrop studies, gravity–magnetic data, at least a few long regional profiles for 2D seismic reflection data, well logs from a few wells, etc.

Questions of Usual Concern in Geophysical Exploration

- What types of geophysical methods are to be used for the given nature of the geological problems?
- What is the nature and magnitude of geophysical anomaly?
- What is the nature of the basement? Is it flat or dipping? Is it highly fractured? The basin deposits are basement-guided or basement-independent?

- What is the nature of local and regional anomalies? What types of processing techniques are to be adopted for separating local from regional anomalies?
- What are the resolution criteria for the type of geophysical method adopted? How to optimize the performance of the given instrument to adapt it for high-resolution data acquisition?
- What are the physical factors responsible for controlling the anomalies?
- What is the nature of noise? What methods are to be adopted for maximizing the signal-to-noise ratio?
- What types of corrections are needed for the raw field data?
- What is the nature of ambiguities involved with the type of geophysical methods used?

Inverse Nature of Geophysical Problems

When some of the fundamental physical properties of a body are known and its effect at some arbitrary point is supposed to be calculated, this is called a forward problem. In forward problems, the observer is interested in the system response for a given input function. So, the system parameters and the input are known, but the output is unknown.

On the other hand, if for a given anomaly curve, that is, the effect, the possible geometry of a causative body for the given value of the density contrast is to be guessed, the possibilities are many, and the problem is called inverse problem.

So, in inverse problems, input and output are known; the objective is to know the system parameters. Here, the mathematically correct possibilities are many. So, the ambiguities in solutions (Figure 3.3) are inherent. However, by making use of geological intuition, the observer is supposed to choose the appropriate solution.

Effect of Size and Depth of the Causative Body

The size and depth of the causative body have a strong influence on the geophysical anomalies. The chances of detection increase with increasing dimensions and decreasing depths. Particularly for natural source-based methods, instead of plotting anomaly versus distance, plotting normalized anomaly

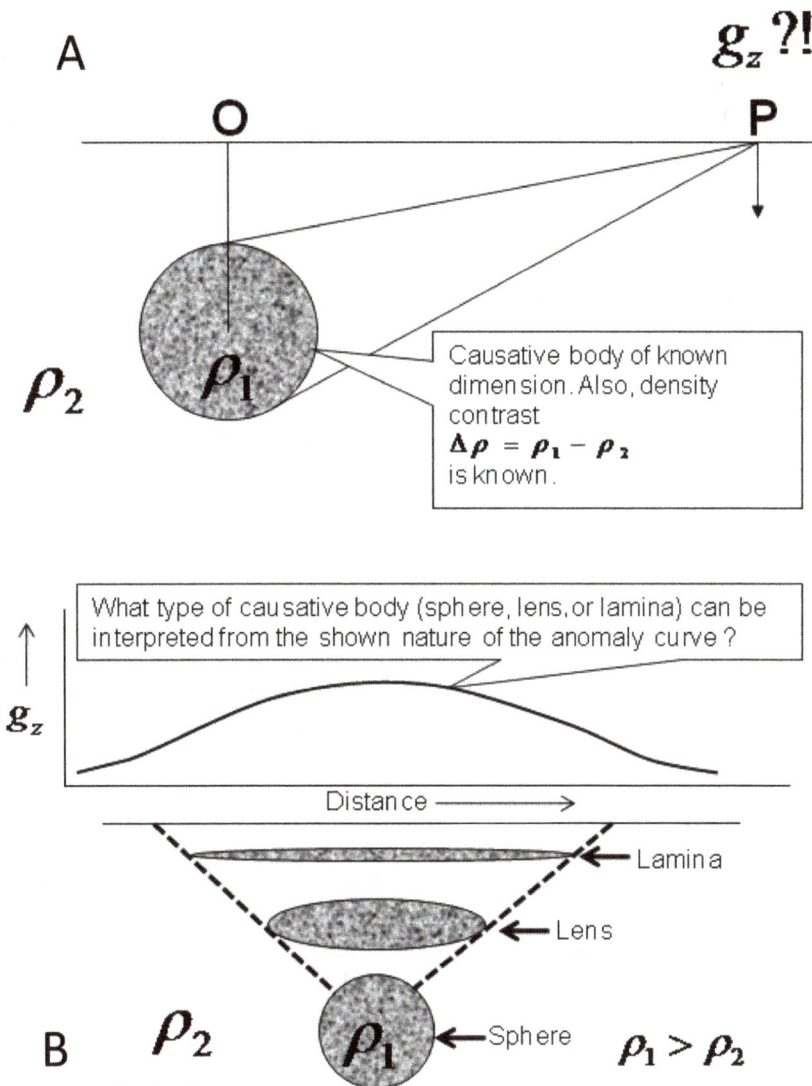

A

g_z?!

O   P

ρ_2 ρ_1

Causative body of known
dimension. Also, density
contrast
$\Delta\rho = \rho_1 - \rho_2$
is known.

What type of causative body (sphere, lens, or lamina) can be
interpreted from the shown nature of the anomaly curve?

g_z

Distance ⟶

←Lamina

←Lens

ρ_2 ρ_1 ←Sphere $\rho_1 > \rho_2$

B

FIGURE 3.3
Qualitative comparison between the forward and the inverse nature of the geophysical
problems.

versus relative distance from the point where the maximum anomaly was
observed, and the overburden thickness yielded a more rational representation.

By 'depth of a causative body', the sense that comes out is the depth to the
centre of mass from the ground surface. But, from the drilling point of view,
it is the top of the causative body that matters. For mineral exploration having a shallow target depth, it is advantageous to calculate the depth to the

centre of mass. However, for knowing the approximate nature of variation in the sediment thickness, where the target depth is much deeper, the depth to the top of the causative body is usually preferred. For example, Peters' half slope method in magnetics makes use of the concept by which the depth to the top of the basement is found.

Planning Geophysical Investigations

In geophysical field operations, either route survey or grid survey is used. What is to be adopted is mostly determined by the nature of terrain, accessibility, level of logistic support available, degree of accuracy needed, and, of course, the method itself. For example, in gravimetric surveys in difficult terrain, conducting grid surveys might appear a very difficult proposition. In such situations, the best alternative is to increase the data density by following, of course, the route survey method that should be substantiated with very accurate GPS data. However, this method cannot be applied to the seismic reflection method. For the 2D seismic reflection technique, terrains with rapid dip variation and high magnitude of dip are not suitable. Sometimes, to carry out the seismic reflection method in hilly terrain, crooked lines are shot, which need much longer computer time and tedious processing of the raw data.

The stages of geophysical investigations depend on many factors. Starting from preliminary measurements in areas where surface or subsurface geology is not known to the final stages of geophysical operations, there are broadly four categories: (i) reconnaissance surveys, (ii) semi-detailed surveys, (iii) detailed surveys, and (iv) prospecting surveys (Bhimasankaram, 1977).

   i. **Reconnaissance surveys**: The fundamental aim is to study the broad-based regional features in the overall tectonic context and to find out the areas where the investigation work is to be focused at a later stage. The mode adopted can be either airborne or ground-based. In airborne surveys, magnetic, gravimetric, or electromagnetic methods are usually preferred. In the ground-based mode of data acquisition, besides gravity and magnetic methods, the magneto-telluric method is also preferred these days.
   ii. **Semi-detailed surveys**: Promising zones outlined by reconnaissance surveys are further detailed. The aim is to geologically map the region in greater detail and demarcate the anomalous zones with the required degree of precision. Almost all types of ground geophysical methods find application in this stage of the investigation.
   iii. **Detailed surveys**: The aim is a complete delineation of particular anomalous zones to establish the possible occurrence of objects of interest; to determine quantitative parameters like depth, physical

TABLE 3.1

Scale Factor Showing the Range of Geophysical
Surveys at Different Stages

Scale Factor in Geophysical Surveys

Stages of Geophysical Survey	Scale Range
Reconnaissance	1:1,000,000–1:100,000
Semi-detailed	1: 50,000–1:10,000
Detailed	1:10,000–1:1000
Prospecting	1:1000–1:500 etc.

properties, size, and dimensions of objects to arrive at the economic potentialities of the expected deposits; and to finally recommend suitable locations for test drilling. In this stage of investigation, apart from the regular grid surveys, a few special profiles are sometimes laid in desired directions with closely spaced stations with a view to precisely trace the anomalous zones of quantitative interpretation.

iv. **Prospecting:** This is mainly meant for exact spatial delineation of the causative body before drilling and also as a lead during mining activities. For example, micro-gravity survey and very close grid dc electrical survey to identify faulted blocks in coal seams (Table 3.1).

'Noise' and the Problems of Enhancing 'Signal'

Any event on the geophysical data record from which information relevant for understanding geology is wished to obtain is a signal.

All unwanted events on the geophysical data record are termed as noise, including coherent events, which interfere with the observation and measurement of signals. Geophysical noise can broadly be divided into three types:

i. Passive noise,

ii. Active noise, and

iii. Instrumental noise.

When the terrain is highly rugged or some shallow localized inhomogeneities are present, the prospective zone becomes masked by those influences which are static and are time-independent factors and thus called passive noise.

Those noises which are time-dependent, for example, natural earth currents of different origins, and microseisms, which influence the field data adversely are called active noise.

Instruments that are supposed to monitor a particular anomalous field suffer from typical inherent problems that usually appear as background noise. At times, these might confuse the observer. For example, as different parts of a gravimeter age, rods and fibres slowly bend, and the springs gradually stretch. These slow mechanical alterations cause the gravimeter dial reading to change with time. This is called 'drift'. Occasionally, an abrupt change of several tenths of a milligal or more may occur in the gravimeters, which is called a 'tare'. Some gravimeters undergo a tare once every few months, whereas others never change in this way.

Resolution

Resolution is defined as the minimum distance between two features so that one can tell that there are two features.

In subsurface studies, the explorer is always highly concerned with vertical resolution and horizontal resolution. In any type of artificial source-based method where the input signal is a wave or a wavelet, the minimum thickness of a bed to see the effects of the top and the base of the bed as distinctly separate is the vertical resolvable limit, and it is about one-fourth of the wavelength (Figure 3.4).

On the other hand, the minimum thickness for a layer to give a reflection is called the detectable limit and is of the order of one-thirtieth of the

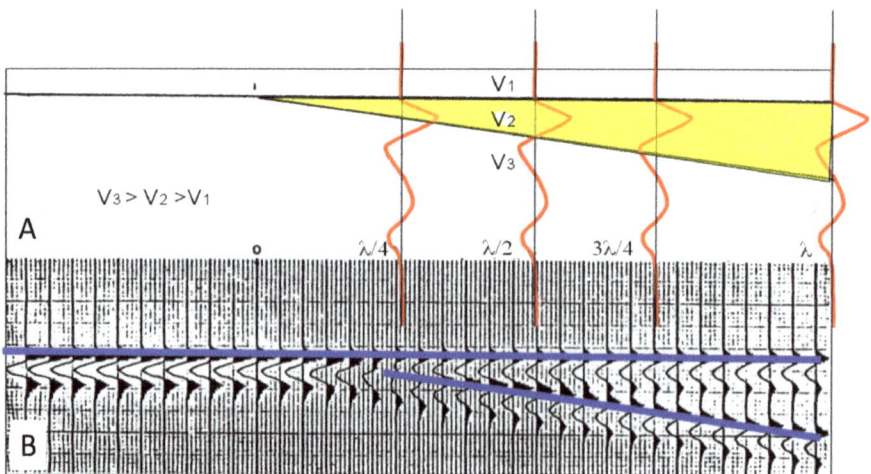

FIGURE 3.4
Vertical resolution.

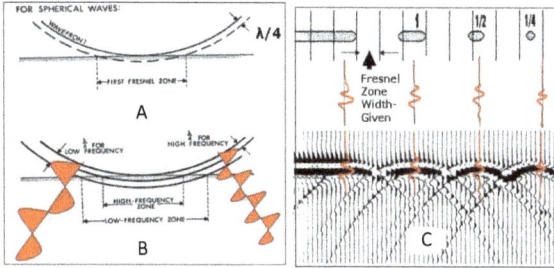

FIGURE 3.5
Horizontal resolution.

wavelength. One important point to remember is that seismic resolving power is only one-hundredth of that of well logging (Sheriff, 1982). For horizontal resolution (Figure 3.5), if the reflector is larger than about one Fresnel zone, the reflection shows the shape of the reflector, whereas for small reflectors, the arrival time patterns are almost identical.

The Fresnel zone is the portion of the reflecting surface from which energy returns to a geophone within a half-cycle after the reflection onset. Energy reflected from the Fresnel zone interferes constructively and builds up the reflection; this is called the Fresnel zone effect. When the objective is to search out stratigraphic traps, a high-resolution survey is a must. The earth acts as a low-pass filter. Accordingly, as the depth increases, the seismic spectrum is mostly dominated by the low-frequency components.

To arrest the fast deterioration of the incident waves in terms of their high-frequency components, explosives are invariably blasted below the weathered zone and at the very initial stage; experiments are conducted for the optimization of various field parameters. These measures help ensure a high-resolution survey at the desired depth.

Data Reduction

Raw field data collected should go through different types of corrections, which is called 'processing'. Data reduction is the conversion of a set of measurements into another set, which is more useful.

Usually, the data reduction involves the steps shown in Table 3.2.

Sometimes, to locate anomalies means – after undertaking different types of corrections – separation of local anomaly (also called 'residual anomaly') from the regional anomaly. The geophysical anomaly obtained is the cumulative effect of several subsurface inhomogeneities. However, for practical

TABLE 3.2

Geophysical Data Reduction Scheme

- Examination for bad data:
    Failure of instruments,
    Variations beyond allowable tolerance,
    Disturbances from outside.
- Resolving location:
    Locating measurement points,
    Adjustments to tie,
    Preparing base map.
- To collate various measurements:
    Application of appropriate corrections,
    To mark line intersections,
    Preparation of profile plots.
- Preparing map:
    Posting of data on map,
    Contouring with geologic sense.
- Selection of anomalies for study:
    Locating anomalies,
    Selection of portions of data free of interference,
    Making appropriate measurements,
    Comparing measurements of same features,
    Attributing significance to anomalies,
    Assigning grading factors
- Preparation of interpretation map:
    Posting data,
    Reconciling the inconsistencies,
    Contouring using geologic sense.
- Estimating reliability and attribute significance:
    Preparation of report with recommendations

purposes, we usually try to assign only two terms, namely, a regional and local anomaly, to the entire spectrum of anomalies. The regional anomaly occupies a relatively large territory with relatively large intensity and reflects the deep-seated large structural feature. On the other hand, local anomalies occupy smaller areas and reflect the shallow localized objects. Thus, if one is interested in the local bodies (such as ore deposits and small structures), it is necessary to identify the local anomalies against the background of large regional trends. On the other hand, if the purpose is to recognize the deeper structures, the local anomalies are to be filtered out.

Data Presentation

For different types of geophysical methods, the mode of data presentation has some common features and differences. It is very common to draw the anomaly *contours* and *profiles* passing over the 'highs' and 'lows'. Depending upon the nature of physical property differences of the causative bodies and the host rocks and the geometry, the profiles present different types of signatures, which are compared with the standard theoretical curves. Sometimes, instead of plotting the anomalies, if the derivatives of the fields are plotted, the causative bodies present a much sharper look.

In seismic reflection methods, the most widely used tool for oil exploration, during data acquisition, two-way travel times and the shot–receiver offset distances are plotted in the raw field records. Later on, on the subsurface reflectors, data for the common depth points are sorted out, corrected, and then stacked, and a section is prepared where reflectors are shown in terms of two-way vertical reflection time values. These are called *time sections*. If the subsurface velocity variations and their exact values are known, and a simple multiplication of two-way time and velocity, divided by '2', would result in *depth sections*. However, the process is not that easy due to many reasons. A very elaborate velocity analysis, detailed monitoring of the density variations from well-log data, preparation of synthetic seismograms, identification of 'marker horizons', and so many other information need to be integrated. It becomes more complex for 3D data. So, time slices are preferred for most of the geological interpretations, except for drilling where there is no alternative to depth sections.

Sometimes, the presentation of certain *attributes* like amplitude, instantaneous frequencies, and phase variations facilitates revealing some of the very interesting features, especially the lateral changes along with the bedding such as those associated with the stratigraphic changes or hydrocarbon accumulations. To have those, the normal seismic data are subjected to complex trace analysis.

Geological Objects and Their Geophysical Generalizations

To simplify the geological problems by adopting easier mathematical calculations for anomalies due to subsurface causative bodies, simple geometric shapes are replaced (See Table 3.3).

The error involved in the aforementioned types of approximations does not usually exceed 5%–10%.

TABLE 3.3

Types of Geological Objects and the Simplified Geometrical Shapes

Types of Geological Objects	Simplified Geometrical Shapes
An isometric ore body having its three dimensions of the same order	Sphere
An ore body with limited thickness and lateral dimensions twice or thrice the thickness	A lens or a limited tabular body
An anticline or a graben	Horizontal cylinder
A vertical intrusive of a limited areal extent or a salt dome	A limited vertical cylinder
Volcanic lavas or sills of a large areal extent	Infinitely extending horizontal slabs
Volcanic lavas or sills of a limited areal extent	Limited horizontal slabs
Dykes or veins	Vertical or dipping sheets

Geophysical Data Integration

Integration of two or more geophysical methods becomes necessary on many occasions. For example, in MT methods for basin evaluation, the usual practice is to proceed from a base station, the subsurface information of which is known. The subsurface information of the subsequent satellite stations is interpreted by referencing the base station. To know very accurate information at the base station, it is essential to start preferably from a known well, log data of which are available. For the interpretation of total field magnetic data by using inverse methods, direct determination of some parameters of the source is possible from the measured data provided certain other parameters are given – for example, source geometry can be calculated by linear programming if the distribution of magnetization is given. On the other hand, if the geometry of the source is given, magnetic distribution can be calculated by using the matrix method. To conduct gravity and magnetic methods jointly is a very old practice. From gravity–magnetic methods, approximate basement mapping is possible, which in turn helps select the field parameters for the regular seismic work. For shallow subsurface studies, it is very common to use refraction and electrical resistivity methods jointly. Integration of geophysical methods does not mean simultaneous use of as many methods as possible. Only a proper understanding of the geological problems and correct appreciation of different geophysical tools can result in a judicious integration. The employment of integrated geophysical methods would be necessary for the solution of geological problems for the following purposes:

  i. To solve different aspects of the problem in full to obtain, as far as possible, a complete picture of the subsurface objects under study;

ii. To eliminate or at least minimize the theoretical and practical ambiguities inherent in the interpretation of data of a single geophysical method;

iii. To pick out useful anomalies belonging to objects of interest from a multitude of anomalies obtained during exploration by using any particular geophysical method and under suitable conditions to recognize the geological nature of the objects based on the different physical properties determined during an integrated geophysical survey; and

iv. To eliminate or at least reduce the effect of geological noises, thereby enhancing the signal-to-noise ratio and picking out weak useful signals.

Gravity Data

During the initial stages of oil exploration, several large oil fields, located mainly under prominent structural highs (anticlines) or associated with structural lows (salt domes), could be discovered by using gravity methods. With time, in comparison to that the seismic methods, the frequency of gravity method use reduced considerably. However, due to the far cheaper rate of data acquisition, processing, and interpretation than its seismic counterpart, as well as its success in the logistically difficult terrains, the method survived. In the virgin areas, where mapping the basement and understanding the gross sediment depositional patterns is the primary target, gravity and magnetic methods are still the first choice. Besides these, with increasing intensity in understanding valley-range problems, where low-angle reverse faults play a very significant role as basin modifiers, gravity methods with enhanced sophistication of interpretational techniques have contributed a lot.

Bouguer Anomaly

Gravity data used for basin analysis and in understanding the problems associated with basin margins are Bouguer gravity data. Field data collected during gravity measurements include cumulative effects due to the gravitational pull towards the centre of the earth, all the inhomogeneities associated with the real earth over a standard datum (maybe the sea level), and the contribution due to the rotational movement of the earth about its axis. When the observed field data are corrected for all the elements (latitude, elevation, and terrain), except the 'anomaly' caused by the lateral inhomogeneity in the

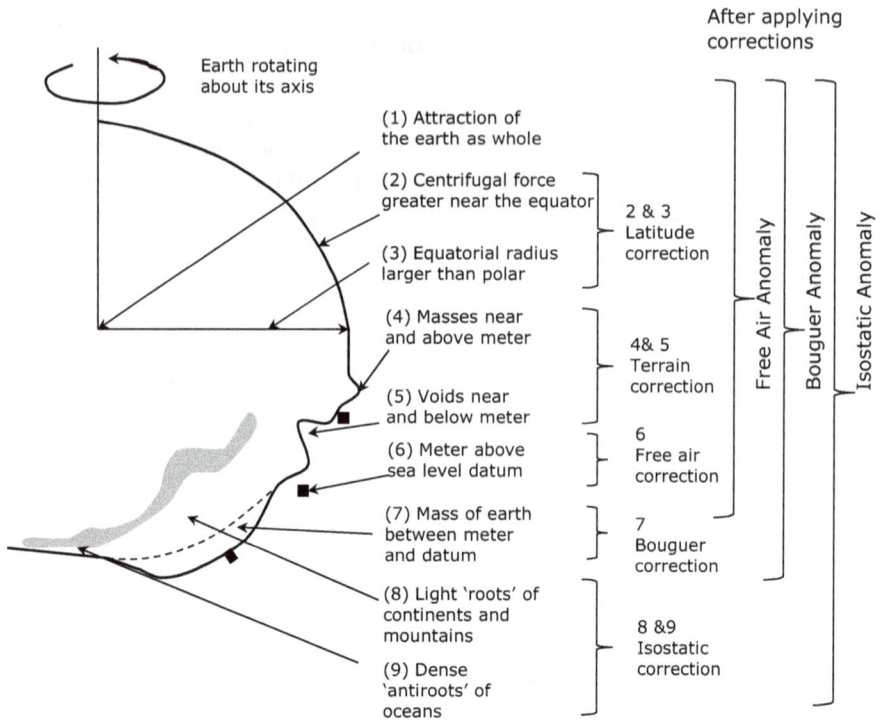

FIGURE 3.6
Different types of gravity reductions (from Sheriff, 2002).

subsurface below the standard datum, the data are called Bouguer gravity data or simply the Bouguer anomaly (Figure 3.6).

Thus,

$$\Delta g_B = g_{obs} - (g_N - \text{Free air correction} + \text{Bouguer correction} - \text{Terrain correction})$$

While making basin-confined observations, we do not bring into consideration the effects due to isostasy.

Here,

Δg_B is the Bouguer anomaly;

g_{obs} is the gravity reading taken by the gravimeter;

g_N is the normal gravity value calculated as per the 1967 Geodetic reference system formula (GRS67) for a given latitude (φ) and is expressed as follows:

$$g_N = g_e + g_e.f(\varphi) = g_e\left(1 + \alpha \sin^2 \varphi + \beta \, \sin^4 \varphi\right)$$

Here, g_e is the gravity value on the equator (= 978.031846 gals);

  $A = 0.005278895$; and
  $B = 0.000023462$.

These constants take into consideration the following standard values for the equatorial radius (R_e), polar radius (R_p), flattening (f), and angular frequency of rotation of the earth (ω):

$$R_e = 6,378,160 \text{ m}$$

$$R_p = 6,356,774.5 \text{ m}$$

$$f = 1/298.247$$

$$\omega = 7.2921151467 \times 10^{-5} \text{ rad/sec}.$$

For the given elevation (h) of a place expressed in meters, the amount by which the point of observation moves away from the centre of the earth as if suspended above the mean sea level (MSL) in free air at the given elevation, the gravitational pull decreases inversely by the square of the distance term, is negative, and is equal to $0.3086\,h$ milligal. This is known as the free air correction.

However, in actuality, there is mass between the gravimeter and the MSL, which strengthens the downward gravitational pull. This additive correction is called Bouguer correction, which is equal to $0.04193\,\rho h$ milligal. For practical purposes, ρ is the density contrast between the causative body and the host rock.

Regional–Residual Separation

The Bouguer anomaly data at a place are the combined effect of anomalies of a different order. Separation of regional from residual components of gravity from the Bouguer anomaly data is of key importance for the types of problems dealt with. According to the needs of the circumstances, the anomaly range is defined and is separated from the observed data.

For example, a basin has a fractured basement and salt domes on it (Figure 3.7). The geometry of the salt domes will not follow the basement configuration because their genesis is independent of each other.

The basement is constituted of hard crystalline rocks having higher density, whereas the incompetent salt is of much lower density than that of its

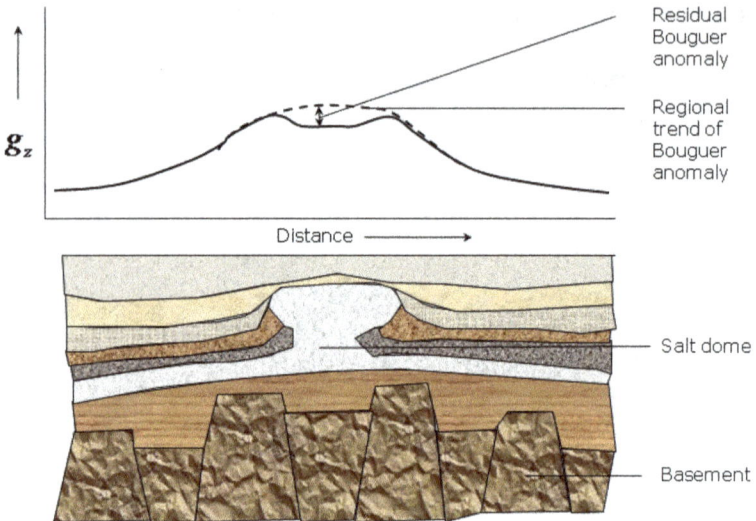

FIGURE 3.7
Bouguer gravity anomaly.

surrounding. The gravimeter placed on the surface will read the combined effect of both. Here, the basement will constitute the regional gravity effect, and the salt will constitute the local or residual gravity effect. Unless these effects are separated from each other, it is not possible to study these phenomena independently.

The Bouguer gravity anomaly data used in the present work are from the seismo-tectonic map of the Geological Survey of India published in 1994.

Seismic data analysis and basic principles of interpretation

The 'Seismic Data'

'Seismic data', in general usage, mean the two-way travel time (TWTT) values of the reflected P-waves at regular intervals on the surface. P-waves (primary or push/pull-type waves) are the faster moving body waves showing properties similar to those of the sound waves with to and fro particle movement, with direction identical to the wave propagation direction, and capable of moving through solids and fluids. The outcome is thus a time versus distance (t-d) record (Figure 3.8).

The artificial seismic signals (input) are being generated at pre-determined points (shot points) by impulsive sources like blasting of dynamites, weight dropping, or simply hammering the earth in the land surveys and 'pressure pulses' in the marine surveys.

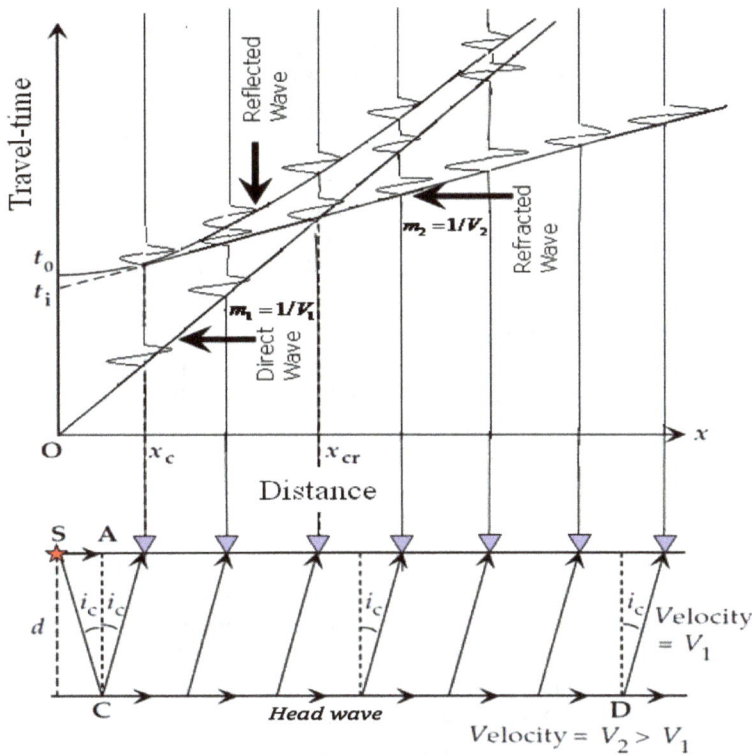

FIGURE 3.8
Seismic wave geometry.

The seismic signals can also be generated in a controlled manner in the form of 'sweeps' by the systems called vibroseis, where the input frequency band is fixed. In theory, to obtain the vertical reflected signals (outputs), the geo-sensors (receiver point) should be placed at the same positions as the shot points. The output signal strength is mainly governed by the reflection coefficients of the stratal boundaries, which in turn are determined by the differences in the acoustic impedances (= Density of the medium × Velocity of the P-waves through the medium). The common assumption for the multi-layered earth is that all the bed boundaries are horizontal. The method is also suitable for those situations where the beds are having a constant dip and the amount of dip is not more than 5°. The method is not suitable for highly variable dip conditions or highly faulted zones, normal faulting, and reverse. For the dipping beds, the usual practice is to shoot the lines along the down-dip direction, which means when the shot points move towards the direction where sediment thickness keep on increasing, receivers trail behind the shots. The arrangement where all receivers lie behind a shot, is called 'end on'. When shots are blasted between the receivers, the arrangement is called a split spread which can either be symmetrical or asymmetrical.

A few cross-profiles are also taken to 'tie-up' the entire data set in the form of 'fencing'. So, we can have data either in the XZ plane or the YZ plane (hence, 2D). As in the earlier days, the custom was to receive only the vertical component of the reflected P-waves, where the receivers used to trail behind the shot points; this data acquisition system is 2D1C. However, by receiving the signals on several parallel lines simultaneously for shots taken successively on an independent line, we get subsurface coverage, not for a common depth point (CDP) but rather for small areas called 'bins' with spatial overlapping; this is 3D1C. For monitoring the status of the reservoirs when we conduct 3D1C from time to time, it becomes 4D1C. Special directional geophones can be used to monitor two other components, namely, SV and SH waves. Then, it can either be 3D3C or 4D3C. For highly specialized problems, at the shot point, three directional shots could be taken. Then, altogether, if the time component is added, it becomes 4D9C.

Noise Reduction

A major problem in the seismic methods is to identify 'signals' from an assemblage of different types of 'noise'. Noises are unwanted signals that can be coherent and incoherent. A great challenge before the data acquisition practices is to either remove or suppress the noise and enhance the signal and thereby enhance the S/N ratio as much as possible. This can be carried out by increasing the number of geo-sensors (Figure 3.9). For any given point in the subsurface, the data are collected several times (multiplicity or fold) for different sets of shot–receiver arrangements having different offsets. Multiplicity is expressed by the following formula:

$$M = \frac{n}{2}\left(\frac{R}{S}\right) \text{ for } R \leq S$$

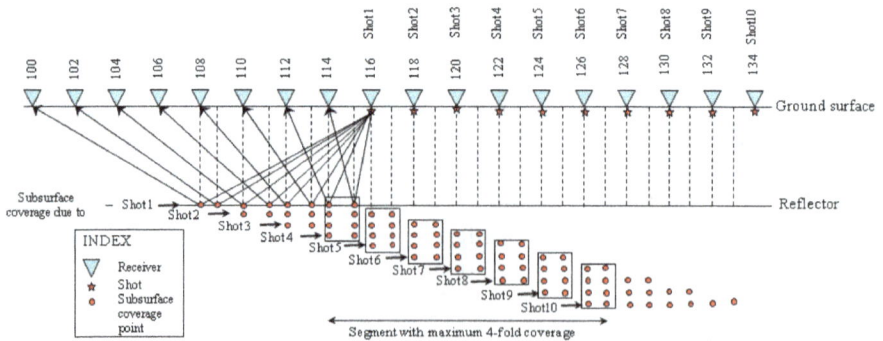

FIGURE 3.9
Multi-fold seismic reflection data coverage of the subsurface.

Where,

    n is the number of channels;

    R is the receiver interval; and

    S is the shot interval.

If the shot interval is less than the receiver interval, multiplicity and, subsequently, the signal quality do not change; rather, the closeness of the sampling points increases.

The S/N ratio improves by a factor $\sqrt{n}$, where n is the number of channels or geophone groups (geo-sensors). For a horizontally stratified multi-layered earth, it can be safely assumed that all the reflection points from various boundaries are aligned below the common midpoint for a particular shot–receiver set having some offset distance in between the two (Figure 3.10). If we can sort out reflection data for a given subsurface point (CDP) obtained by using different shot–receiver offsets (Figure 3.11) and then subtract the additional time (normal moveout correction, NMO) spent by different ray paths over the normal reflection (i.e., zero-offset reflection), then all the data can be treated as equivalent to normal reflection.

Additive effect (stacking) of the NMO-corrected data for a given subsurface point helps cancel out-of-phase and incoherent noises and strengthens the signal, thereby bringing a huge improvement in the S/N ratio.

Two-way zero-offset travel time can be very easily read from the t-d curve. Additionally, NMO correction helps straighten the reflected events which are hyperbolic in nature and thereby treat the information derived from the seismic data as equivalent to the lithological boundaries.

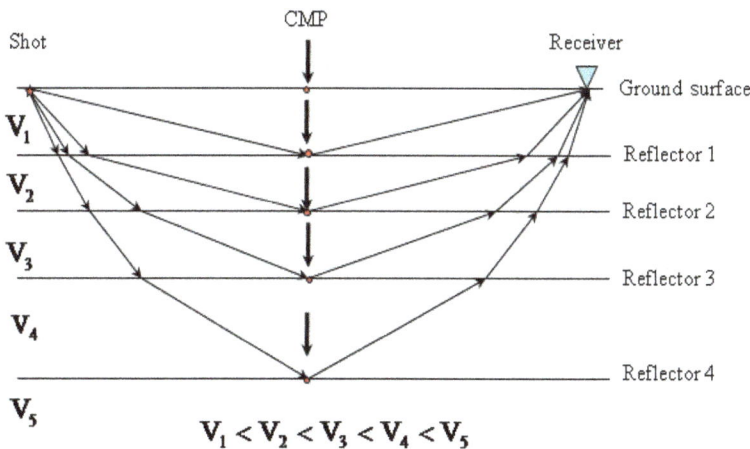

FIGURE 3.10
Understanding of common midpoint data coverage.

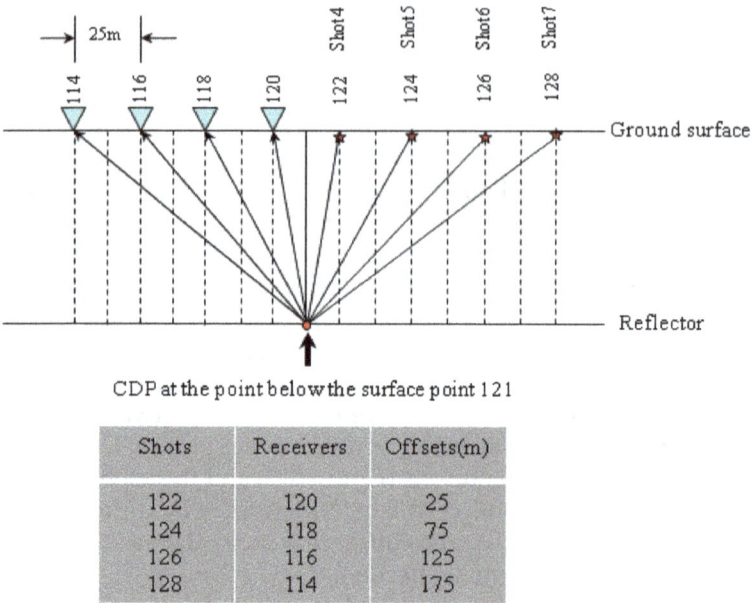

FIGURE 3.11
An example showing four-fold CDP data acquisition. CDP, common depth point.

The processing of data involved up to this stage where the NMO-corrected data sets are stacked is called 'routine'. The introduction of the CDP technique brought a dramatic change in the history of seismic methods of geophysical surveys. People have started comparing seismic facies as almost equivalent to lithofacies, which may, of course, lead to serious misgivings. For the dipping beds and the situations where out-of-the-plane reflections influence the data, the generally 'assumed' reflection points differ from the 'actual' reflection points. As a result, a fictitious migrated surface gets constructed, which needs to be 'relocated' by a special class of special processing techniques called 'migration'.

2D Seismic Data Interpretation

The processed 2D seismic data are similar to a series of virtual deep drillings carried out vertically at short regular distance intervals chosen based on the horizontal resolution needed for the size of the causative body to be identified with the difference that the subsurface locations of different interfaces are mostly expressed in terms of TWTT (time section), rather than depths. If the velocity analysis is very accurate, the 'time section' can be converted to the 'depth section'. But the general practice in the industries is to work with the 'time sections' along with actually drilled logs procured at strategic locations.

In the aftermath of the late 1970s, with the publication of the *AAPG Memoir 26* (Payton, 1977), seismic stratigraphy developed as a very important tool to understand sediment deposition characteristics in the subsurface. Besides locating structural traps, it became possible to identify stratigraphic traps from the seismic sections. Some of the general but crucial aspects of reconstructing sedimentary facies from the seismic facies are as follows:

1. From the law of original horizontal continuity, it is generally expected that the amplitude 'standout' along a horizontal or near-horizontal boundary, representing the meeting surface of two lithological units can be traced regionally by joining either the 'crest' or 'trough' of the reflected wave cycle. But in many situations, this condition does not remain true. The reasons are usually related to either sedimentology or some of the post-depositional structural changes. For example, if the nearshore stratal surfaces thin out, these may look like suddenly terminating surfaces top lapping against another surface in seismic sections, which in actuality is not true. This is simply due to the problem of vertical resolution. On the other hand, a highly faulted basement will show several discontinuous segments. Thus, 'marking' a horizon is always guided by some a priori understanding of the genetic factors associated with the basin evolution.

2. A 'marker horizon' is always preferred as the first choice to mark that has already been confirmed either from the outcrops or from the drilling information, as well as the open-hole wireline log indications.

3. 'Intervals' with similar geometrical characteristics are identified, which are sometimes guided by the 'lap-out' patterns of a set of beds against a surface (Figure 3.12).

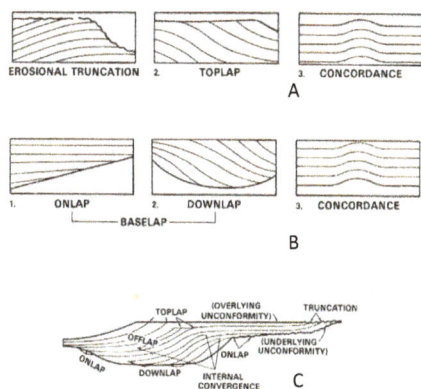

FIGURE 3.12
Different types of lapping characteristics of stratal geometry.

4. The help of 'synthetic seismograms' is taken to map prominent bed boundaries and the conversion of 'time sections' into 'depth sections'. A synthetic seismogram is the vertical response graph of a representative wavelet for a 'model earth' constructed from the borehole information of the density and sonic log data. Placing the synthetic seismogram over the processed seismic section at the place of the drilled borehole where the logging was performed, the lateral continuity of a bed boundary can be traced very efficiently.

5. The next stage in the conventional lithostratigraphic studies is to map thickness variations of different layers (or 'intervals'). Accordingly, 'isopach' maps are prepared.

6. Extending the studies on a regional basis, sometimes different potential reservoirs like fans, channel bars, and pinch-outs, provided proper capping mechanisms exist, are mapped. These mappings have two tasks – identification of the lateral extent of the causative body and the spatial variation of the thickness of the same body.

7. It is usually possible in the seismic sections to differentiate marine sediments from the fluvial ones. Marine sediments thin out towards the coast. Accordingly, ancient coastlines can be mapped.

8. Besides mapping siliciclastic sediments, limestone beds and different types of diapirism can also be identified in seismic stratigraphy. For limestone beds, there are situations where lithofacies differ considerably from the seismic facies.

9. As the displaced beds (about a vertical or a near-vertical plane) can be mapped independently, the fault lines can be drawn as a natural corollary. Comparing the bed thickness variation from the bottom to the top about the fault plane, growth faults and blind thrusts can be identified.

Shallow Subsurface Geophysical Field Data Samples

Uphole Data

The uphole survey is conducted for shot hole depth optimization before the routine seismic data acquisition for a deeper investigation of the subsurface so that the explosive charges can be loaded below the low-velocity layer (LVL) to maximize energy penetration and retain the high-frequency components as much as possible. From uphole survey data (Figure 3.13), the objective is to find the thickness of the LVL and the velocities of the weathered and sub-weathered zones (Figure 3.14) at regular intervals and thereby mapping the lateral thickness variation of the LVL. Most of the oil exploration companies

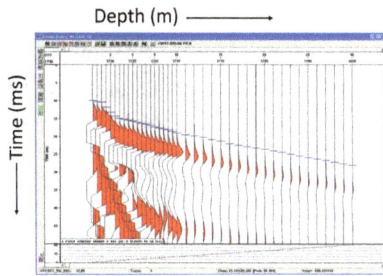

FIGURE 3.13
Picking of signal in the uphole data.

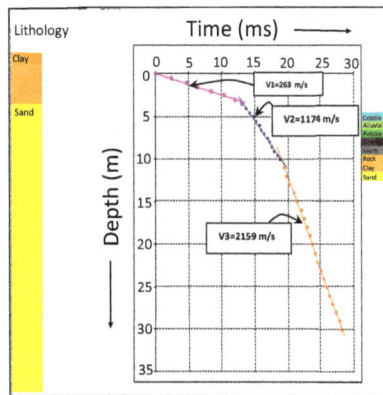

FIGURE 3.14
Plotting of the uphole data.

preserve the uphole data because the uphole data are also needed for the static correction that involves placing the shot and the receiver on a standard datum plane. Uphole survey data can be a good tool for shallow subsurface stratigraphy. By mapping the lateral variation of velocity of the LVL and the sub-weathered layer, the degree of alluvial consolidation can be separated, which in turn can help distinguish older flood plain deposits from the channel fills (Table 3.4).

Shallow Refraction Data

In shallow refraction investigations, several geophones are placed at different offset distances from the shot point. For the present purpose, the *'reversed-catching-up time–distance plot acquisition method'* was used with an irregular pitch between the channels. The aim of thickening at the beginning and the end of the spread was to more accurately determine the velocities of the first layer (Figure 3.15).

TABLE 3.4

Channel Distribution Pattern Followed for the Shallow Refraction Survey

Channels	1	2	3	4	5	6	7	8	9	10	11	12	13	14	15	16	17	18	19	20	21	22	23	24
Distance between channels (m)	0	2	4	9	14	19	25	29	34	39	44	49	54	59	64	69	74	79	84	89	94	99	101	103

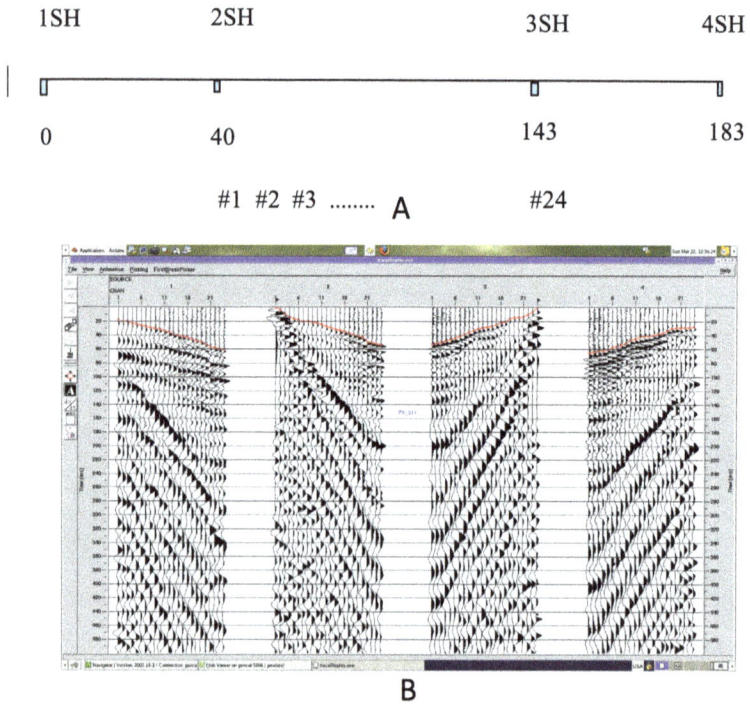

FIGURE 3.15
Picking of refracted events from the field records.

Remote Sensing and GIS

To monitor and measure the temporal changes in the landscape and comparison of the topography of different times, we could collect topographic maps of two different times. The earliest topographic map available was prepared during the 1912–1926 seasons having a scale of 1:253,440 (i.e., 1 inch=4 miles). The second one is the 1976 topographic map of the Survey of India having a scale of 1:2,50,000. We could also collect IRS-P6-LISS-3 images acquired on 15 December 2005 with a spatial resolution of 23.5 m and IRS-P6-LISS-3 image, taken on 27 January 2007 having the same resolution. The standard methods of digitization, image-to-image registration by selecting proper ground control points, and ground verification were followed using ERDAS IMAGINE

8.5 software for co-registration of all data. Different thematic maps including geomorphology and structural aspects of the study area were integrated with images and topographic maps in a GIS environment by making use of ArcView software.

Data Used

As the study area chosen was quite large and the principal objective was to integrate fragmented studies into a broader framework of basin scale, we had to depend on the subsurface data bank of one of the premier oil companies, OIL.

For understanding the temporal changes in the landscape, we had to depend on secondary sources like topographic maps, satellite images, and Shuttle Radar Topographic Mission (SRTM) data.

For shallow subsurface stratigraphy and the study of neotectonics in the foredeep areas of the south bank of the Brahmaputra valley, shallow refraction and uphole data sets were used extensively for the first time in four numbers of strategic windows (Table 3.5).

TABLE 3.5

Types of Data Used in the Present Work

Sl. No.	Data Type	Details	Scale/Resolution
1.	a. Topographic maps (1912–1926) b. Topographic maps (1976)	83M,83I,83J 83F,83B,78N,78J,78G, 78H,78D collected from Office Library and Records, London, UK 83M,83I,83J 83F, 83B, 78N, Survey of India map from the Dept. Of Applied Geology, Dibrugarh University	1: 253,440 (that is, 1 inch=4 miles) 1:2, 50,000
2.	Satellite imagery	IRS-P6-LISS-3 images acquired on 15 December 2005 IRS-P6-LISS-3 image, taken on 27 Jan 2007	23.5 m resolution
3.	Elevation data	SRTM data	Spatial Res-90 m Vertical Res-1 m
4.	Gravity data (Bouguer anomaly map)	Seismo-tectonic atlas prepared by Dasgupta, S., Narula, P.L., Acharyya, S.K., Banerjee, J., 2000. Geological Survey of India.	
5.	2D seismic sections	Prepared by the Oil India Limited, Duliajan, Assam	
6.	Uphole survey data	From the Oil India Limited, Duliajan, Assam	
7.	Shallow refraction data	From the Oil India Limited, Duliajan, Assam	
8.	Shallow subsurface lithology data	Department of Applied geology, Dibrugarh University, Assam	

Data Integration

The planning of the data integration is shown in Figure 3.16.

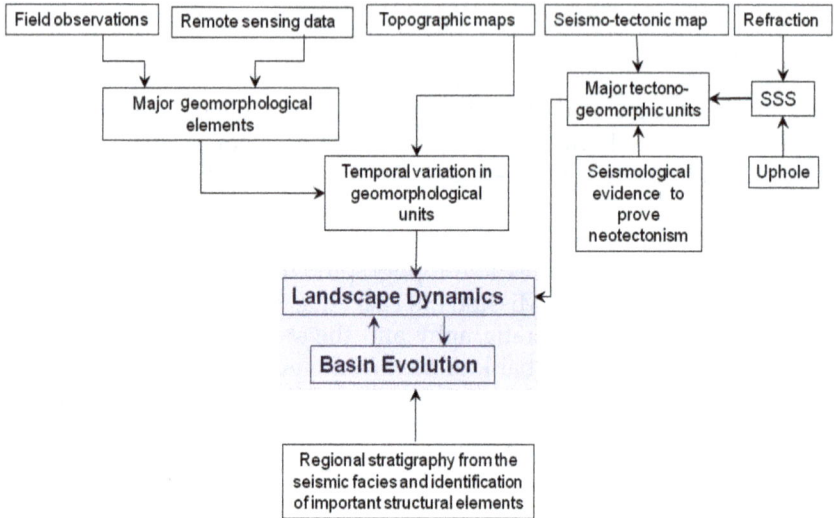

FIGURE 3.16
Data integration as it is planned for the present work.

4

Fluvial Dynamics and the Changing Interfluves

Rivers are the real architect (Miall, 1991, 1985), the principal dynamic agency to sequestrate sediments and assign definite geomorphic signatures to the valleys. Changing landscape and fluvial dynamics in the Himalayan foreland basins have proved the major role played by the mountain-fed rivers (Sinha and Friend, 1994). While exploring the origin of rivers and fluvial megafans (Gupta, 1997), different forms on the surface (Guiseppe and Heller, 1998) were appreciated as products of processes that could also be extended to understand ancient environments and their causation (Gregory and Schumm, 1987). This is one of the major reasons behind the emphasis given on a thorough understanding of recent sediments, particularly in the petroliferous basin studies, which helps process-based analysis of sedimentary basins and basin evolution on different scales. With the availability of satellite data, it has become easier to monitor short-term geomorphic changes in the landscape. To link up earth processes on the surface, shallow subsurface, and then the deep subsurface, a major task is to identify the changes in geomorphic parameters and then understand different controls behind the changes. In the study of earth surface dynamics, the meaning of 'dynamism' is related to the temporal form consistency of geomorphic features like a channel on the surface relief. In this particular sense of usage, if a river flows through the same path for the last 10 years or so, we may say that the river does not show any dynamism on the decadal scale. But the same river might show a remarkable shift from the course through which it used to flow hundred years back. Thus, the river showing no decadal-scale dynamism might be highly dynamic on the century scale. Moreover, for big braided rivers, the principal channel within the overall channel belt may keep on changing, as described by Coleman (1969) as 'wandering of thalweg'. So, the overall channel belt may not show any temporal variation, but there might be intra-channel dynamism. For the multi-channel big braided rivers having a high slope, twin sources of discharge, namely, the monsoonal precipitation and the glacial melt, sediment supply from high mountains brought by different tributaries as well as the incised older flood plain deposits of the valleys, the landforms inside the channel belts are usually of three types: (i) *bars*, which are not so steady, sometimes ephemeral, with a very less vegetative cover and mostly sinking underwater during high flood seasons; (ii) *islands*, having a dense vegetative cover and various levels of soil consolidation, with more or less consistent geometry and less affected by the alternate degrada-tion – aggradation process in the immediate neighbourhood; and (iii) *relict islands*, which used to be the interfluves earlier between two major streams, comparatively much elevated above the local base level, entered the present channel belt by river capture mechanism or some avulsive processes. These islands constituted of much older flood plains, and their vegetative covers are also of much older origin.

Among the largest tropical rivers (Latrubesse et al., 2005; Tandon and Sinha, 2007) of the northern hemisphere, the place of the Brahmaputra is very special. This international river passing through the three most populous countries – China, India, and Bangladesh – is a conduit through which a huge mass trans-fer of sediments takes place into the Bay of Bengal, the *sink*, from the *source* that represents broadly the active zone of the continent-to-continent collision between the Indian and the Eurasian plates. The catchment processes respon-sible for supplying vast quantities of sediment include erosion of actively uplifting mountains of the Himalayas, slope erosion of the Himalayan foot-hills, and movement of alluvial deposits stored in the Assam valley (Thorne et al., 1993). On a global scale, the sediment yield of the Brahmaputra (Goswami, 1985) within the Indian territory alone (804 tons/km^2/year) is ~5.5 times higher than that of the Amazon (≈146 tons/km^2/year) – the largest river in the world in terms of the catchment area, number of tributaries, and volume of water discharge. With a mean annual discharge of 21,200 m^3/s, measured at Bahadurabad, Bangladesh, (Latrubesse, 2008), the Brahmaputra River is the seventh-largest river in the world (Hovius, 1998; Tandon and Sinha, 2007) and has created a thick and extensive valley fill in its alluvial reach. The large-scale dynamics of the Brahmaputra River have fascinated geomorphologists across the globe for over 3 decades, and a large number of studies, particularly in the lower reaches of the river in Bangladesh, bear testimony to the international attention this river has received (see Coleman, 1969; Bristow, 1987; Curray, 1994; Goodbred and Kuehl, 1998, 2000, Goodbred et al., 2003).

By contrast, the upper reaches of the river in Assam have received inad-equate attention so far (Goswami, 1985; Sarma, 2005; Kotoky et al., 2005; Sarma and Phukan, 2004, 2006; Singh, 2006; Singh and Lanord, 2006; Lahiri and Sinha, 2012, 2014). We can mention at least five major reasons why researchers were fascinated to study the Brahmaputra River system. (i) The huge mass transfer of sediments orchestrated by it influences the Quaternary evolution of several basins and sub-basins in amazingly diverse ways. This necessitated studies relating sediment budgeting, sediment provenance, and relative contribution of different sources to this effect. (ii) Being located in a tectonically active triple junction of three plates, namely, the Eurasian, the Indo-Burman, and the Indian plate, even subtle changes in the intraplate and interplate relationships are supposed to be manifested in the river dynamics and the landform changes. Thus, understanding the temporal variability of the fluvial dynamics comes as important evidence to identify and establish structural elements, which are mostly *blind*. (iii) The biggest repositories of sediments like the Bengal basin, forming under the marine conditions, rich in incised valley deposits are directly connected to the big continental river systems, like Brahmaputra and Ganges, and generate excellent petroleum systems (source- reservoir- cap rocks and migration mechanisms) during different periods. By studying sediment sorting patterns, types of clay minerals, diagenetic transformations, and cementation characteristics of the recent sediments, the ideas can be extended to the palaeo-environments, which, in turn, give a lead to oil exploration. (iv) Dynamic river systems like the Brahmaputra show a general tendency to redefine the landforms quickly, not only on a decadal scale but sometimes even every year by rapidly eroding the bank materials, avulsions, anabranching, river capture, etc. We are still far away from accurately predicting the incoming changes to be caused by the river. However, these elements of uncertainty affect the socio-economic life of the land immensely. This called forth closer observations and detailed scientific studies of the river system. (v) The *fifth* cause for the increased emphasis on the study of the Brahmaputra River system is the fast increasing human intervention in the catchment areas of the Brahmaputra and its tributaries by constructing big dams for generating hydroelectricity. Unless the existing hydrodynamics is understood properly, proper river simulation studies are not possible to understand and predict the consequences of massive changes in water and sediment budgets of the river regime.

The Brahmaputra River divides the Upper Assam valley into two distinct geographic zones, the north and south banks. Concerning the direction of flow, the right bank stands for the north and the left bank stands for the south segment of the valley. In addition, this reach of the river hosts one of the largest alluvial islands in the world, Majuli Island, which is known for its significant cultural heritage (Sarma and Phukan, 2004) apart from its geomorphic significance. The Brahmaputra River shows significant geomorphic diversity in this region, which is strongly manifested in the morphodynamics of the river at a historical time scale. Moreover, the tributaries joining the main river from these two banks are quite different in terms of their morphometric characteristics and temporal dynamics.

That the rise of the Himalayas and influx of the eroded materials from the mountainous reach to the valley have greatly influenced the fluvial dynamics of the Brahmaputra River is a well-known fact. However, for a dynamic geomorphic regime, obliteration of the earlier imprints of changes due to the cultural noise and scarcity of shallow subsurface data makes it difficult to trace the Quaternary history of the basin evolution linearly.

Available studies (Mathur and Evans, 1964; Bhandari et al., 1973; Dasgupta and Nandy, 1982; Ahmed et al., 1993; Dasgupta and Biswas, 2000; Kent and Dasgupta, 2004) suggest that the present-day Brahmaputra valley, an NE-SW trending intermountain alluvial relief, was earlier a part of the Assam–Arakan basin, and to be more precise, it constituted mainly the shelf part of the basin. For the sake of convenience, the Brahmaputra River and the valleys through which it passes can be divided into at least three reaches.

The upper reach of the Brahmaputra (Figure 4.1a) contains a northeast–southwest flow that extends from the confluence of three rivers – the Siang, the Dibang, and the Lohit – wherefrom the river gets its name the Brahmaputra to the portion adjacent to the Mikir Hills.

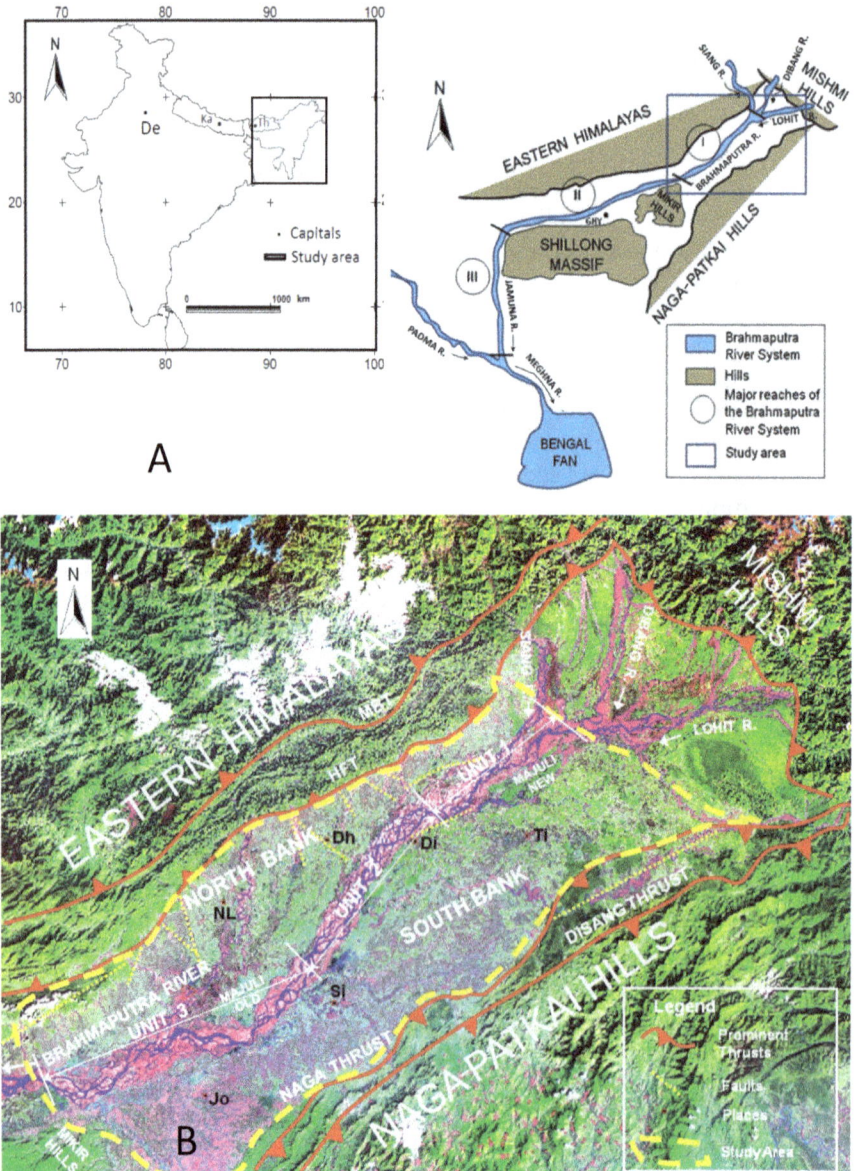

FIGURE 4.1
Three major reaches of the Brahmaputra River.

In the middle reach, the westbound, near-parallel continuation of the Brahmaputra with the eastern Himalayas extends from the Mikir Hills up to the western tip of the Shillong massif. The Brahmaputra, in its lowest reach; takes an L-turn; flows in the southward direction; and enters Bangladesh. There it extends from the upper tip of the western boundary of the Shillong massif up to the confluence with the Ganges. Afterwards, the river is known as the Meghna. The term 'Brahmaputra valley' is used jointly for the upper and middle reaches. Interestingly, basins undergoing active tectonic adjustments are not considered suitable for hydrocarbon prospects; however, the Brahmaputra valley, despite being a seat of large-scale seismic activities, is also an excellent preserver of hydrocarbons. It is pertinent to remember in this particular context that there is evidence where geomorphology provided tools for finding oil (Lattman, 1959; Leeder, 1993).

In the present study, attention was focused on the upper reach of the Brahmaputra valley, where most of the old and new oil fields, as well as highly prospective oil fields, exist. The older topographic maps suggest that three rivers – the Siang, the Dibang, and the Lohit – used to meet at a place called Kobo to form the Brahmaputra in 1915. This confluence point shifted by ~16 km downstream to a place called Laikaghat by 1975. By 2005, a further downstream shift of 19 km occurred, as observed in the satellite imagery. Although the Brahmaputra has been described as a braided river, the conventional definitions for braided rivers (Leopold et al., 1957; Lane, 1957; Bristow, 1987) are as anastomosing channels or the presence of several bars and islands having intertwining association with channels are not adequate to describe it. In the study reach of the Upper Assam area, the Brahmaputra appears to be a multi-channel and multi-pattern river that tends to generate an anabranching (Latrubesse, 2008) pattern very frequently.

Data and Approach

For the present study, the IRS-P6-LISS-3 images were acquired on 15 December 2005 with a spatial resolution of 23.5 m, and older topographic maps of 1:253,440 scale corresponding to 1912–1926 and 1977 (Scale: 1: 2,50,000) have been used. Digital image processing of the satellite images obtained from the National Remote Sensing Centre, Hyderabad, was carried out to enhance the geomorphic features for mapping. Shuttle Radar Topographic Mission (SRTM) data with a spatial resolution of ~90 m and vertical resolution of ~1 m were used to find out point elevations and computing slope. All temporal data were georeferenced and registered on a common platform for investigating the temporal variability in bank line, channel width, and planform parameters of the Brahmaputra and its tributaries for three different periods (1915, 1975 and 2005). Reach I was divided into three units (Figure 4.1b). The basis of the morphological divisions was the presence of exceptionally large river islands termed locally as the 'Majuli', the literary meaning of which is landlocked between two rivers. 'Majuli' islands (described earlier as the relict island) differ from the other sandbars in the sense that the latter features develop directly as the consequence of the sediment load redistribution, whereas the former features representing much older flood plains are due to the sudden diversions like anabranching or avulsion bringing older flood plains inside a braided regime. The formation of Majuli-like landforms is thus a part of river dynamics that might be related either purely to the variability in the sediment dispersal pattern or to neotectonic influences, and of course, there might be an interplay of both. The uppermost unit 1, where the Brahmaputra of the 21st century is widest, starts from the confluence of the Siang, Dibang, and Lohit rivers; extends to 51 km downstream; and contains a newly formed large alluvial island (Dibru-Saikhowa Island, locally called 'new Majuli'). The 68-km-long unit 2 is the narrowest, and two important tributaries, the Burhi Dihing and Disang join the Brahmaputra on the southern bank of this unit. Unit 3 is Majuli Island, the largest alluvial island in the world, which has been under serious threat due to severe bank erosion (Sarma and Phukan, 2004). Unit 3 is 121 km long, and three important tributaries join in this unit, the Subansiri from the north and the Dikhau and the Dhansiri from the south bank. Each unit was further subdivided into second-order (smaller) reaches (Figure 4.2), altogether 37 in number (unit 1: 9;

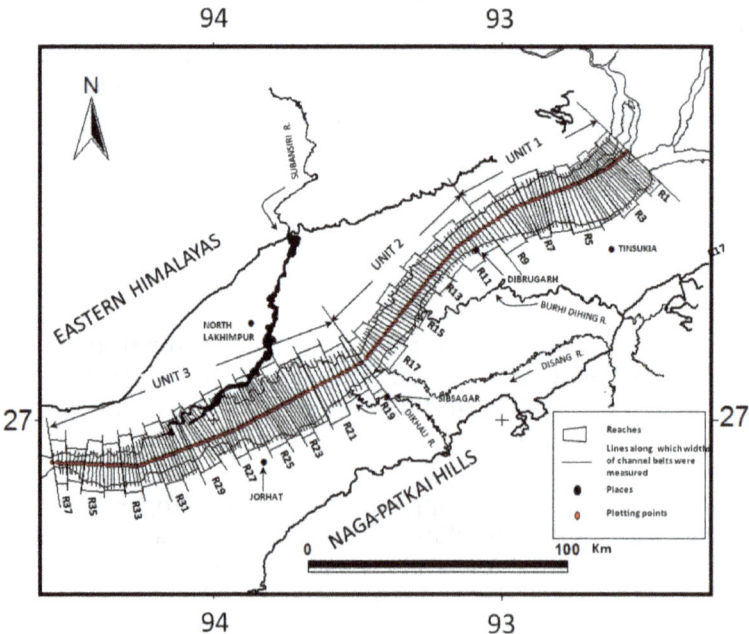

FIGURE 4.2
Second-order reaches for detailed measurements.

unit 2: 9; and unit 3: 19 reaches), 4.5–9 km long, to measure bank-line shifts and all planform parameters such as braid channel ratio (after Friend and Sinha, 1993) for three different periods.

All data were integrated into a GIS environment to document the morpho-dynamics of the Brahmaputra and its tributaries flowing through the study area and to understand the causative factors.

It is general knowledge that when the bank lines of a river shift, lateral erosion occurs. Additionally, if the width of the river increases at a place, then the erosion of banks occurs. However, changing bank lines do not mean a change in width. Similarly, changing channel belt areas (CHBs) can be due to the changing channel area (CH), bars, or both. Thus, several questions arise. What is the trend of bank-line shift? Is it uniform for both the banks? What is the trend of width variation? Is there any relation between the bank-line shift and the width variation? Are braid bar (BB) areas increasing or decreasing? Is there any change in the BB/CH ratio in different reaches?

Again, monitoring absolute widths at different places may not reveal any-thing about the place-specific changes concerning the overall planform changes in the river. Do these changes have any correspondence with variable slopes in different reaches? In addition to these issues related mainly to geomorphol-ogy, there are, of course, other fundamental questions: Can geomorphological parameters alone explain the fluvial dynamics and the changing landforms? How recent seismic events could have influenced the valley topography?

To get concrete answers to the previous questions, a few geomorphic param-eters, such as the existing sinuosity of the rivers, temporal changes in sinuos-ity, sinuosity–slope relationship, basin asymmetry factor, temporal variation in channel–belt area, channel–belt width, sandbar area, and the ratio between the channel and channel–belt areas, have been calculated. Moreover, studies were conducted on aspects related to meandering, bank migration, thalweg shift, avulsion, and anabranching. In this chapter, discussions will be about the Brahmaputra River, Majuli Island, the Dibru-Saikhowa reserve forest (the new Majuli), and some of the important tributary rivers coming down from the eastern Himalayas (north bank rivers) and the Naga-Patkai hills (south bank rivers) to join with the Brahmaputra.

Based on the data generated and incorporating observations of other work-ers, a morphotectonic zonation of the study area will be performed. Moreover, with simple models, the effect of the great earthquake of 1950 on the valley topography and its influence on the fluvial dynamics will be explained.

Dynamics of the Main Brahmaputra River

Valleys dividing big rivers into tectonically active areas showing basin asymmetry are indicative of a distinct tendency of the valley tilt (Keller and

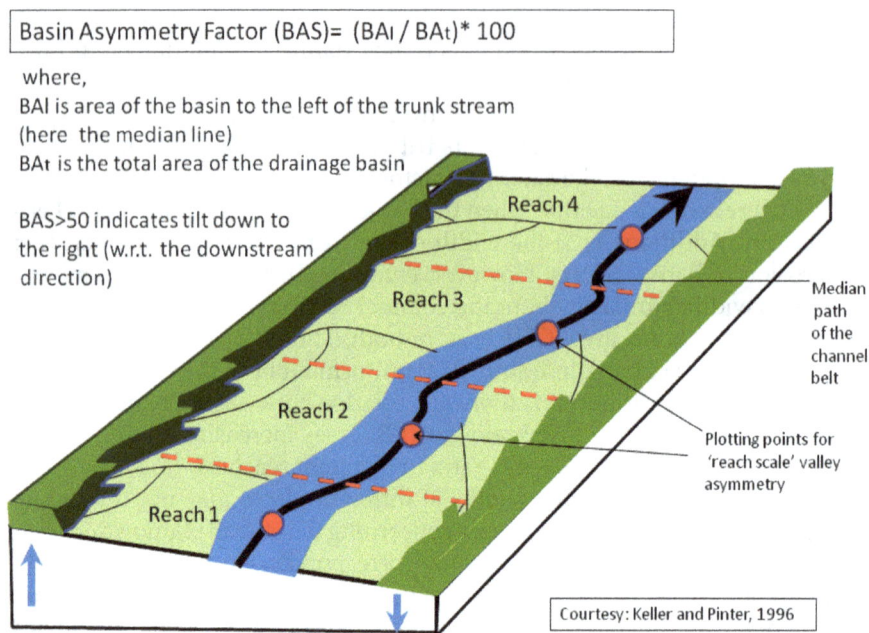

Basin Asymmetry Factor (BAS)= (BAI / BAt)* 100

where,
BAI is area of the basin to the left of the trunk stream
(here the median line)
BAt is the total area of the drainage basin

BAS>50 indicates tilt down to
the right (w.r.t. the downstream
direction)

Reach 4

Reach 3

Reach 2

Reach 1

Median path of the channel belt

Plotting points for 'reach scale' valley asymmetry

Courtesy: Keller and Pinter, 1996

FIGURE 4.3
Basin asymmetry factor.

Pinter, 2002). In this scheme of basin asymmetry measurement, a value of 50 stands for perfect symmetry and hence no tilt. The Brahmaputra River asymmetrically divides the upper reach of the Brahmaputra valley, with the overall basin asymmetry factor (BAF) of 59. The calculation of the BAF and the plotting points for different reaches are shown in Figure 4.3 (modified from Keller and Pinter, 2002). The overall valley tilt is along the eastern Himalayan margin. However, mean basin asymmetry for different units shows different values with a fast decreasing trend from 72 to 62 and then 52. For the three segments studied the magnitude of asymmetry varies considerably (Table A2.1 in the appendix and Figure 4.4). The maximum asymmetry (76) is observed in unit 1 and the minimum asymmetry (38) in unit 3.

A stretch of about 54 km, located mostly within unit 3, shows <50 asymmetries, indicating the superseding influence of the Naga-Patkai thrust (NPT) over the eastern Himalayan frontal thrust (HFT) on the basin-scale tilting. The median path of the Brahmaputra River (excluding Majuli Island) in the overall stretch (Figure 4.3) is almost straight with slight temporal variation in sinuosity from 1.09 to 1.1 in the span of 90 years (1915–2005). Although small, all three segments show some changes in the sinuosity of the median path of the river (Table A2.2 in appendix).

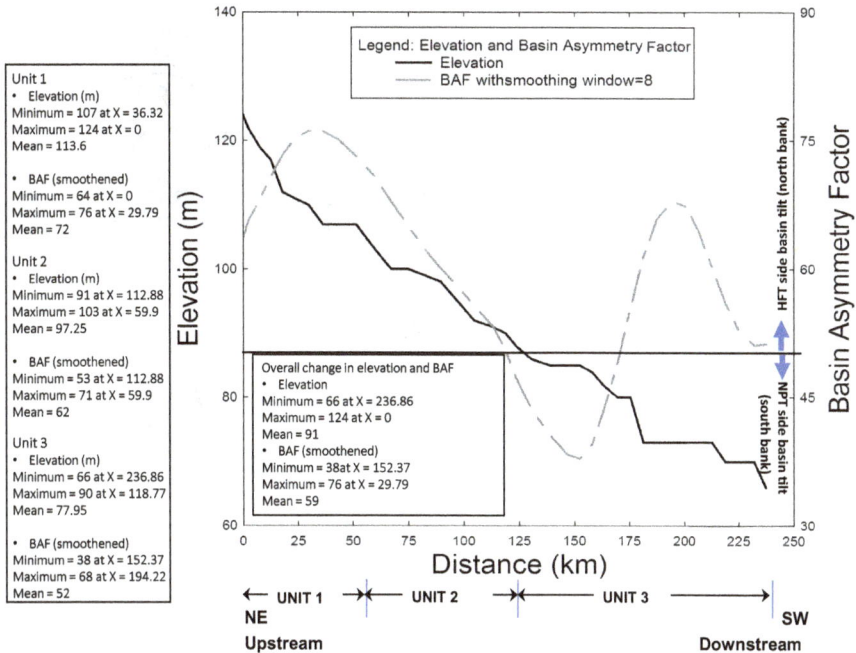

FIGURE 4.4
Elevation and basin asymmetry factor plotted together.

Bank-Line Shift

For the Brahmaputra River, the bank-line migration does not always follow unidirectional characteristics. It is interesting to observe different segments of the Brahmaputra River: when one bank shifts, the other may not. If the migration direction of the modern bank line is away from the historical bank lines, stretching occurs, whereas if the migration is towards the direction of the historical median line, there is narrowing. When both the bank-line shift, the direction of shifting may or may not remain the same. If the direction of migration is opposite, it is a positive lateral stretch that is widening.

If the migration direction is the same, depending upon the degree of migration, the effective change in width might be positive, negative, or simply no change at all (Figure 4.5a–h).

Figure 4.6a–c show the positions of both the north and south banks of the Brahmaputra River for different periods for all three units. A careful analysis of the maps shows that the nature and extent of the bank-line shifts are quite different in the three units. Unit 1 shows the maximum shift (Figure 4.6a), and the south bank has been much more mobile in recent times than the

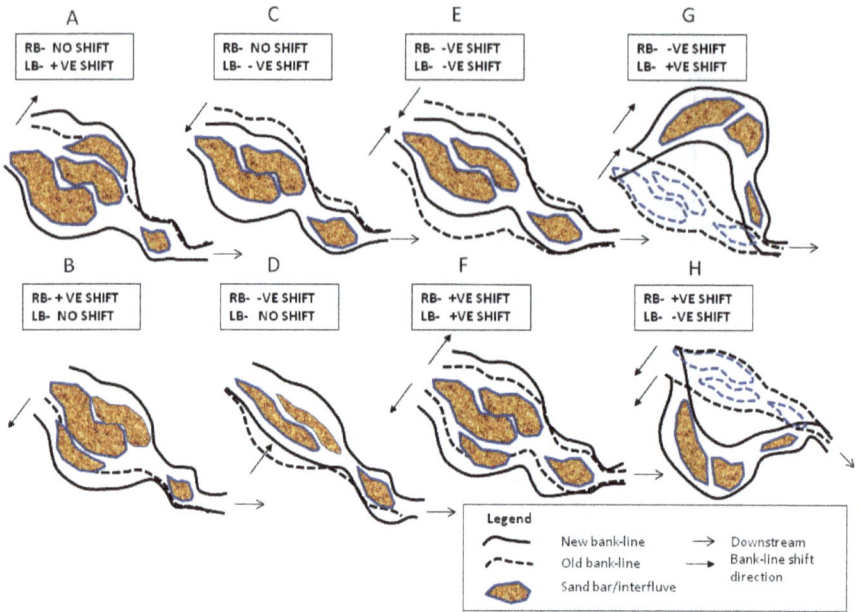

FIGURE 4.5
Different possibilities of bank-line shifting.

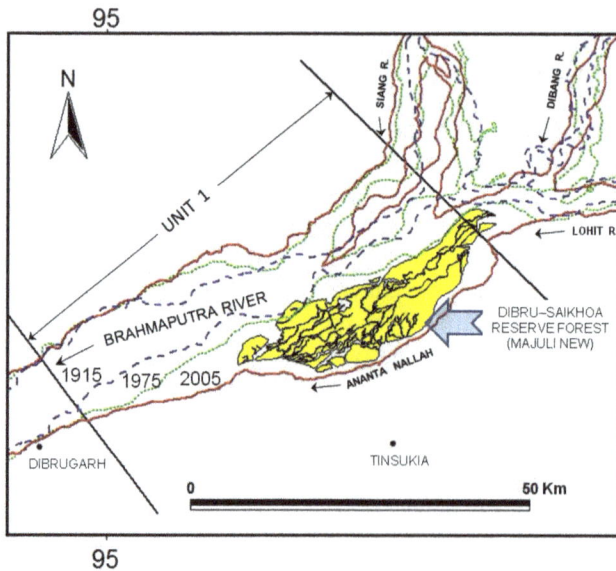

FIGURE 4.6
Planform bank-line shifting of the Brahmaputra River during 1915–2005.

(Continued)

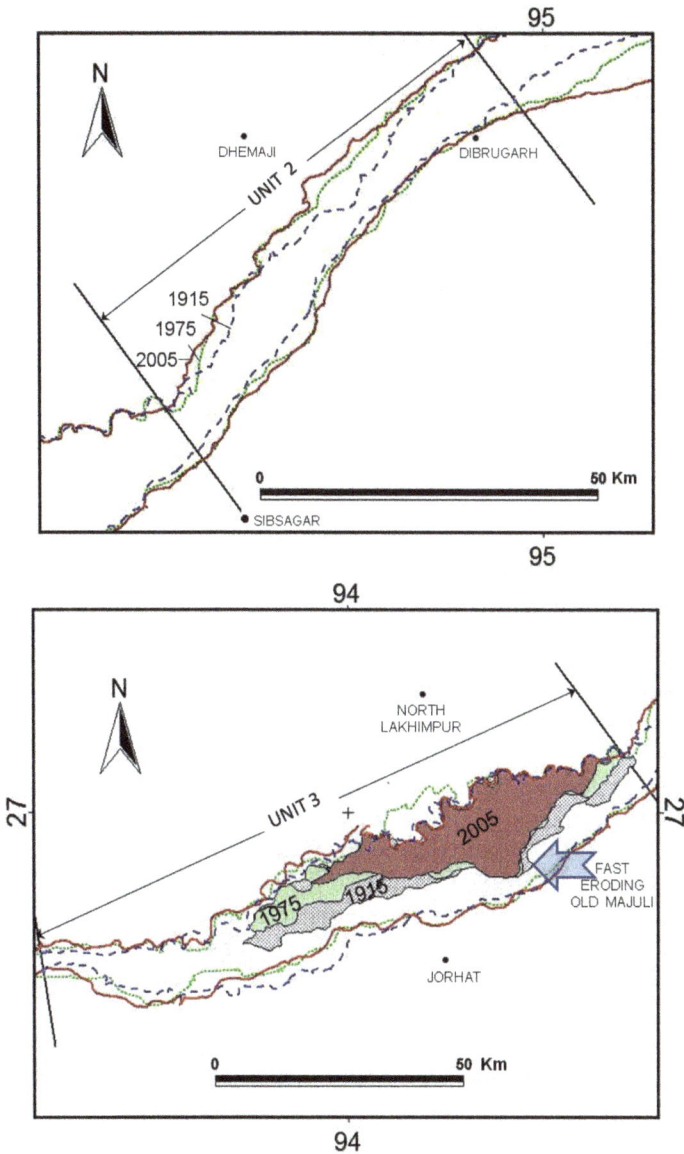

FIGURE 4.6 (*CONTINUED*)
Planform bank-line shifting of the Brahmaputra River during 1915–2005.

north bank. However, the bank-line shifts are not uniform during the period of study for both north and south banks.

For example, the mean shift in the north bank for the period 1915–1975 was 1.45 km, whereas a shift of about 0.7 km (approx. half) was recorded for the south bank during the same period. However, the mean shift in the north bank during the period 1975–2005 was just 0.06 km, while the mean shift in the south bank was 2.05 km (approx. 34 times) during the same period.

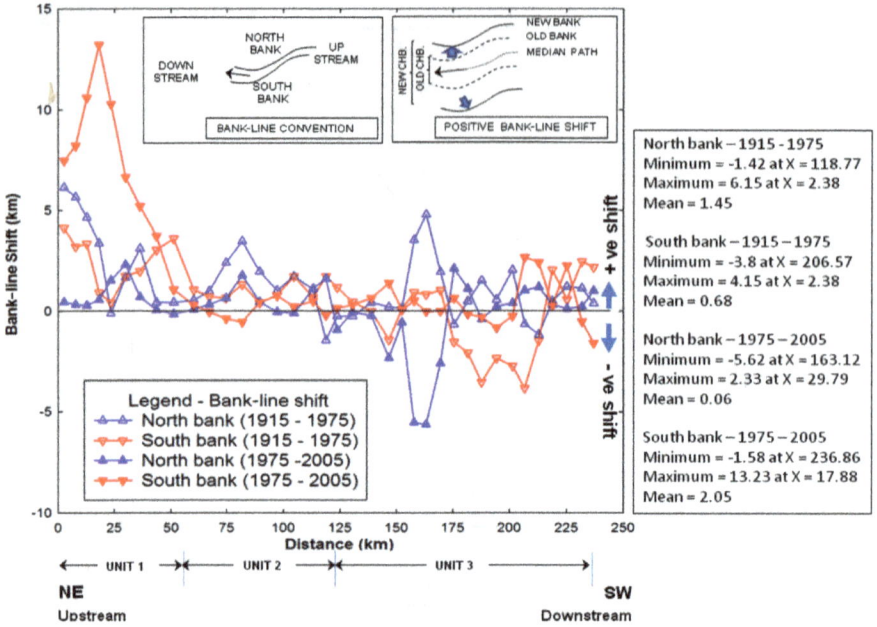

FIGURE 4.7

Graphical representation of bank-line shifting of the Brahmaputra River during 1915–2005.

Furthermore, the reaches showing maximum shifts in both north and south banks are located in the most upstream part (unit 1), whereas the minimum shifts are recorded in the downstream reaches (Figure 4.7) (unit 2 or 3). Although the overall tendency of shifts of the riverbanks was positive, there was a significant magnitude of negative shifts in selected reaches at different times (see Figure 4.7). For example, the minimum south bank shift was 3.8 km (at $X = 206.57$ km, unit 3) from 1915 to 1975, whereas the minimum north bank shift was 5.62 km (at $X = 163.12$ km, unit 3) from 1975 to 2005 (Figure 4.7).

Basin Asymmetry and Bank-Line Shift

When we compare the differences in the bank-line shift between the two periods – 1915–1975 and 1915–2005 – and thereby identify the erosion-prone (EP) zones with basin asymmetry, it is observed that the north bank-line (right) shift (Figure 4.8) conforms with the basin asymmetry.

The point to be noted is that the positive excursion of the north bank line emphasizes a tilt towards the eastern Himalayan side of the basin, whereas the negative excursion indicates a tilt towards the Naga Patkai side. The things are just the opposite for the south bank-line: negative excursion should emphasize

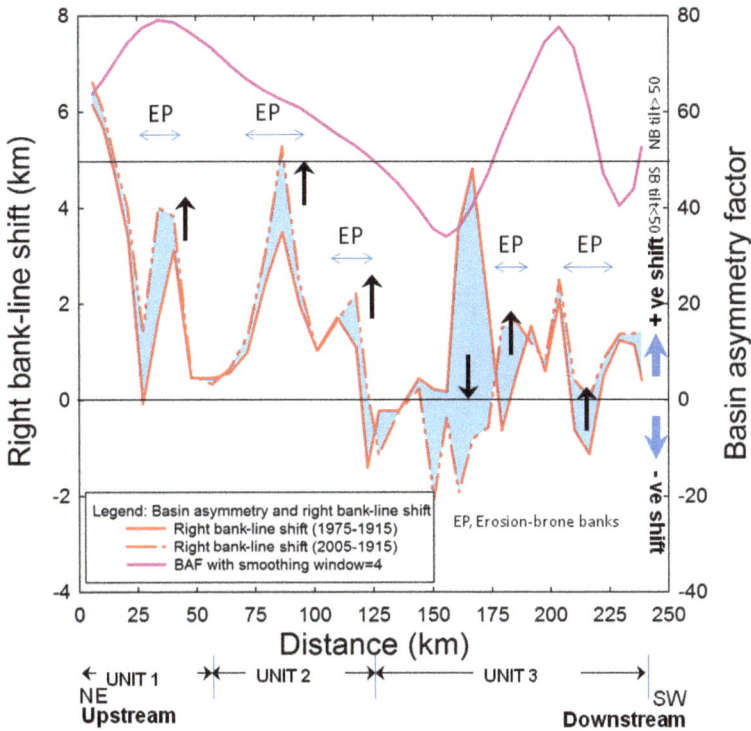

FIGURE 4.8
Basin asymmetry and right bank-line shift of the Brahmaputra River.

a tilt towards the eastern Himalayan side, and a positive excursion a tilt towards the Naga-Patkai side. The pattern of the left bank-line (south bank) shift (Figure 4.9) is in partial conformity with the trend of basin asymmetry.

Planar and Temporal Variability of Width

As a consequence of the overall positive bank-line shift in both the banks, the widths of the channel belt of the Brahmaputra River have changed significantly over the years. Measured at several points along the median course of the belt, the average channel width in the study area increased steadily from 9.74 km in 1915 to 11.6 km in 1975 and then further to 14.03 km in 2005 for the 240-km-long stretch of the river. However, the individual reaches showed varied patterns of increase. Figure 4.10a shows absolute width variation (Tables A2.3 and A2.4 in appendix). There was a continuous increase in the width in unit 1. A similar trend is observed in unit 2, albeit with a lower rate,

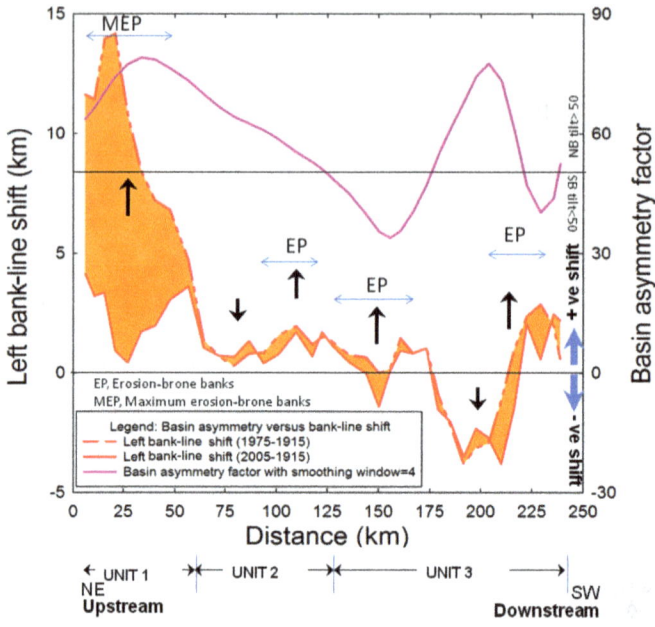

FIGURE 4.9
Basin asymmetry and left bank-line shift of the Brahmaputra River.

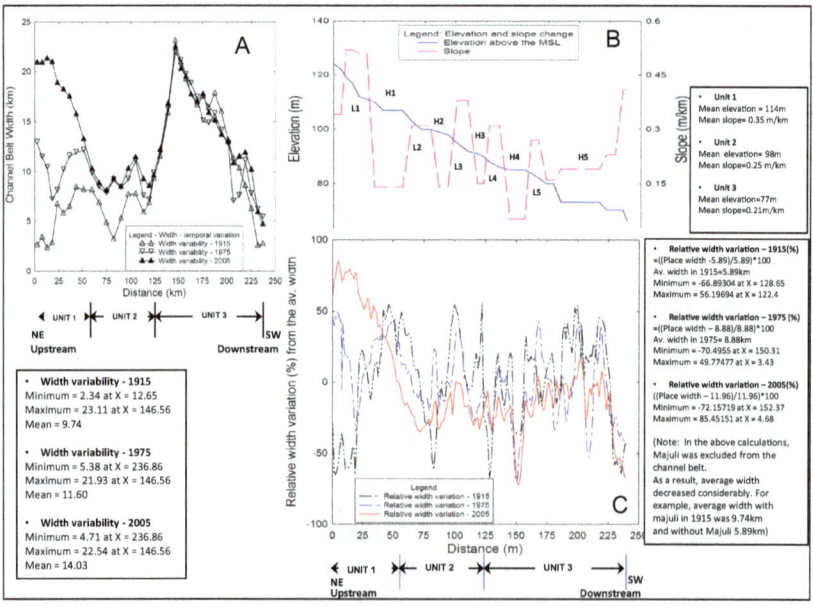

FIGURE 4.10
Planform variation in the widths of the channel belt of the Brahmaputra River.

and unit 3 does not show much change. In addition to knowing the absolute change in width at different locations (Table A2.3 in appendix), we are also interested to know how the river as a whole behaves at different reaches. In other words, what is the relative change a particular reach is experiencing over time concerning the average geomorphic characteristic of the river? This can be known by measuring the relative width variation (Figure 4.10c and Table A2.4 in appendix). One important point to note in this context is that the consequence of the inclusion and exclusion of Majuli Island in calculating the average width of the overall channel belt in a particular year is tremendous. For example, the average width with Majuli in 1915 was 9.74 km and without Majuli was 5.89 km for an about 240-km-long channel in the study reach. Since for most of its northern bank, Majuli is practically bounded by a very narrow flow-less stream called Kherkutia Suti, which was closed by anthropogenic intervention on the upstream side of the island at a place called Tekeliphuta; practically, the Brahmaputra River in its entirety is flowing along the southern bank of Majuli.

A relative change in width gets masked when Majuli Island is included in the calculation. This can be clearly seen in Figure 4.11. Qualitatively, in unit 1 and unit 2, the relative change in width seems to show similar characteristics with the difference that in 1915 and 1975, relative widths with Majuli show lower values than the average width of the river.

A plot of the longitudinal slope derived from elevation data from SRTM DEM (Figure 4.10b and Table A2.1 in appendix) shows that the upper part of unit 1 with an average slope of 0.38 m/km shows the maximum temporal variability in width (250% during the period 1915–2005). A dramatic change occurred between 1915 and 1977, during which the average width of the channel belt

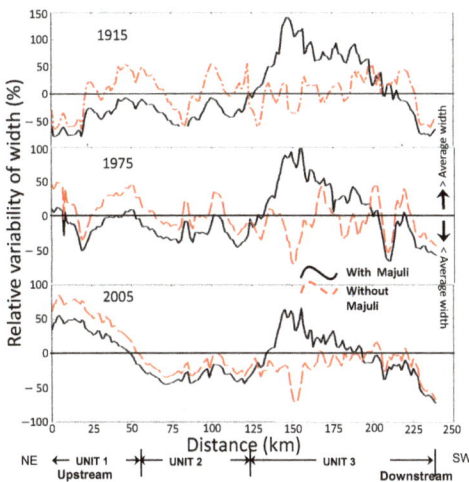

FIGURE 4.11
Relative variability of width (%) from the average width of the channel belt.

of unit 1 doubled (Table A2.7 in appendix). Most of unit 2 with an average slope of 0.21 m/km shows the least variability. Table 3.8 shows that the average width of unit 2 was 6.25 km in 1915, which was more than the average width of unit 1 in the same year (5.28 km). In contrast to the dramatic change in unit 1, the average channel–belt width of unit 2 increased to 8.8 km in 1977 and then to 9.42 km (a total of 51% increase from 1915 to 2005). The most downstream unit 3 has variable slopes (average 0.14 m/km) and shows moderate variability in width over the years. This unit has remained relatively stable during the 90 years (Tables A2.7 and A2.8 in appendix). A minor decrease in the channel–belt width between 1915 and 1975 (~2%) is noted, followed by a ~3% increase in 2005. Therefore, there is no net increase in the average channel belt width in this segment during the period of observation (1915–2005).

Moreover, when we plot the effective changes in the relative width variation for the period 1975–2005 (Figure 4.12), the masking effect due to Majuli is more clearly visible. So, to magnify the relative change in width, it is better to exclude Majuli from the calculation. Figure 4.13 shows the effective width variation for different reaches during different periods. In unit 1, widening continues for the entire segment. However, there are areas where erosion has

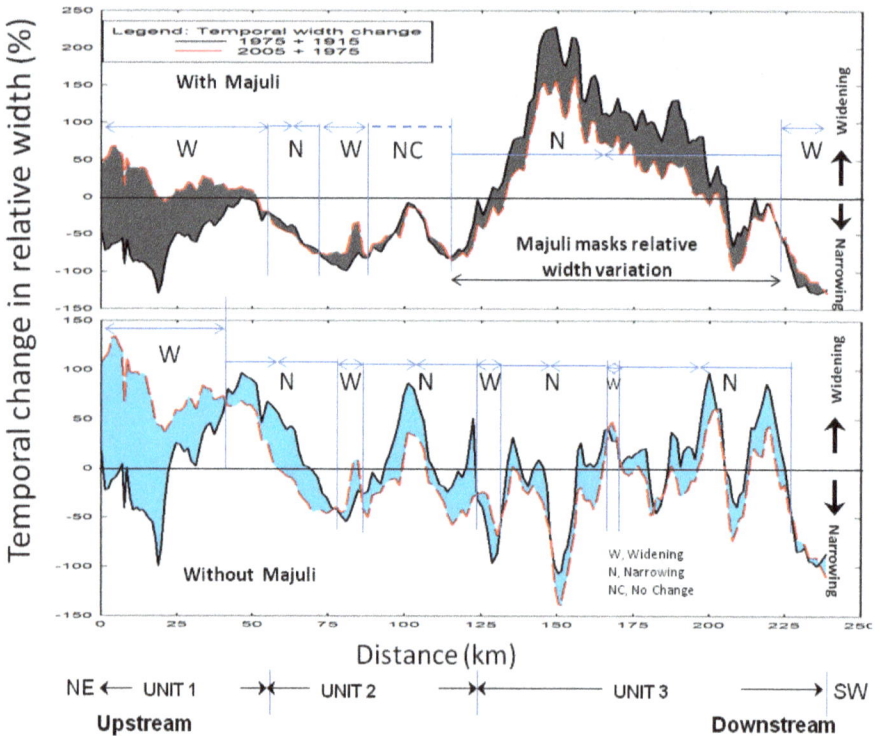

FIGURE 4.12
Masking due to Majuli Island.

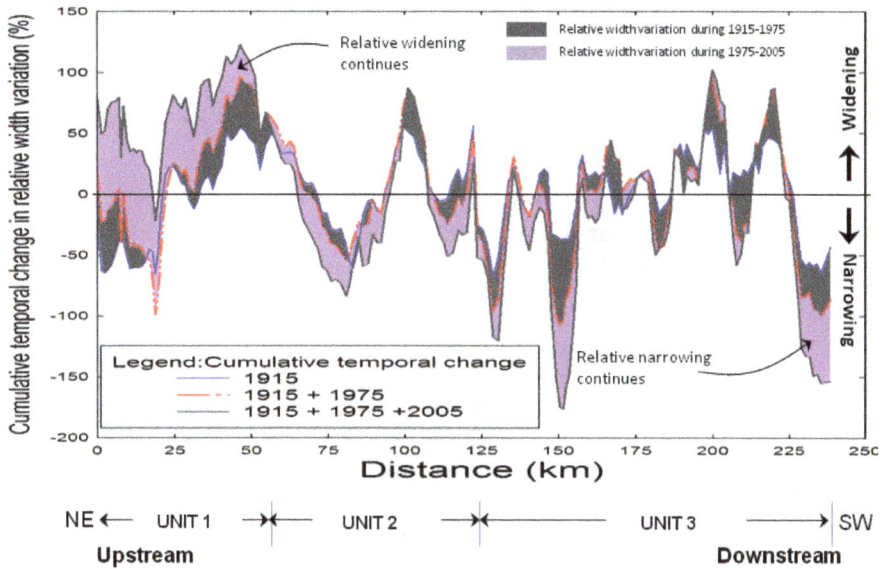

FIGURE 4.13

Temporal change in relative variability of widths (%) from the average width excluding the presence of Majuli Island.

accelerated very recently. The overall scenario for unit 2 is narrowing, but the middle segment shows older widening that was not activated in the last 30 years. Unit 3 shows alternate zones of narrowing and widening and mixed up conditions, some portions were highly active throughout the period being considered, there are places where the spurt in bank erosion was initiated very recently, and then there are also places where substantial recent activities seem to be arrested completely (probably due to anthropogenic intervention) (Box 4.1).

BOX 4.1   METHODS OF MEASURING WIDTH VARIATION

Absolute width variation: In reach-scale measurement at a given time, this is the planform variability in the average width of the channel belt for each of the successive reaches.

Relative width variation: This is the percentage variability of the width of a reach at a given time with respect to the average channel belt width of the entire study length of the stream at that time. This improves the visibility of widening and narrowing stretches in the normalized sense and additionally provides a means for temporal comparability.

Thus,

$$\text{RWV} = \left(\frac{\left(\text{Reach scale channel belt width} - \text{Average channel belt width} \right)}{\text{Average channel belt width}} \right) \times 100$$

Effective width variation: This is the cumulative effects of relative width variations of the channel belt for a given reach at different times.

$$\text{EWV} = \left\{ \frac{(R_1 - A_1)}{A_1} + \frac{(R_2 - A_2)}{A_2} + ... \right\} \times 100$$

where R_1 is the average channel belt width for a given reach (e.g., in 1915); R_2 is the average channel belt width for the same reach at different time (e.g., in 1975); A_1 is the average channel belt width, for example, in 1915, for the entire length of the channel; and A_2 is the average channel belt width, e.g., in 1975, for the entire length of the channel.

Trend Analysis of the Bank-Line Shift and Width Variation

The bank line of a river flowing over a tectonically active landscape can be assumed to represent a superposition of 'controls' having different orders expressible in the form of waves of variable lengths and frequencies. Thus, large-scale tectonic controls having regional dimensions will represent long-wavelength low-frequency waves, and the more the control is of the localized origin, the wavelength will be shorter and the frequency will be higher. Although the mathematical operation of Fourier transform (FT) is usually meant for a transform of functions from the time domain to the frequency domain, and the same principle is applied here to replace discrete cumulative time with discrete cumulative distance, and the amplitude term is replaced by the magnitude of bank-line shift (A_b). Thus, the FT will give a different kind of frequency content (cycles/unit length) of the bank-line shift [$B(f_n)$], which can provide us with a scheme for classifying the nature of forcings operational in shaping the bank line. Moreover, this helps discriminate whether the principal cause of change observed at a particular reach is the fall out of the local control or regional control. Additionally, whether changes observed in the bank-line shifts of two banks of a large river are caused due to the same 'forcings' or not can be compared to (Box 4.2).

BOX 4.2   DEFINITION OF FAST FOURIER TRANSFORM (FFT)

The fast Fourier transform (FFT) is used for computing the frequency content of the bank-line migration. Given a variable function of amplitude of bank-line migration with distance $A_b(d_k)$ with N consecutive sampled values, the FFT implementation by DPlot software for the present purpose calculates the corresponding function in the frequency domain as follows:

$$B(f_n) = \frac{1}{N} \sum_{k=0}^{N-1} A_b(d_k) e^{2\pi i k n / N}$$

where

$$d_k = k\Delta$$

$$\Delta = dis \tan ce\_int erval$$

$$f_n = 0, K, \frac{1}{2\Delta}$$

It is observed (Figure 4.14) that the trend of bank-line shifts can be divided into at least three classes. Class I represents the short-wavelength ($\lambda = (1/f)$, f is cycles/unit length) forcings with wavelengths equal to or less than 15 km. These are responsible for bank-line shifts of the order of less than or equal to 300 m. Class II represents the medium-wavelength forcings with wavelengths greater than 15 km and less than or equal to 51 km. These are responsible for the bank-line shifts of the order of greater than 300 m and lesser than or equal to 870 m. Class III represents the long-wavelength forcings with wavelengths greater than 51 km and lesser than 260 km. These are responsible for bank-line shifts greater than 870 m and lesser than or equal to 2,570 m (Lahiri and Sinha, 2015).

A significant difference is observed in the *frequency content* of the bank-line shift of the north bank and the south bank of the upper reach of the Brahmaputra River. Some of the prominent peaks are shown for both the north and south bank-line shifts (Figure 4.14). The changes taking place in two different phases, respectively, 1915–1975 and 1975–2005 seem to have a uniform influence on the north bank-line shift. However, the same thing cannot be seen to have influenced the changes in the south bank. The influence of class II forcings is almost absent in the south bank-line shift.

Another interesting thing observed is the relative influence of bank-line shifts on the width variation (Figure 4.15). The widening of unit 1 seems to be controlled by the south bank-line shift (SBS), and the widening of unit 2 seems to be controlled by the north bank-line shift (NBS). Unit 3 presents a

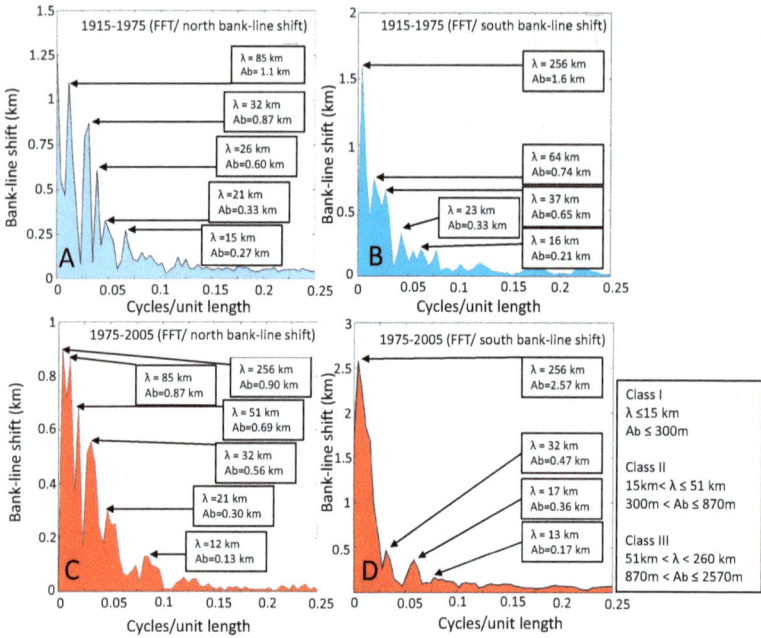

FIGURE 4.14
Fourier transform of the bank-line shifts.

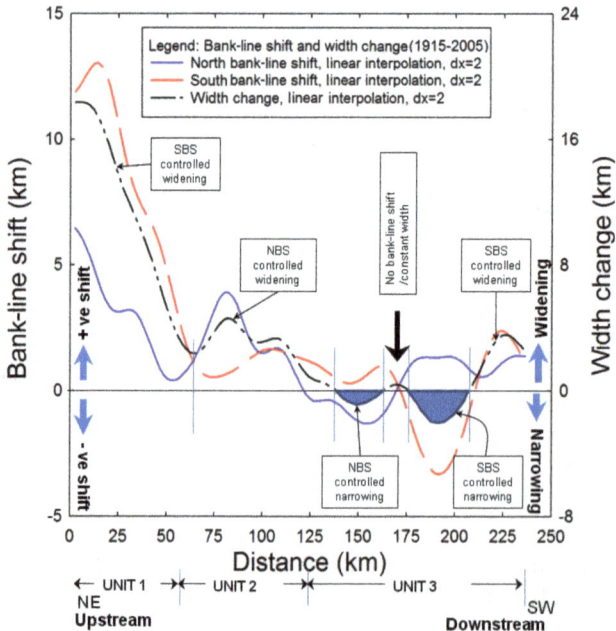

FIGURE 4.15
Relationship between bank-line shift and the width variation.

complex situation. Firstly, there is a narrowing controlled by the NBS, followed by a stretch that practically shows no change, and then there is again a narrowing controlled by the SBS, and in the end part, there is a widening controlled by the SBS.

Channel Avulsion and Migration

In unit1 of the Brahmaputra channel belt, the Lohit River, as well as the Dibang River, shows a prominent avulsive character. The channel capture mechanism is also seen. In the topographic map of 1975, both the channels are observed to flow along the western side of the Dibru-Saikhowa reserve forest. However, from 1995 onwards, the Lohit River started to divert its partial discharge along the Ananta Nallah, and very soon, the Dibru River channel was captured. This caused the Dibru-Saikhowa reserve forest to become an island (Gogoi et al., 2022). Currently, the entire discharge of the Lohit River, partial discharge of the Dibang River, and several lesser southern streams are flowing through the previously existing Ananta Nallah and joining near Rohmoria, is causing unprecedented bank erosion.

We have observed earlier that the median path of the Brahmaputra channel–belt is having a slight temporal swing in the eastward direction and westward swing in unit 3, whereas the intermediate unit 2 is more or less steady. Thus, the channel belt as a whole does show neither avulsion nor migration. However, there is a very high degree of intra-channel variability. These are mainly of five types: (i) changes in the anastomosing pattern, (ii) flow diversion, primary channel becoming secondary, and vice versa, (iii) moving meanders, (iv) channel capture, and (v) anabranching.

Morphodynamics of the Brahmaputra River

If the sediment supply increases for some reason (e.g., tributary contribution and deforestation in the catchment) for mountain-fed rivers flowing through comparatively young valleys, the vertical incision diminishes, and the average thalweg depth decreases. To accommodate the average discharge, lateral erosion dominates in the alluvial reaches where channel banks are composed of unconsolidated sand. For situations where all planform parameters, such as channel belt width, CH, sandbar area, and CHB, are continuously increasing, the temporal variability in the ratio of CH to CHB can give additional information about the aggrading tendency. A highly negative change may

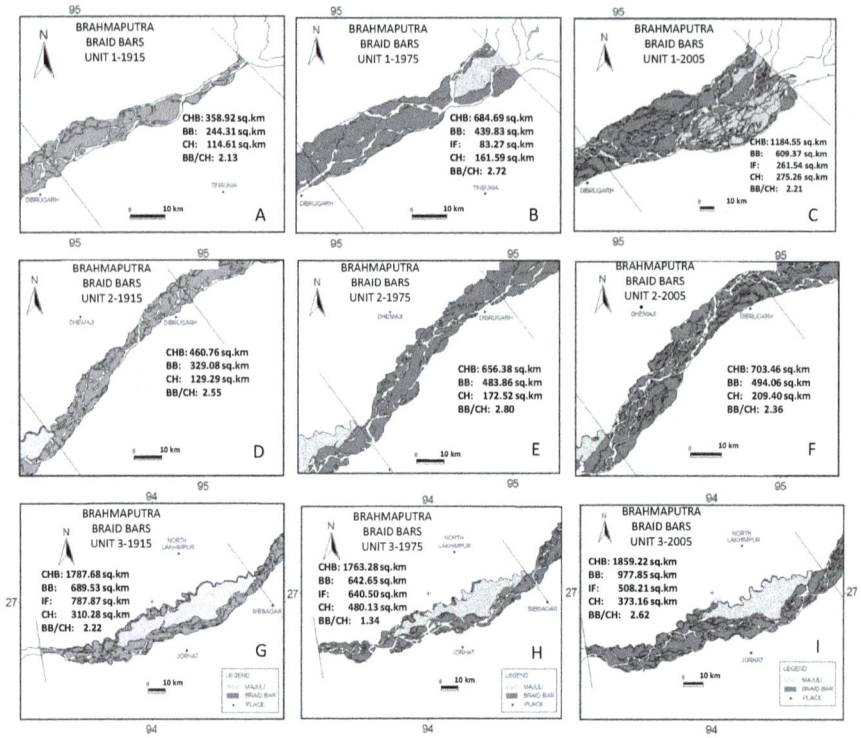

FIGURE 4.16
Changing morphology of the BBs during 1915–2005. BB, braid bar.

indicate aggradation, and a positive trend should be manifested in degradation. Similarly, a large positive change in the ratio of the bar area to the CHB should be associated with aggradation, and the negative change with degradation.

Figure 4.16 shows the major morphodynamic changes in three units of the Brahmaputra River. We have also measured various planform parameters, such as CHB, sandbar area, and CH, for three different years (1915, 1975, and 2005) to quantify the morphodynamic changes (Tables A2.5 and A2.6 in the appendix).

Figure 4.17a–d shows the planform variability of channel belts, BBs, channels, and the BB/CH ratio of the Brahmaputra River for two different periods (1915–1975 and 1975–2005).

During these periods, reach 4 in unit 1 shows a sharp rise of 21 km² in the CHB (19 km² in 1915 to 40 km² in 1975) and 101 km² (from 40 km² in 1975 to 141 km² in 2005) caused by the avulsive characteristic of the Lohit River that brought the Dibru-Saikhowa reserve forest (new Majuli) within the Brahmaputra channel belt. Unit 2 shows a much lesser propensity to change in the CHB, except for a few erosion-prone sites. In unit 3, the CHB seems to be fairly stable during the period of study (Figure 4.17a), but the bar areas have changed significantly. In addition to this, a reversal of the trend is

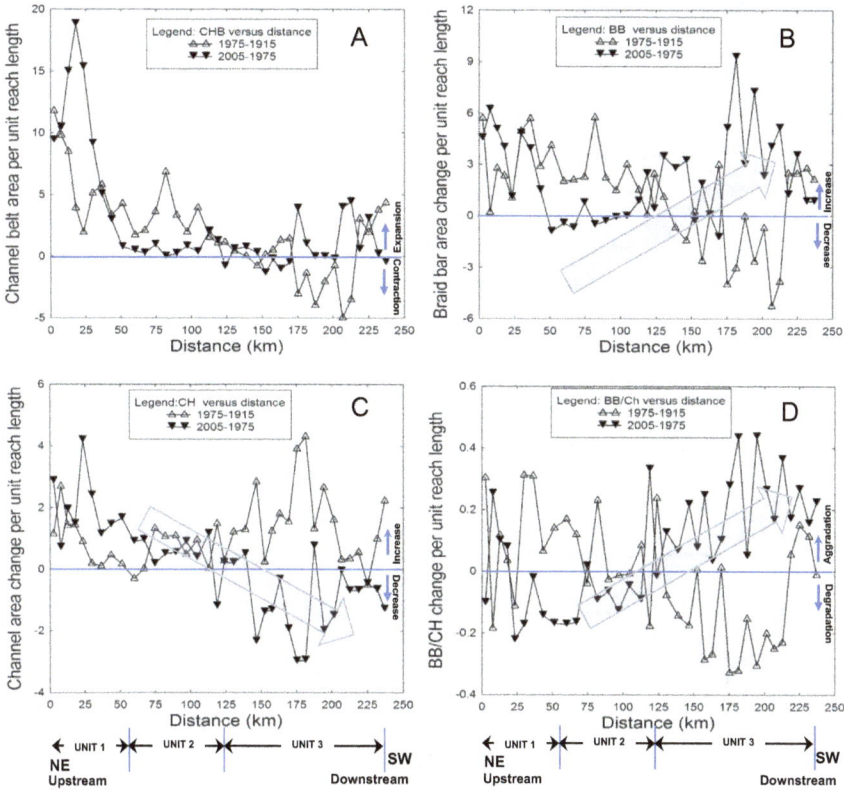

FIGURE 4.17
Temporal variations of morphologic parameters like areas of channel belt, BBs, channels, and BB/channel ratio. BB, braid bar.

observed – the reaches where the channel belt decreased earlier were widened later, and vice versa. This is most likely related to the anthropogenic intervention because Majuli Island as well as the riverbanks in the adjoining reaches have been protected through embankments (existing embankment length is about 160 km; District Disaster Management Plan of Jorhat, 2011) that are often breached during the flood season. These embankments constrain the bank erosion and lateral shifting for a few years. However, sediment load is not distributed freely in the adjacent flood plains, and the river bed rises rapidly. Subsequently, the river breaches through the embankments and crevasse channels develop, adding to channel multiplicity. Units 1 and 3 show a fluctuating trend over the two periods mentioned earlier in terms of the bar area, CH, and their ratios (Figure 4.17b–d). In general, unit 1 showed an aggrading trend, and unit 3 showed a degrading trend from 1915 to 1975. From 1975 to 2005, these trends were reversed. Moreover, although the channel belt was widening in unit 1 from 1975 to 2005, unit 3 was undergoing a major aggradational phase.

Major Changes in River Islands

Majuli Island

Majuli is the largest riverine island in the world having human habitation (population: 0.16 million), and the abode of Vaishnavite spiritual centres called 'satras' (around 64) is located about 150 km away in the downstream direction from Kobo, the place of origin of the Brahmaputra River (Figure 4.1b).

The length of Majuli Island is at present about 64 km, and the maximum width is about 20 km. The elevation contour map (Figure 4.18) shows that from the upper tip of the island to the lower tip, the elevation varies from 92 m above the mean sea level (AMSL) to 72 m AMSL, which means a difference of 20 m. The elevation contour map of the island (Figure 4.18) also shows a prominent high in the central part.

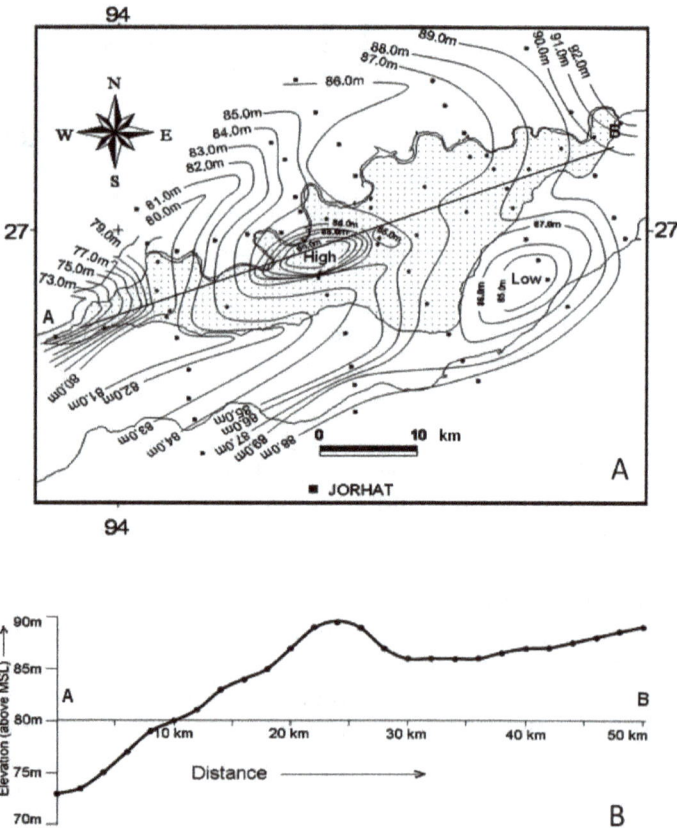

FIGURE 4.18
Elevation contours in and around Majuli Island.

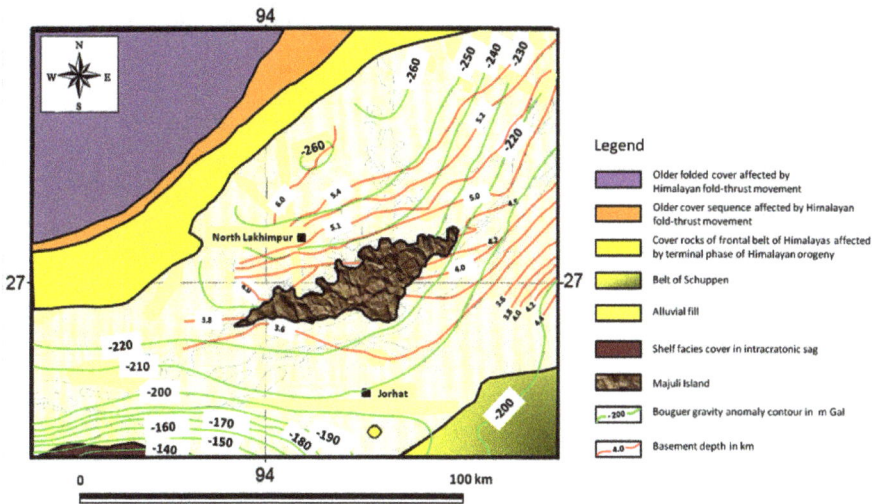

FIGURE 4.19
The geo-tectonic setting of Majuli Island.

The geo-tectonic setting of Majuli Island (Figure 4.19) shows that the place is situated between the Bouguer gravity anomaly contours of 220–240 mGal and first-order basement depth contours of 3.6–5.0 km (Dasgupta et al., 2000). We have investigated in detail the possibilities of influences of geomorphic parameters like CHB, channel belt width (W), BB area, CH, and bank-line migrations on the trend of erosion of Majuli Island and making use of the 'method of least squares' found out the relations of second-order 'best-fit curves' (Lahiri, 2013).

We have also compared the pattern of thalweg migration during different periods with the rate of erosion of the island.

Geomorphology, History, and Mechanism of Erosion of Majuli Island

Majuli is a relict island because its age is older than the age of the Brahmaputra River at its present location flowing around the island. Approximating a spindle shape, despite the recent surge in the rate of erosion, Kotoky et al., 2003. Majuli is a fairly steady landmass within the Brahmaputra River that does not submerge completely even during maximum flooding like other smaller islands located in the adjacent areas. After entering the channel belt of the Brahmaputra River, it has remained an important question about the

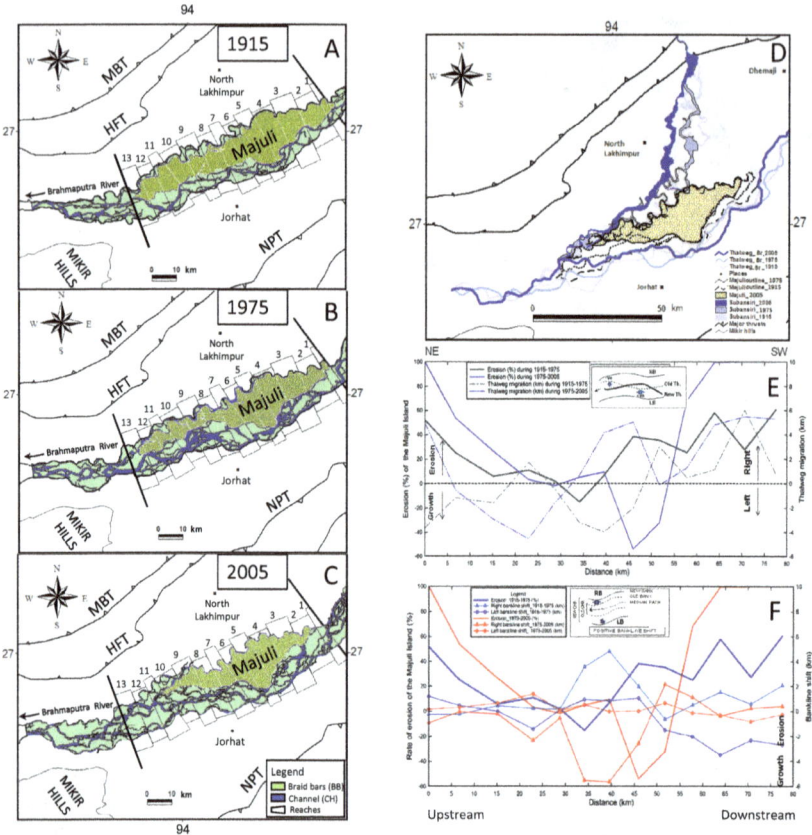

FIGURE 4.20
The mode of erosion of Majuli Island.

mechanism of erosion (Figure 4.20) of the island. Is it the head-on impact of the high-stream power river, the lateral impact of the multi-channel river system, or both, principally what is causing the erosion of the island?

Geomorphic mapping of Majuli Island (Figure 4.21) using multi-date satellite images and toposheet covering a period of about 90 years reflects a highly dynamic regime and a very high rate of erosion of the island. Between 1915 and 1975, the surface area of the island reduced from 787.87 to 640.5 km² (18.7% reduction) and then to 508.2 km² by 2005 (35.5% reduction as compared with 1915). The average rate of erosion in the last 30 years or more has increased considerably from 2.46 km²/year (1915–1975) to 4.40 km²/year (1975–2005). The length of this spindle-shaped island has also reduced greatly from 79.7 km in 1915 to 75.16 km in 1975 and then 63.33 km in 2005 (about a 20.5% reduction compared with 1915). In addition to this, a major tributary of the Brahmaputra, the Subansiri, has shifted significantly towards the SW;

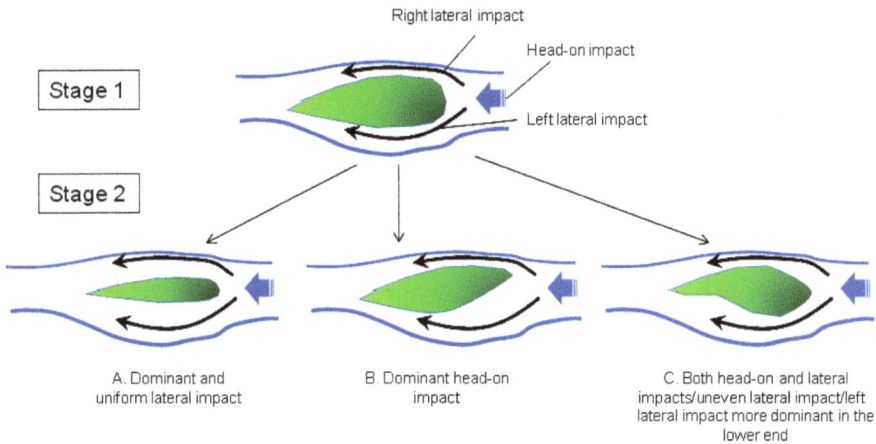

FIGURE 4.21
Plano-temporal variation in the Majuli bearing the Brahmaputra during 1915–2005.

in the past (as seen in the map of 1915), before joining the Brahmaputra, it used to divide into the number of secondary trunks, and its impact on Majuli Island was divided. Lately, after the 1950 earthquake, there is a narrow secondary channel left, and the main trunk, practically the only one, joins the Brahmaputra near the western part of the lower Majuli (Figure 4.21f). Thus, the migrating thalweg and the shifting confluence both acted in unison, and consequently, the rate of erosion in the end part of the Majuli was so high that two of the lowermost reaches eroded completely, as observed in the imagery of 2005. If we compare changing geomorphic parameters in 1915–1975 (Figure 4.22a) with those in 1975–2005 (Figure 4.22b), we have a few interesting observations (Table 4.1):

Based on the rate of erosion, Majuli Island can distinctly be divided into three parts: upper, middle, and lower Majuli. Middle Majuli is fairly stable compared with upper and lower Majuli.

As expected, the trend of variation in the CHBs and the average widths (W) shows a close similarity, and the reduction of Majuli Island for the period 1915–2005 can be correlated with the second-order approximation of the polynomial fit as follows (Figure 4.23a and d):

$$ME = 36.84 - 0.7604\ CHB + 0.154\ CHB^2$$

$$ME = 33.45 + 0.668\ W + 0.3538\ W^2$$

We also observe that the changes in the CHB and the average width of different reaches (W) are not influencing the erosion of Majuli directly because of expansion or shrinkage, or even when there is no change in the channel belt, erosion continues unabated. However, in those reaches where the channel

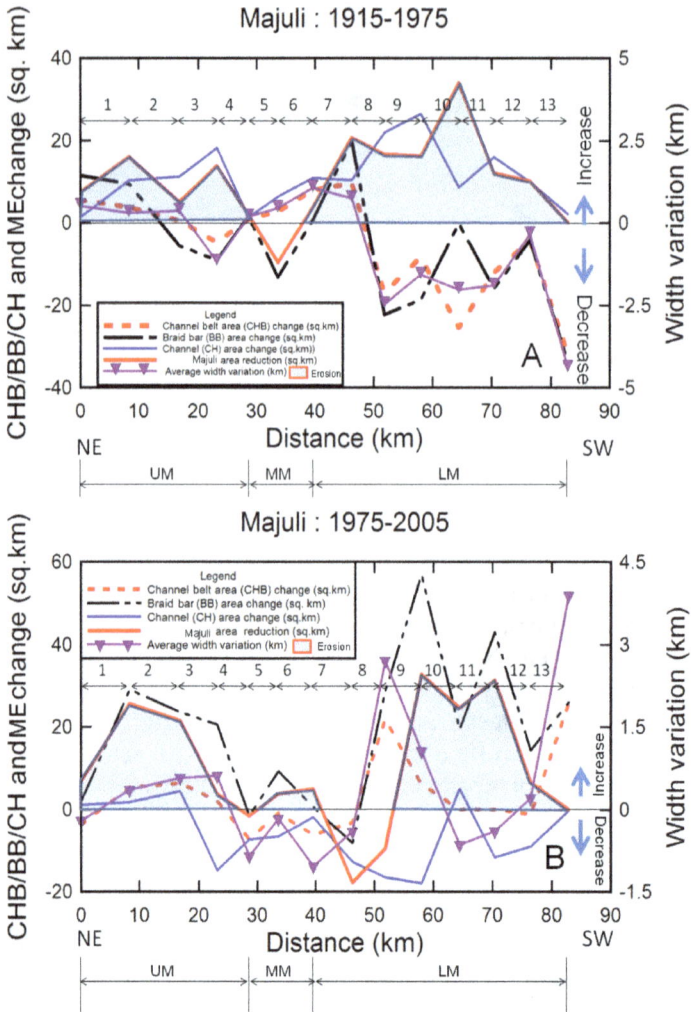

FIGURE 4.22

Quantitative analysis of the geomorphic parameters observed about the temporal variability of planform erosion of Majuli Island.

belt shows a narrowing tendency (due to human intervention in the form of construction of embankments, etc., or otherwise), the rate of Majuli erosion increases much faster.

BBs formed due to the river processes show different types of changes over time like emerging new ones, waning older ones, and some stable braidbars, which remain for a much longer period. The cumulative BB areas in different smaller units (excluding the area covered by Majuli Island) show an overall decreasing trend of about 9.3% from 1915 to 1975. However, from 1975 to 2005, the trend was not only reverse, there seems to be a vigorously progressive tendency of BB formation with an overall growth rate of 53% and a

TABLE 4.1

Temporal and Planar Trends of Erosion of Majuli Island for the Period 1915–2005

Time Period	Upper Majuli (28 km Wide)	Middle Majuli (10 km Wide)	Lower Majuli (45 km Wide)
1915–1975	About 44 km² area eroded, which was about 11.4% of the landmass during 1915. The Brahmaputra CHB and obviously and the average width (W) of the channel belt were showing a crest and trough with a decreasing trend towards the downstream direction. The BB area was showing the same trend with a mild increase of about 6.37%. CH, on the other hand, was showing a sharp increase of 59.8%, indicating an overall degradation trend. **Formation of new BBs much lesser than the erosion rate of the older flood plain areas of Majuli Island.**	This portion does not only show absence of erosion; about 9.6-km² area (about 15% more than 1915) was added. This was probably a direct fall out of the massive embankment projects undertaken. CHB, W, and CH were showing an increasing trend. BB was showing a reducing trend of about 31%. Thus, in spite of the expansion of the island in this unit, the degradation trend was common to the upper Majuli. **Check in erosion and reduction in new BB formation.**	A severe erosion of about 113-km² area, that is, about 1.74 km²/year (>34% of the 1915 landmass). CHB and W were showing a contraction trend. BB decreased by about 12%; CH, on the other hand, increased by 86.7%. This again shows degradation. **Severe erosion of Majuli Island as well as reduced tendency of new BB formation.**
1975–2005	Erosion aggravated in absolute terms (about 56 km²) as well as in percentage (about 16% of the landmass of 1975). CHB and W increased. BB was showing a tremendous increase of about 56.6%. With a crest and a trough, CH decreased by 13.4%. A clear case of aggradation. **Severe erosion of Majuli Island and the simultaneous increase in the new bed bar formation.**	Small but positive erosion; about 5% of the 1975 landmass. CHB and W decreased showing contraction of the Brahmaputra in this patch. CH decreased by about 31%, and BB increased by 32.6%. Aggradation. **Mild erosion of Majuli Island and substantial increase in the new BBs.**	Unabated massive erosion – about 32.3% of the land mass existing during 1975 was eroded; at the increased rate of 2.4 km²/year. Frontal and end parts of this zone show expansion of the channel belt, with most of the middle region unchanged. CH decreased by 28.4%. BB increased massively (53.5%), showing vigorous aggradation. **Severe erosion of Majuli Island and vigorous formation of the new BBs.**

CHB, channel belt area; BB, braid bar; CH, channel area.

simultaneous increase in severity of erosion of Majuli Island. Thus, a growth rate of 38.8% observed for BB during the period 1915–2005 is not a continuous growth phase, rather it includes a complete cycle of high-low-high. The second-order polynomial fit equation of reduction of Majuli Island concerning the changing BB areas can be expressed as follows (Figure 4.23b):

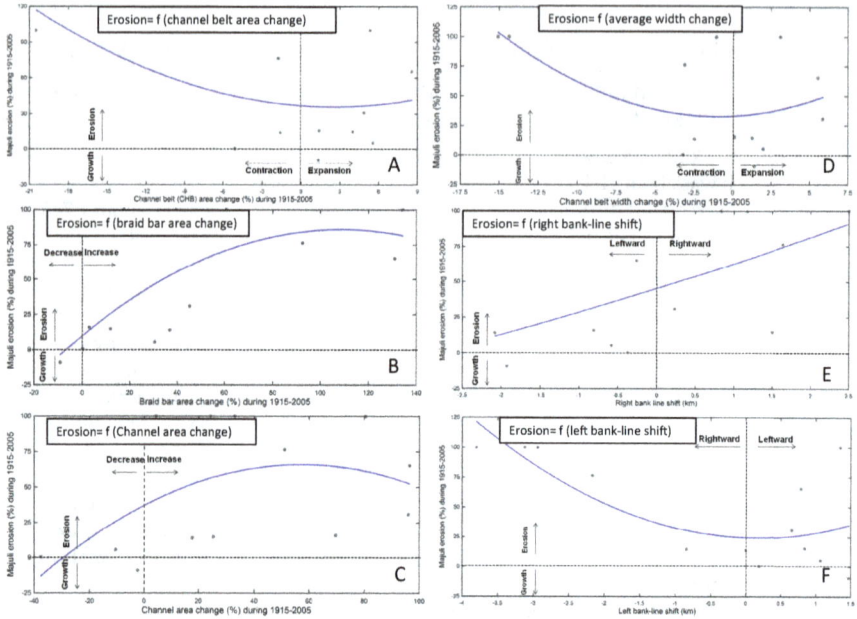

FIGURE 4.23
Application of the method of least squares and polynomial curve fitting for the erosion of Majuli Island during 1915–2005.

$$ME = 9.777 + 1.405\ BB + 0.006463\ BB^2$$

The increasing rate of erosion might facilitate increasing cumulative areas of the BBs. However, looking at the event in other ways, if the BBs are forming due to the greater volume of sediment supply from the much higher upstream side, changing sites of aggradation, and the creation of more accommodation space in the immediate neighbourhood of Majuli Island, the increasing surge in the new BB formation can also induce increasing erosion of Majuli Island.

There is a remarkable shift in the comparative trend of the changing CH and the BB area. From 1915 to 1975, they show a more or less opposite trend, whereas the overall CHB was following a similar trend to the BB. Also, the average amplitude of the CH was higher than both the BB and the CHB. This shows that wherever BB deposits increased, the actual CHs decreased. Moreover, the increase in CHB was principally due to the increasing area of BBs. However, from 1975 to 2005, the CH and BB were showing a similar trend, and the BB was maintaining consistently much higher amplitude all along Majuli Island. As we move downstream, the amplitude difference between the BB and CH continues to increasing, showing a rising trend in aggradation. The BB/CH ratio reduces drastically from 2.34 in 1915 to 1.24 in 1975 and then again to 2.49 in 2005. Thus, the overall trend of aggradation has increased from 1915 to 2005. The second-order polynomial fit equation of reduction of Majuli Island concerning the changing CHs can be expressed as follows (Figure 4.23c):

$$ME = 37.24 + 1.006 \ CH - 0.008772 \ CH^2$$

The bank-line shifts, right and left with respect to the downstream direction of flow, can be correlated with the erosion of Majuli Island as follows (Figure 4.23e and f):

$$ME = -45.42 + 17.16 \ RB + 0.4711 \ RB^2$$

$$ME = 24.7 - 1.99 \ LB + 6.189 \ LB^2$$

The effect of the right bank-line shift (Figure 4.23e) and the left bank-line shift (Figure 4.23f) shows an opposite trend in the erosion of Majuli. The right bank is very close to the western border of Majuli. Over the years, due to human intervention and migration of the Subansiri River in the further downstream direction, the right bank channel flow has practically stagnated, and erosion has also decreased rapidly. Thus, higher discharge accelerates erosion in the right bank, and vice versa. On the other hand, when the left bank line moves closer to Majuli Island, the force per unit area of discharge cross-section increases and that accelerates erosion of the island. Thus, there is no doubt about it that the erosion of Majuli Island is principally caused by the river laterally.

Thalweg Migration

As mentioned earlier, a major question related to the trend of erosion of Majuli Island is whether the place is eroding mainly due to the frontal impact of the highly fluctuating discharge of the Brahmaputra or the lateral impact of its different channels. Like other mountain-fed rivers, during selective short phases in the year, the Brahmaputra transports the major bulk of water and sediments. The fluctuating characteristic of impounding streams on a land-mass will loosen the resistance of its frontal part if it is flat. When we compare the nature of erosion in different parts of Majuli Island, it is observed that there is indeed erosion in the frontal part, but from 1975 to 2005, the rate of erosion in the lower Majuli (~32.3%) was just double the upper Majuli (~16.3%). The Brahmaputra adjacent to Majuli Island is changing in different ways. 'Thalweg migration' is a very significant characteristic (Coleman, 1969) of the Brahmaputra River. In some of the reaches, it becomes very difficult to identify the main trunk. On the other hand, over time, secondary trunks might become the principle one, or the principal trunk can itself shift else-where. Yet, assuming the widest trunk is also the deepest one, when we map the median path of the widest channel, the thalweg, a distinct temporal pattern emerges (Figure 4.21d). In the upper and the middle Majuli, the thalweg shows an eastward migrating trend away from the island, whereas in most of the lower Majuli, the thalweg is moving closer to Majuli, and proportionately, the erosion rate is escalating in the lower fringe (Figure 4.21e). Thus, the main erosion is due to the lateral impact. This observation is substantiated by the nature of the thalweg migration mentioned in the previous section. As seen in Figure 4.21d, thalweg is migrating away from Majuli Island in its upper part, and just after crossing the middle stretch, in the entire lower part the thalweg is migrating towards Majuli Island. Thus, thalweg migration may not be the main cause of erosion for the upper Majuli, but it is so for the lower Majuli.

Subansiri Effect

Another major cause behind the higher rate of erosion of the lower Majuli seems to be related to the shifting characteristic of the Subansiri River (Figure 4.21f). Before the 1950 earthquake, the impact factor of the river discharge on the western boundary of Majuli Island was substantially lower because the river was having several branches. During the 1950 earthquake, probably due to the co-seismic subsidence of the Subansiri depression (Lahiri and Sinha, 2012), the river shifted further in the SW direction, its sinuosity was reduced, and it preferred non-branching. Thus, the lower part of Majuli Island experienced a double impact. Advancing thalweg of the Brahmaputra and enhanced thrust of the Subansiri acted in unison to accelerate the rate of erosion of Majuli Island.

Effect of Erosion of Majuli Island on BB Formation

This is a general expectation that the erosion of Majuli Island is influencing the nature of BB formation in the adjoining channel belt of the Brahmaputra. Accordingly, when we investigate the relation in the form BB=f (IF) and try to fit a straight line, for different time durations, we get straight-line equations as follows (Figure 4.24a):

$$BB = -17.78 + 0.7167 \ ME (1915 - 1975)$$

$$BB = 48.07 + 0.3975\ ME\,(1975 - 2005)$$

$$BB = 14.38 + 0.6631\ ME\,(1915 - 2005)$$

Here, we observe a jump in the distribution pattern of the points from the status during 1915–1975 and 1975–2005. The broader temporal change during 1915–2005 is found to be lying somewhere in the middle. All these curves show a near-parallel and increasing general trend, showing erosion and new BB formation have a positive correlation, and more erosion results in an increase in the cumulative areas of the BBs. However, at zero erosion of Majuli, during 1915–1975, we observe a decay in the changing areas of the

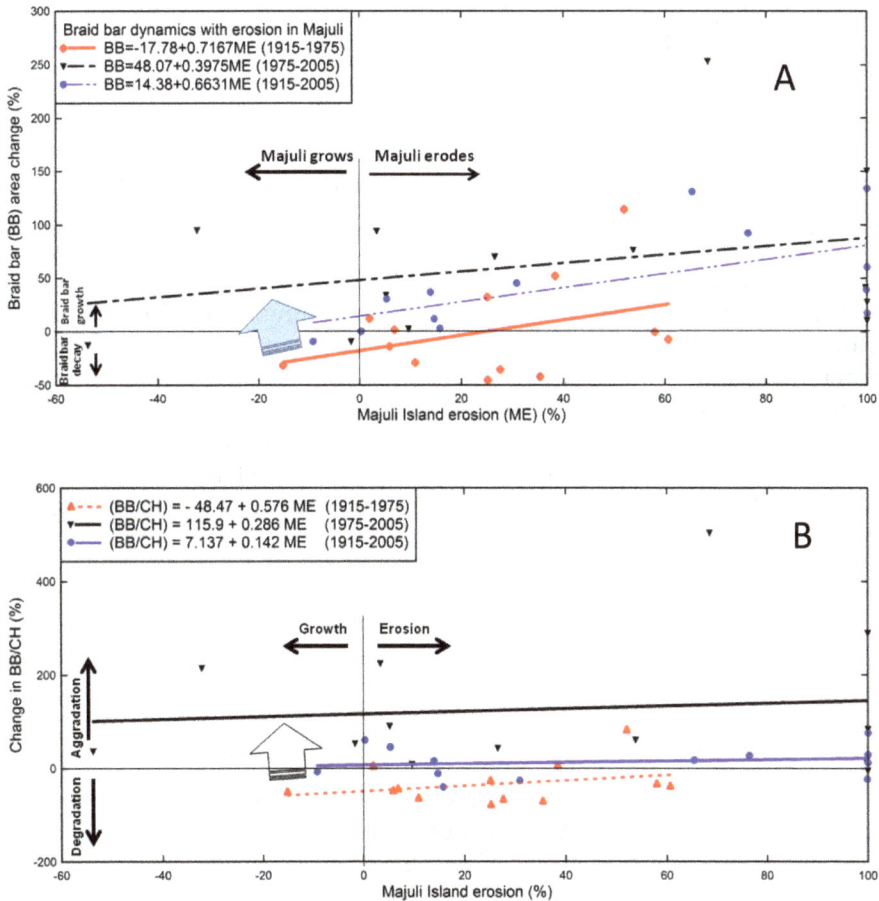

FIGURE 4.24

Linear minimum least square relations.

BBs, and during 1975–2005, for the identical condition, there was growth. Thus, there must be other factors responsible for the BB formations.

Erosion of Majuli Island and the Aggradation Tendency

When the percentage change in the BB/CH ratio is plotted as a function of the percentage erosion rate of Majuli Island per unit area, the linear curve fit method gives us the following equations:

$$(BB / CH) = -48.47 + 0.576 \, ME(1915 - 1975)$$

$$(BB / CH) = 115.9 + 0.286 \, ME(1975 - 2005)$$

$$(BB / CH) = 7.137 + 0.142 \, ME(1915 - 2005)$$

We observe a remarkable jump in the aggradation tendency of the Majuli bearing channel belt of the Brahmaputra (Figure 4.24b) from 1915–1975 to 1975–2005. Secondly, the slope of the curves indicates that there is a positive correlation between the increase in the rate of erosion of Majuli Island and the aggradation tendency in the corresponding channel belt reaches. However, the BB/CH over the period 1915–2005 shows a minimal change, albeit with an upward tendency, with an increasing rate of erosion. Thirdly, in the places where there is no erosion of Majuli Island, aggradation continues. This shows that there are other causes of aggradation. Fourthly, the low slope angle of the straight line indicates even a very high increase in the rate of erosion brings just an incremental change in the BB/CH. Thus, the aggradation is influenced only marginally by the erosion of Majuli Island.

Existing Hypothesis for the Genesis of Majuli Island and Related Questions

The existing explanation for the genesis of Majuli Island (Sarma and Phukan, 2004) puts the old Brahmaputra about 250 years back flowing through the northern side of the present-day Majuli Island as a considerably low-energy meandering course named Lohit, the width of which was comparable with one of its present-day tributaries, the Burhi Dihing. The model shows the presence of one Dihing River passing through the present course of the Brahmaputra River along the southern part of the present-day Majuli Island.

The great flood of 1750 is identified as the agency that converted the earlier low-energy meandering river Lohit into a very high-energy braided river, the present-day Brahmaputra. The model also suggests that a sudden rise in the magnitude of discharge necessitated the development of anabranch, in line with the suggestions made by Richardson and Thorne (2001) and Jain and Sinha (2004). This anabranch flowing through the earlier course of the Dihing formed a landlocked area called Majuli. So, the explanation is given not only about the making of Majuli island but also about the formation of the Brahmaputra River in this part of the land. The model adds further that in the last 250 years, the discharge through the main Brahmaputra continued decreasing very fast, and consequently, the anabranch flowing through the southern flank of Majuli Island continued flourishing. This model raises several questions. (i) Why 250 years back, Brahmaputra was flowing as a meandering river? (ii) As the stream power of a meandering course is much lesser than the straight braided channel, why was the stream power of the Brahmaputra so low? (iii) Was it due to a much lesser volume of water discharge from the catchment area? (iv) The catchment area of the Brahmaputra is mainly located in the Eastern Himalayas and although ice-melt water constitutes a significant component of the discharge into the Brahmaputra, it is mainly the monsoonal precipitation in the catchment which is responsible for the large volume of water and sediment transport. Did the catchment area for the Brahmaputra valley witness a drastic change in the monsoonal precipitation 250 years back? We have dealt with most of these issues in Chapter 6 from a new perspective and explained the surface–deep subsurface connection to understand the 'Majuli'-type situations for the valley having a high degree of tectonic control.

Dibru-Saikhowa Island – the New Majuli

Segment 1 of the upper reach of the Brahmaputra channel belt has underwent some major changes in the last 2 decades, which gave birth to a major island having an area of around 300 km². The location of the Dibru-Saikhowa Island is shown in Figure 4.25 about the major geological elements of the surrounding area.

The Brahmaputra River in this segment acts as a multi-channel river system where the individual presence of three major tributaries – the Lohit, the Dibang, and the Siang, emerging from different catchment conditions with different stream powers can be felt very distinctly. A comparative study of the temporal changes (Figure 4.25) in fluvial dynamics shows three phases: firstly, bifurcation of the Siang River and its north-westward shift: secondly, confluence shift and positive stretching of the Brahmaputra channel belt; and thirdly, avulsion of the Lohit River by channel capturing and formation of

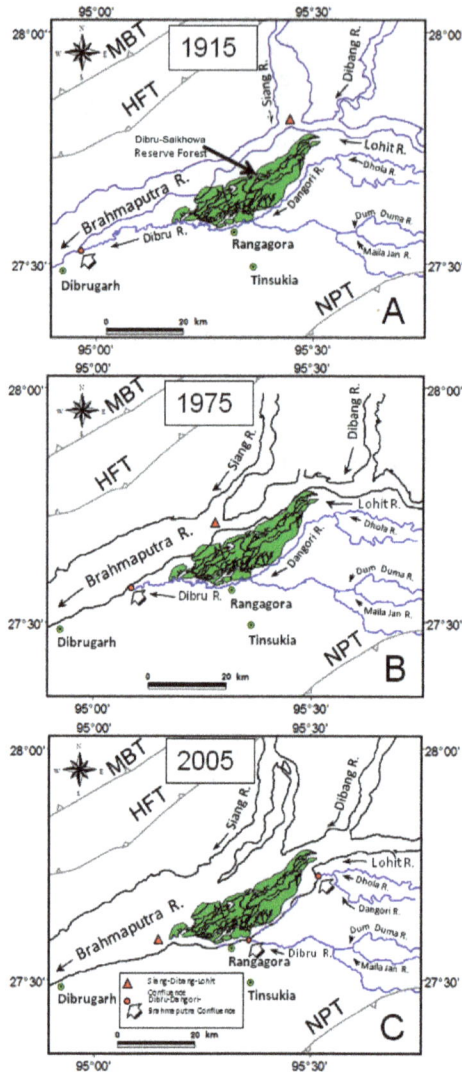

FIGURE 4.25
Recent evolution of Dibru-Saikhowa Island.

the Dibru-Saikhowa Island, accompanied by further shifting of the confluence (Gogoi et al., 2022).

Avulsion of the Lohit River from the northern fringe of the Dibru-Saikhowa reserve forest towards its southern side has reduced the place to an island. Two confluences are found: At the first confluence C1, Dibang and the Siang rivers meet each other. At the second confluence C2 below the Dibru-Saikhowa Island, discharges from the Dibang and the Siang meet with the Lohit, along with many other smaller streams of the south bank of the valley.

A maze of thin channels is present on the Dibru-Saikhowa Island, most of which are parallel to those of the NE-SW-bound flow direction of the Brahmaputra River. However, a few criss-crossing channels are also there.

Bank erosion in Rohmoria

Geographically, as can be seen in the topographic map of the Survey of India prepared during 1912–1926 (scale 1:253,440, i.e., 1 inch = 4 miles), a large village located around 25 km away from the Dibrugarh town towards the upstream (southward) direction of the Brahmaputra River was known by the name 'Rohmoria'. With unabated bank erosion, the most affected adjoining areas extended in the downstream direction from Nogaghuli (where earlier the Dibru River used to meet the Brahmaputra River) and up to the Balijan tea garden. The further upstream direction was associated with the name 'Rohmoria'. The first decade of the 21st century saw a much quicker validation of the 'Rohmoria' in Assam synonymous with 'the place of fastest riverbank erosion in the entire Brahmaputra valley' (Kotoky et al, 2005; Sarma, 2008). Rohmoria reduced in area from 236.54 km² in 1915 to 160.53 km² in 1975. If we assume that the bank erosion was consistently continuing throughout, this amounts to an erosion rate of 1.27 km²/year. In 2005, the remaining area of Rohmoria was measured as 74.57 km² only, and that means the rate of erosion attained a significant jump, an average of 2.86 km²/year in the last 30 years (Figure 4.26).

There used to be a road called 'Tamuli Ali' – a very old one believed to be constructed during the days of the Ahom dynasty, which connected Dibrugarh with the Tinsukia town. Rohmoria used to be a place lying in between. Later on, this was rechristened as the Dibrugarh–Rongagora–Tinsukia (DRT) Road. The old map of the pre-1950 earthquake shows clearly that the DRT Road, running more or less parallel to the Dibru River, divided the Rohmoria area into two halves. In the 1976 map, the rapidly advancing frontier of the Brahmaputra River had already engulfed a portion of the DRT Road. In 2005, the road was cut off.

Avulsion of the Lohit River from the western part of the Dibru Saikhowa reserve forest to the eastern flank was the first important cause of the rapid rate of erosion of Rohmoria. Secondly, the rapid shifting of the confluence point of the Lohit, Dibang, and Siang rivers in the further downstream direction, as discussed earlier, increased the pressure tremendously on the bank at Rohmoria. Thirdly, the intra-channel–belt flow characteristics of the Brahmaputra, the major channel becoming minor, and vice versa, continued changing. This is most probably due to the unevenness of the sediment dispersal and the locally raised flow diversion structures built by the state departments.

Currently, an advancing frontier of the main channel, in the form of a 'bow' (Figure 4.27), is directed towards Rohmoriâ, making it highly erosion-prone

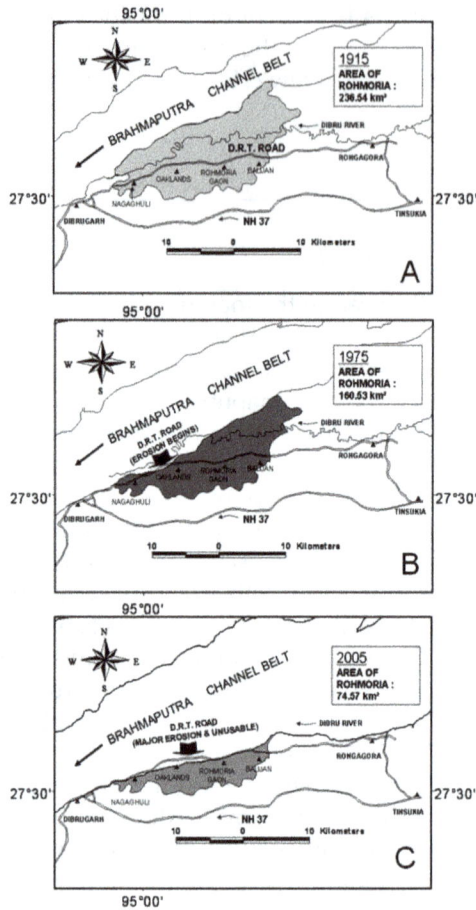

FIGURE 4.26
Bank erosion in Rohmoria.

(Lahiri and Borgohain, 2011). The fourth factor is the presence of very loose sands belonging to the older flood plain deposits, just below the clayey top-soil. This facilitates rapid toe-cutting of the banks and consequent slumping of the bank (Lahiri, 2018). Visibly, these are the four major factors responsible for the rapid bank erosion at and near Rohmoria.

Dynamics of the Brahmaputra Tributaries

The Brahmaputra River has tributaries along the north bank originating in the eastern Himalayan watershed, as well as along the south bank draining

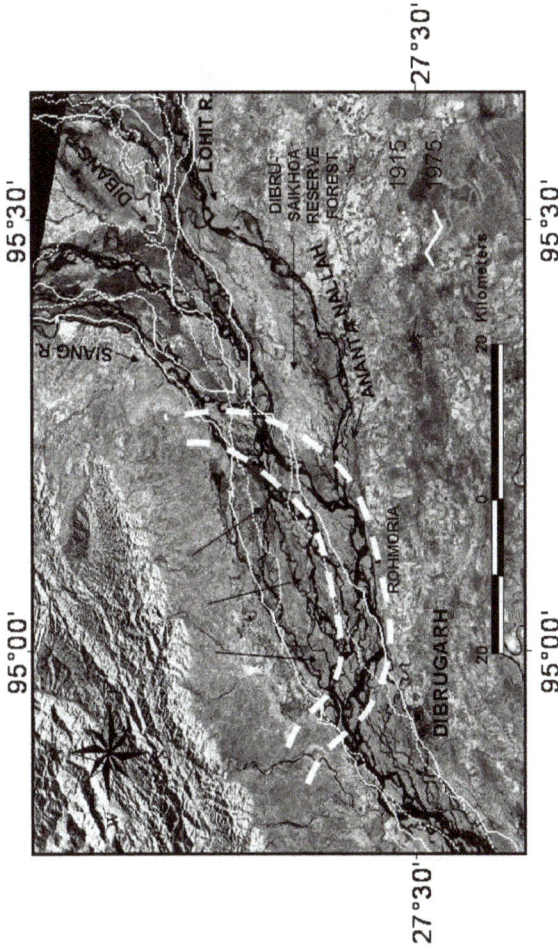

FIGURE 4.27
Advancing 'bow-like' river dynamics.

through the watershed of the Indo-Burman Range. Significant differences are noted in the fluvial dynamics of the northern and southern tributaries.

Most of the major tributaries joining the Brahmaputra from the north have distinct palaeochannels, as revealed from a comparison of historical maps and recent satellite images (Figure 4.28a). This suggests that avulsion is the dominant mechanism of fluvial dynamics on a historical scale. For example, three rivers (viz., the Jiya Dhol, the Sisi, and the Simen) have avulsed westwards between 1915 and 2005, and palaeochannels are marked to the east of the modern channels. The other tributaries have remained fairly stable during this period. A comparison of sinuosity indices of the older and new channels reveals that except for the Simen, the older channels have distinctly lower sinuosities than the new channels. This observation is contrary to the general belief

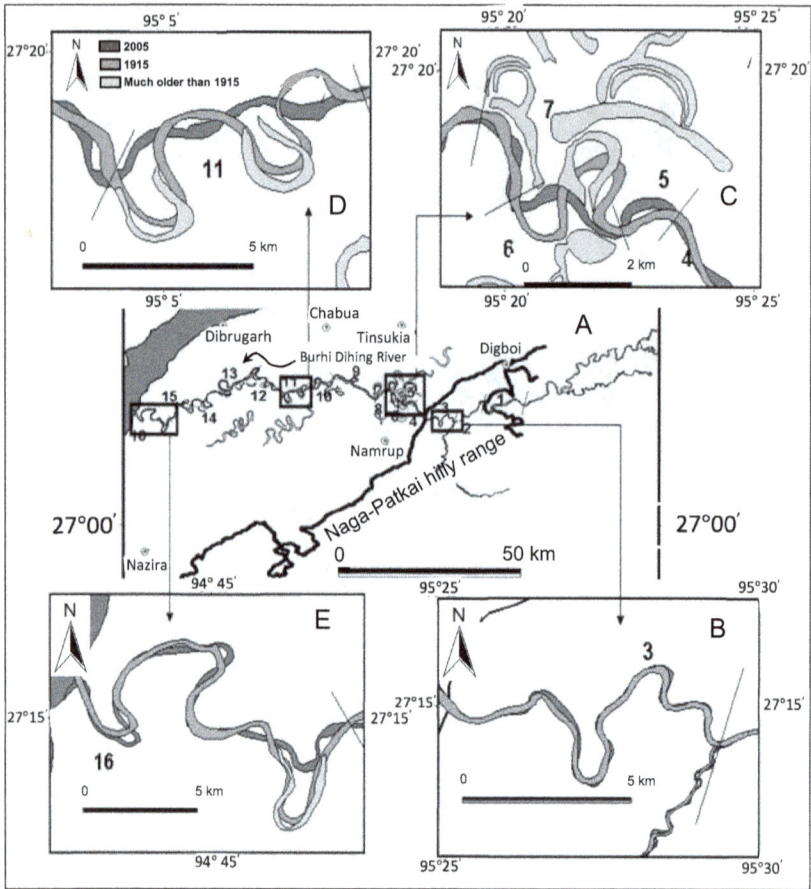

FIGURE 4.28
Dynamics of the Burhi Dihing River.

that the new avulsion channels have morphological characteristics similar to those of the older channels (Allen, 1965). Such morphological changes owing to avulsion are fairly common in the Gangetic River and have been reported by previous workers (Sinha, 1996; Richards et al., 1993). The unidirectional (westward) shifting of all the northern tributaries is certainly remarkable.

South Bank Tributaries

Three important southern tributary rivers originating from the Indo-Burman hilly range are the Burhi Dihing (Figure 4.28), Disang (Figure 4.29), and Dikhau.

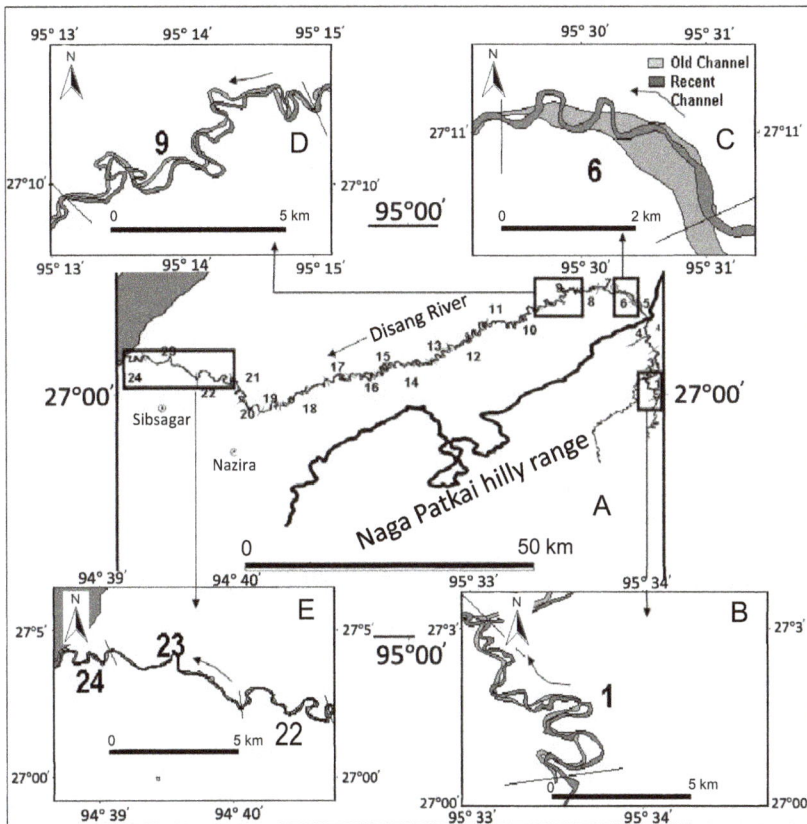

FIGURE 4.29
Dynamics of the Disang River.

Out of these, the Disang has a much longer flow path (by about 20 km) than the other two before they join the Brahmaputra. Furthermore, a series of subtle impressions of palaeochannels are observed on the imagery along the southern bank of the Brahmaputra River. These palaeochannels were not found on the historical topographic maps; therefore, they must be older. This implies a larger avulsion frequency of these channels. Historical maps show, however, that several oxbow lakes (Figure 4.28) formed during the last 90 years (1915–2005), and this suggests that meander migration has been one of the common mechanisms of channel shift in southern tributaries.

These three prominent rivers of the south bank (viz., the Burhi Dihing, the Disang, and the Dikhau) have average sinuosities of 1.37, 2.06, and 1.96 (Tables A2.9–A2.11 in appendix), respectively (Figure 4.30a–c), within the valley after they emerge from the Naga-Patkai hills and then joining the Brahmaputra River. Figure 4.30 shows the reach-scale sinuosities for these

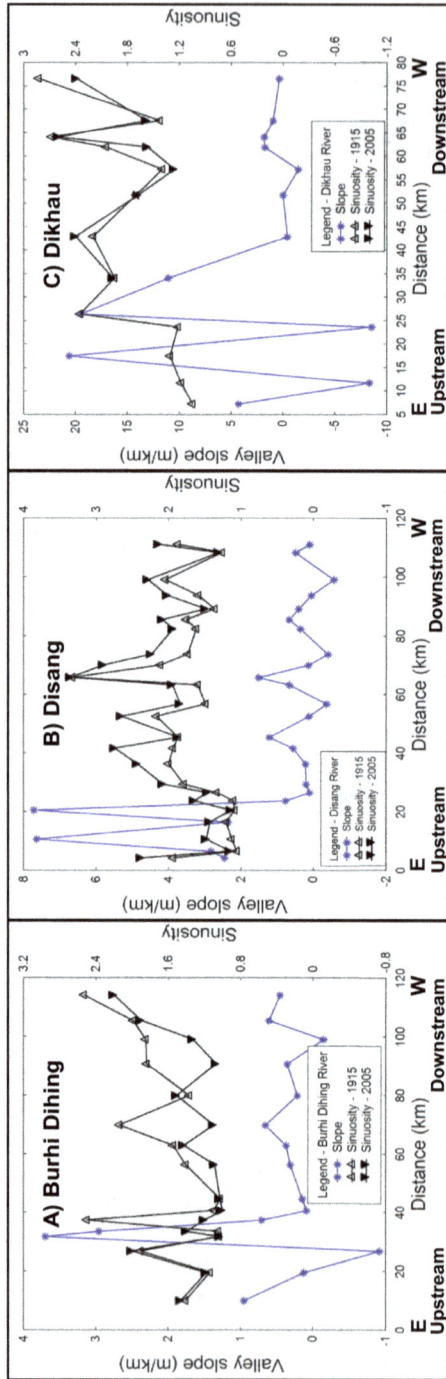

FIGURE 4.30
Slope–sinuosity relationship of three south bank tributary rivers.

rivers for 1915, along with the slope of the individual reaches. The Disang River shows a continuous increase (~ 50%) in sinuosity over a period of 90 years in all reaches with high valley slopes. The Burhi Dihing and Dikhau Rivers also show an increase in sinuosity in reaches of high channel slopes, although we note an overall decrease in sinuosity temporally.

North Bank Tributaries

North bank tributaries originate from the eastern Himalayan watershed. In the north bank, the Subansiri is the most prominent tributary, but in addition to the Subansiri, we have considered seven smaller channels: Jiya Dhol, Sisi, Simen, Dikari, Depi, Silli or Lekojan, and Remi (Figure 4.31a—j and Tables A2.12–A2.19 in appendix). Most of the northern tributaries of the Brahmaputra River flow through the Subansiri depression (discussed in the next section), and sediment loading in the Subansiri depression has likely resulted in a south-westward tilt of the adjoining parts of the Himalayan foreland, which, in turn, caused unidirectional avulsion of the tributaries of the north bank (Figure 4.31a and b). It is also interesting to note that the avulsive shifts in these rivers diminish as we move north-eastward away from the Subansiri depression.

Morphotectonic Zonation of the Upper Reach of the Brahmaputra Valley

Seismic data (discussed in detail in Chapter 6) indicate that the thickness of the alluvium along the Himalayan Frontal Thrust in the Brahmaputra valley reaches about 2 km (Das Gupta and Biswas, 2000), while the average surface elevation in these parts of the valley is ~100 m AMSL. Accumulation of thick alluvial fills requires not only a high sediment flux from the hinterland but also continuous creation of accommodation space in the foreland. In the active foreland basins such as the Himalayas, crustal readjustments through tectonic movements (subsidence) are obvious mechanisms for the creation of large accommodation space, particularly in the frontal parts. In such tectonically active regions, a common manifestation is a very dynamic river regime, but it is often difficult to relate the two without sound geophysical data and characterization of major tectonic domains. Based on the tectonic map of the area interpreted from the seismic data acquired by the oil companies (discussed in Chapter 6), we have attempted a first-order tectonogeomorphic zonation of the valley: (i) the central uplift, (ii) the slopes, and (iii) the depressions (Figure 4.32). This zonation has helped us explain the spatial variability in fluvial dynamics.

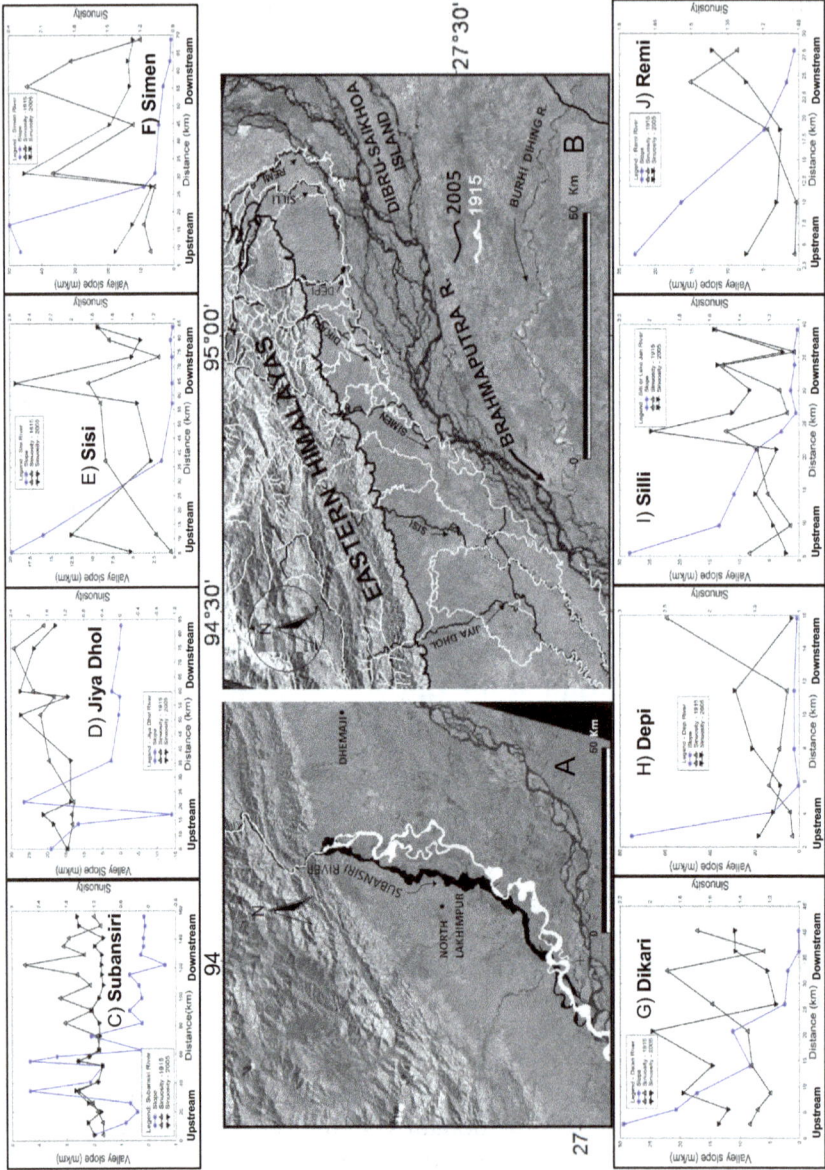

FIGURE 4.31
Tributary rivers of the north bank of the Brahmaputra River.

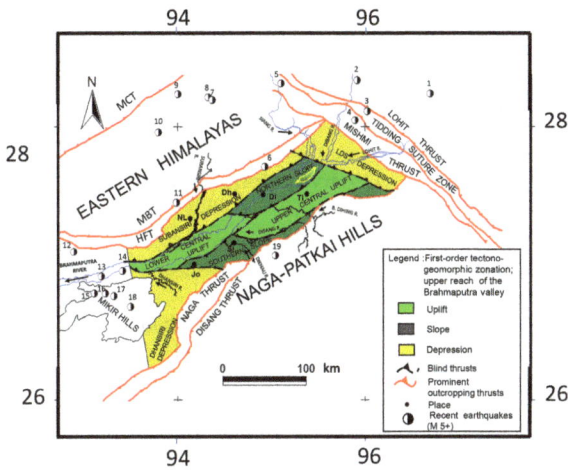

FIGURE 4.32
Morphotectonic zonation.

The lower uplift zone with an area of ~3,258 km^2 is closer to the Himalayan front, and Majuli Island is located in this zone. The northern boundary of this zone is a blind fault that runs parallel to the HFT, and the southern margin runs approximately parallel to the Naga thrust. The north-eastern end is marked by another blind fault parallel to the Mishmi thrust. Of the two slopes, the northern slope is principally controlled by the Himalayan orogeny, and the southern slope is controlled by the Indo-Burman plate dynamics. Furthermore, a thick accumulation of sediments has been recorded in three major depressions: the Subansiri, the Dhansiri, and the Lohit, in the seismic sections.

The central uplift is an NE-SW trending block, around 7,466 km^2 in an area that can be divided into upper and lower uplift zones. The upper uplift zone with an area of ~4,286 km^2 is closer to the Naga-Patkai thrust, and it seems to have influenced the southern bank line of the Brahmaputra River.

Currently, unit 1 and most of unit 2 of the Brahmaputra channel belt are mostly confined within the northern slope. Unit 3 falls in the lower part of the central uplift zone and is flanked by the southern slope.

Based on the measured data on the morphodynamics of the different units (Table A2.1 in appendix), the interpreted response for the period of study (1915–2005) is summarized in Table 4.2.

Unit 1 with a high valley slope shows the maximum bank-line shift and channel belt widening, albeit with varying rates during the two time periods (1915–1975 and 1975–2005). Sediments eroded from the banks, mostly through slumping during falling stages (Coleman, 1969; Goswami, 1985), are deposited in the channel belt and are manifested as the overall increase in the sandbar area in this unit (Table A2.1). Therefore, unit 1 is characterized as a major zone of aggradation. Unit 2 shows a much lower but increasing trend in channel belt widening and minimal bank-line shift and continuous

TABLE 4.2

Differential Response of the Brahmaputra Channel Belt as Interpreted from the Morphodynamics during the Period 1915–2005

Units	Tectono-Geomorphic Setting and Terrain Characteristics	Observed Morphodynamics (1915–2005)	Interpreted Response
Unit 1	Northern slope, average channel slope 0.35 m/km; mean elevation 114 m	Maximum bank-line shift South bank more mobile in recent times Continuous increase in channel belt width and CH (CH/CHB = −34%) Sandbar area increased drastically (BB/CH = +76%)	Zone of major aggradation and channel widening; highly dynamic in response to higher sediment supply from the northern slopes
Unit 2	Northern slope, average channel slope 0.25 m/km; mean elevation 98 m	Minimum bank-line shift Increasing trend in channel belt width and CH but with a lower rate (CH/CHB = +3.57%) Sandbar area decreased (BB/CH = −7%)	Zone of aggradation (with some reaches of minor degradation) and minimal dynamics; gradual diminishing of the effect of slopes
Unit 3	Central uplift zone, average channel slope 0.21 m/km; mean elevation 77 m	Moderate shift in bank line, pockets of significant negative shift (narrowing) Channel belt width stable, some increase in CH (CH/CHB = +15%) Sandbar area nearly unchanged (BB/CH = −18%)	Zone of degradation, channel narrowing (and deepening) in response to uplift

CH, channel area.

increase in the sandbar area. Therefore, we have characterized this unit also as a zone of aggradation, except that some minor degradation in some reaches may have occurred in the last ~20 years, as reflected by a negative change in the ratio of the sandbar area to CH. We argue that the morphodynamic response of units 1 and 2 is a manifestation of northern slopes that resulted in a higher sediment supply.

Higher sediment supply, in turn, results in the 'wandering of thalweg' (Coleman, 1969), undercutting of banks, and slumping. As a result, unit 1 has shown a very high degree of morphodynamics during the last 90 years, and the effects have gradually diminished in unit 2 because of the distal position of this unit.

By contrast, unit 3 has a significantly lower average channel slope and shows a distinctly different trend in morphodynamics. The most significant temporal changes observed in this unit are significant channel narrowing (and deepening) in certain reaches, as evidenced by negative bank-line shift (Figure 4.7).

Although we observe an increase in the CH, the sandbar area decreased by ~14% during 1915–1975, but the trends reversed during 1975–2005 and nearly reverted to 1915 conditions. We characterize this unit as a zone of degradation and argue that this is the response to the uplift in this zone. However, this unit

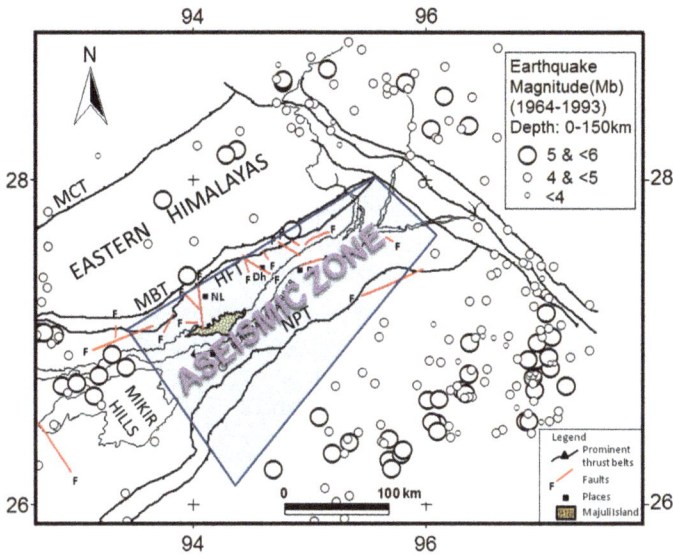

FIGURE 4.33
Recent seismicity.

has seemingly started experiencing aggradation in recent times. A recent large-magnitude earthquake ($M=8.7$) in 1950 and subsequent readjustment of the valley (post-1950 inter-seismic changes), as evidenced by several earthquake events of $M > 5$ after the 1950 earthquake (Tandon, 1954), may have been responsible for such reversals in morphodynamics (Table A2.20 in the appendix). Furthermore, the reach-scale degradation–aggradation in each unit is probably caused by the second-order tectonic differentiation. Variable sediment dispersal and variable geologic control may also generate variable stress patterns.

As a result, the Brahmaputra is crossing several second-order faults that are manifested as 'highs' and 'lows' (Figure 4.10c). The 'damming' owing to the regional highs (Schumm et al., 2000) increases aggradation, lateral erosion, and a relative increase in the width of the river, which we observe when the temporal and planar trends of relative width variation are compared with the valley elevation and slope (Figure 4.10b and c). The exact relationship between these second-order faults and the first-order major faults is not clear yet, and this needs further investigation using high-resolution subsurface data.

Recent Seismicity and Fluvial Dynamics

Some workers have identified the Upper Assam valley as a 'seismic gap' (Khattri, 1987) between the epicentres of two great earthquakes of 1897 and 1950; others called it 'aseismic' (Nandy and Das Gupta, 1991). Mukhopadhyay

(1984) endorsed the 'aseismic' hypothesis and opined from his seismotec-tonic studies that the tectonic stress gets distributed within the thrusts of the Himalayan and Burmese arcs lying on either side of the Upper Assam valley, and that this keeps the valley aseismic. This observation is in line with the study of the Coda waves (Hazarika et al., 2009) for the smaller earthquakes (magnitudes varying from 1.2 to 3.9). A quality index of the coda waves 'Qc' is supposed to have higher values for the lesser decay. Unconsolidated mate-rials, highly fractured or otherwise, are supposed to cause a greater degree of attenuation of the waves and that is why will show a lower value of 'Qc'. The eastern part of the Upper Assam valley closer to Majuli Island represents highly unconsolidated materials, thereby causing a higher degree of seismic energy attenuation in all ranges of frequencies (1–18 Hz).

The area along with the major thrusts and the Mikir Hills is much more active than the Naga-Patkai thrust, as reflected by the distribution of epicen-tres of recent moderate earthquakes ($M > 5$) (Figure 4.33). The presence of blind faults (interpreted from the deep seismic data) within the valley suggests that although the thrust planes at the valley margin are the major zones of stress release, stress is also partially transmitted to the valley. As a result, we expect co-seismic subsidence and inter-seismic uplift in the valley (Burbank and Anderson, 2012), which will continue uplifting some of the regions and gener-ating simultaneously accommodation space in the depressions.

Morphotectonic Model for Morphodynamics

Figure 4.34 presents a morphotectonic model for the study area summariz-ing the co-seismic and inter-seismic forcings influencing the morphody-namics of the valley. Figure 4.34a shows the pre-1950 earthquake scenario depicting the highly dynamic and aggradational regime in the upper two units. Frequently avulsing tributaries along the north bank in contrast to the older meander bends of the south bank tributaries reflect the influence of two different tectonic domains, namely, the Himalayan thrust belt and the Naga-Patkai thrust. Figure 4.34b shows the morphodynamic changes caused by the 1950 earthquake ($M = 8.7$), which resulted in co-seismic subsidence in the Subansiri and Lohit depressions and generated accommodation space. Figure 4.34c shows the post-1950 earthquake (present-day) scenario.

Aggradation in the Lohit depression tremendously increased channel insta-bility and lateral erosion in unit 1. The site of effective aggradation shows a swi-tchover from unit 1 to unit 3. Subsidence in the Subansiri depression continued the north bank tributaries to avulse in the southwest direction. The south bank tributaries remained mostly unaffected during the entire period. Large tracts of highlands called Majuli, lying between two rivers and subsequently becoming part of the main river system of the valley, are very interesting components of

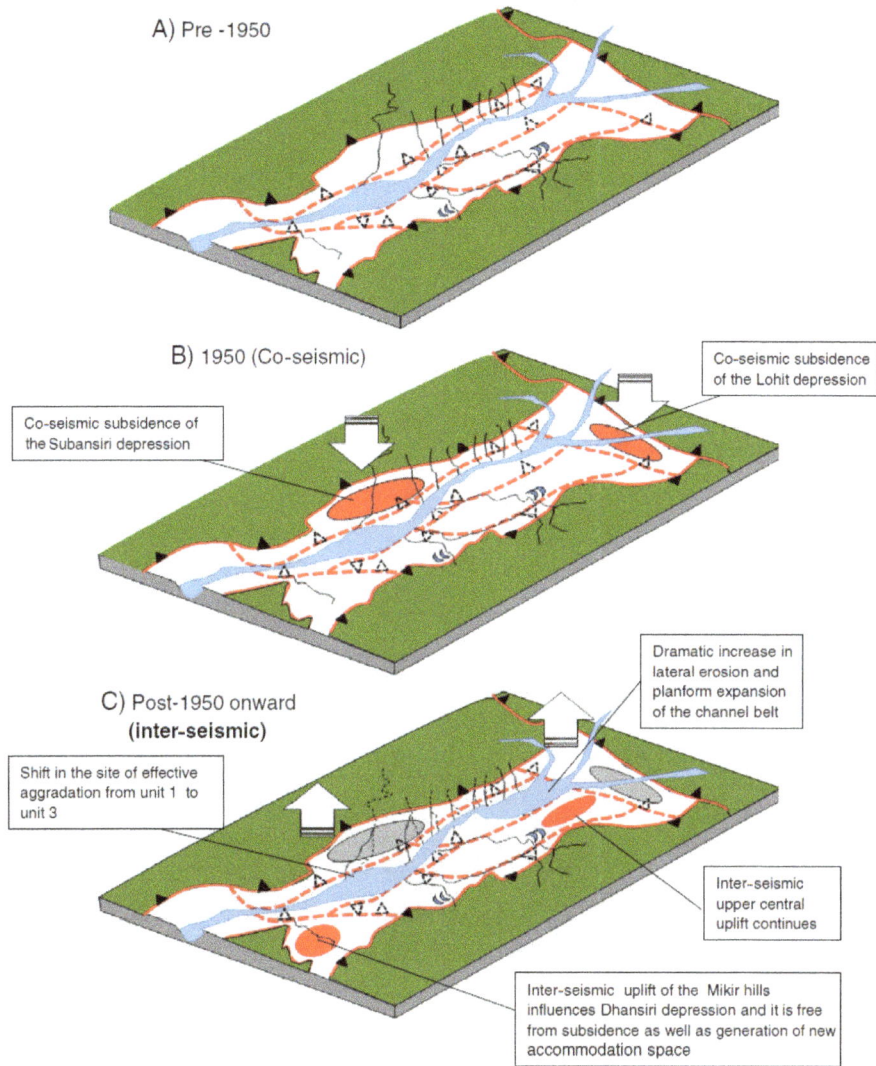

A) Pre -1950

B) 1950 (Co-seismic)

Co-seismic subsidence of the Lohit depression

Co-seismic subsidence of the Subansiri depression

Dramatic increase in lateral erosion and planform expansion of the channel belt

C) Post-1950 onward
(inter-seismic)

Shift in the site of effective aggradation from unit 1 to unit 3

Inter-seismic upper central uplift continues

Inter-seismic uplift of the Mikir hills influences Dhansiri depression and it is free from subsidence as well as generation of new accommodation space

FIGURE 4.34
Models summarizing co-seismic and inter-seismic changes influencing the river dynamics.

the Brahmaputra River system. The present study shows that the conversion of these landforms into islands is not necessarily a fallout of episodic events like co-seismic changes. Basins having strong structural controls and propensity to develop highly uneven accommodation spaces as well as selective uplifts influence the landscape–riverscape duality in much more surprising combinations. Particularly, the Majuli-type landforms have major structural bearings, which we have described in Chapter 6.

5

Quaternary Geomorphology, Shallow Subsurface Stratigraphy, and Neotectonics

The dynamic processes involved in the evolution of geological structures, visible on the surface in outcrops and interpreted in the subsurface by geophysical data, play an important role in controlling geomorphology. For example, the alignment of prominent oil fields of the Upper Assam valley seems to be controlled by the Naga-Patkai Thrust (NPT) belt. In addition to the outcropping frontal thrusts, the en echelon thrust front seems to be advancing underneath towards the central axis of the valley in the form of blind faults (Verma and Mukhopadhyay, 1976, 1977; Murthy, 1983; Ranga Rao, 1983; Ranga Rao and Samanta, 1987; Bally, 1997; Sinha et al, 1997; Dasgupta et al., 2000; Kent et al, 2002; Kent and Dasgupta, 2004; Goswami and Goswami, 2007). Blind fault-related folds provide hydrocarbon reservoir conditions. Moreover, these structural features influence river dynamics and the associated landform evolution. We have seen in the previous chapter that in the last 100 years, prominent tributaries of the Brahmaputra, like the Burhi Dihing, Disang and Dikhau, flowing through the south bank of the valley have been much less dynamic than the north bank tributaries. However, the presence of rich palaeochannel signatures on the landscape, as could be seen in the satellite imageries, indicates the dynamic characteristics of these rivers. The directional changes in the river dynamics, the preferences, are most of the time guided by neotectonics associated with the frontal thrust belt. Some of the palaeochannels are apparent on the surface, and the remaining are concealed below a thin veneer of overburden. Hence, it is very difficult to disentangle the network of palaeochannels in a chronological order from the surface signatures alone. As a primary assumption, palaeochannels can be differentiated from the older flood plain deposits, which they incise due to the lateral variability of physical properties. Such distinction can, in turn, be used to compare different stages of shifting for a given channel from layer discrimination; for example, the higher the number of layers, the older the age of the channel under consideration. Subsequently, a comparison of the preferred direction of river dynamics can help understand the structural control and their consequences like oil migration and probable sites of traps.

Of all the geophysical data, seismic data sets are considered highly reliable to understand vertical and lateral variations of the lithofacies. However, irrespective of the reflection or refraction method, it is usually the boundary between two formations (not the formation as such) that is mapped. Unless

DOI: 10.1201/9781003302353-5

there is contrast in the physical properties, particularly the velocities of the seismic waves (for the present purpose, it is the *P*-waves) through the media and their density values are sharp, the magnitude of reflection coefficient does not become high, and the amplitude standout (differential amplitude over the background noise amplitude by which a 'signal' due to reflection or refraction is identified) does not become prominent enough to mark boundaries. The boundary definition for marine sediments is much stronger than that for continental sediments. Marine sediments are well sorted due to the low-energy environment and remain homogeneous and isotropic at the regional scale compared with the high-energy, poorly sorted fluvial deposits. Sea-level rise–fall and transgression–regression about the coastline bring a drastic change in the sedimentological characteristics of a given place. For example, a place that was earlier shallow shelf margin having mostly the coarser channel deposits of a constructive delta may subsequently become a part of the pro-delta with predominantly finer clays mixed up with carbonates and some turbidites as well. The boundary between these two sets of depositions will be very well defined in the seismic sections. However, the continental sediments, orchestrated mainly by fluvial processes, do not help have well-defined formation boundaries, unless there is a drastic climatic change massively affecting the sediment supply. During seismic investigations due to reflection, we depend mostly on two-way travel time data. Near-surface low-velocity zones (involving more time for seismic waves to travel) influence the deeper data (high-velocity faster zones involving much lesser time to travel a given distance than the shallower horizons). Thus, in oil exploration, the shallow subsurface (within a depth range of 100 m from the surface) acts mostly as a 'noise-enhancing horizon', and this layer is mostly sliced off to implement 'static corrections' as accurately as possible so that the deeper horizons (time surfaces) can be picked up with better resolution and confidence. The topmost layer of the shallow subsurface, also called the weathered zone, is also commonly referred to as the low-velocity layer (LVL). For the seismic data processor, the uniform thickness of the LVL makes matter simple in terms of putting the plane of observation on some standard datum below the actual surface. However, the thickness of the LVL is mostly variable, and that is why meticulous monitoring is carried out by conducting shallow refraction and uphole surveys. Uphole survey data are also mandatory for the regular seismic reflection survey for depth optimization of shot holes (the most common source of artificial energy is dynamite sticks and detonators blasted below the LVL) to facilitate deeper penetration of seismic energy and for maintaining rich frequency content of the input signal. The point that we intend to raise is that once the deeper subsurface boundaries are identified accurately, the shallow subsurface data (velocity and thickness variations obtained from the refraction and uphole methods) become practically useless for oil industries. The basic objective of this section is to show the great utility of these shallow subsurface data sets in understanding at least three major aspects: (i) deciphering palaeo-landscapes by extrapolating surface features to the shallow subsurface, (ii) shallow subsurface stratigraphy and basin evolution during late Quaternary, and (iii) tectonic influence on the sedimentation pattern that can be extended subsequently to seismic micro-zonation. Moreover, shallow subsurface studies can also help realize some of the similar objectives pursued during hydrocarbon exploration activities to obtain deeper subsurface information but sometimes meet with failures. As a typical example, we may refer to mountain front zones, which are greatly influenced by thrust belt tectonics. The leading edges of the thrust belts, involving complex strains, give rise to numerous types of spatially variable dip situations for the layer boundaries. Seismic data quality deteriorates for variable dip situations. Additionally, the presence of boulder beds in the mountain fronts causes the seismic energy to scatter and severely restricts the depth of penetration of energy and thereby drastically cutting the targeted depth of investigation. Under such circumstances, shallow refraction and uphole data can be immensely helpful to understand subsurface stratigraphy and controls of sedimentation patterns including tectonics.

The use of shallow refraction data is very old. That waves can pass along the planes of the elastic solids was first observed by Lord Rayleigh (1885). Although significant literature did not come up until the end of the First World War, the utility of the method was more or less established (Green, 1974). Pioneering articles by Barton (1929) and Heiland (1929) established the importance of the refraction method. Muscat (1933) explained the theory of refraction shooting in detail. The fan shooting technique, which was used in the earlier days for shallow structural traps of oil, was mainly based on the refraction method. Later, this method was used for mapping subsurface structures (Gardner, 1939), water prospecting, and rock investigations (Hasserlström, 1969; Bachrach and Nur, 1998; Grelle and Guadagno, 2009). In addition to these, workers have used this method to map ancient channels (Pakiser and Black, 1957) and in exploring Quaternary deposits (Burke, 1973). With improved instrumentation, processing techniques, and interpretational software, there is also an instance where the method was used to understand groundwater contamination (Zelt et al., 2006). Compared with the shallow refraction methods, the published literature on uphole data is very scanty (Woodward and Menges, 1991).

In this chapter, uphole data were mostly used over shallow refraction seismic data due to at least five reasons: (i) Uphole data are generated from one-way travel time, whereas the shallow refraction data are the two-way travel time having increasing offsets of geophones from the shot points that reduce the data accuracy; (ii) uphole data are the outcome of almost a near-vertical ray path, and this helps maintain good bed boundary resolution; (iii) as uphole data acquisition is carried out inside a borehole having known shot depths that keep on changing for a fixed geophone on the surface, the average velocity calculated is also much more accurate than that of the refraction method; (iv) in refraction and the uphole methods, the observer is supposed to pick up the 'first break' of the geophone response. With increasing offset, the amplitude of the waves suffers, and there is also the effect of phase change. Thus, refraction data sets, with increasing offsets, are bound to be distorted. Uphole data, on the other hand, because of single geo-sensor use (sometimes, an additional geophone is used for precaution), is free from amplitude and phase distortion, and (v) lastly, in the refraction survey, we assume to have subsurface information below the point where the shot hole is blasted. Furthermore, all the information about velocities and thicknesses of the layers are plotted on the map (as a 1D problem) below the shot point. However, the geophones, which are planted at different offset distances from the shot point, pick up signals only after the waves travel laterally for a considerable depth and distance away from the shot hole. Thus, the travel time incorporates more information about earth materials from the adjacent part of the vertical path below the borehole than the borehole itself. This is the reason, for dipping beds, different apparent velocities are obtained in the up-dip and down-dip directions for the second, third, and any deeper layers, and we are supposed to calculate the realistic average velocities again from these values. There is no such problem with uphole data. Since the ray path is vertical, the effect due to the dip of the beds is practically ruled out. Moreover, we receive information strictly for the earth material lying vertically below the shot point on the surface (1D data acquisition in its true spirit) that we plot on the map.

There are, of course, logistic advantages with shallow refraction studies. The uphole seismic survey needs actual drilling of holes, 50–100 m deep, depending upon the extent of shallow subsurface information we aim to gather. Using hammer-based energy source, a shallow refraction survey does not need practically any 'preparation' and is much more cost-effective than the uphole survey.

Some of the major research questions investigated in this chapter are as follows:

1. What are the specific patterns of lateral variability of the shallow subsurface in the south bank of the Brahmaputra valley that constitutes the frontal part of the NPT belt?
2. How did the landscape in the frontal part of the NPT belt evolve during the late Quaternary period?
3. How does the NPT belt in the margins of the upper reach of the Brahmaputra valley influence the modern landscape?

Theoretical Aspects

Lateral variability in the geophysical properties of the layers below the ground surface is contradictory to Steno's principles (1669) regarding original lateral continuity and original horizontality and the superposition principle, whereby older layers are superposed by the younger layers. However, the fact of the matter is in the ideal source–sink relationship, particularly in the deep marine environment (excluding, of course, the seafloor spreading zones), these principles are very much valid, and for falling sea level and regressive conditions, when these marine depositional successions (without the intervention of any structural modifiers) become part of the continents, the outcrop study is bound to endorse Steno's principles. Thus, Steno's observations are valid for most of the continental stratigraphy which is genetically related directly to the marine depositional environments. However, in the high-elevation continental basins with the huge mass transfer of sediments, the landscape is mostly constituted of older flood plains incised by channels having different stream powers.

Recently deposited sediments within the channels and flood plains are usually poorly consolidated and have lesser seismic velocities than the older flood plains. But tectonic processes can bring up the older rocks which may have much higher seismic velocities, thereby generating a drastic lateral variation.

Then, there can be regional 'lows' having much thicker low-velocity sediments. Thus, we are bound to get (i) different types of lateral discontinuity, (ii) lack of horizontality of recently deposited strata, and (iii) several complex associations of older strata with the younger ones. This apparent deviation from Steno's principles is the subject of this chapter, which has a tremendous influence on surface landform evolution.

A 'Layer'

For any geophysical method, it is common to treat sedimentary facies as multi-layered earth, each layer being homogeneous and isotropic. However, when we probe the lateral inhomogeneity of 'one layer' as a three-dimensional problem by data acquisition at several points on the ground surface, the assumption is that all the individual 1D observations are made on several non-interfering smaller blocks (homogeneous and isotropic), each infinite in lateral extent, and are juxtaposed to form ultimately a single layer (Figure 5.1).

Velocity as an Attribute for Basin Analysis

Shallow refraction surveys for a given place provide highly reliable data on the velocities of the topmost layer (weathered layer or simply the LVL) and the sub-weathered layer below the ground surface. The lateral variation in the velocity of a layer can be divided into several classes. To make the

Geological situations   Equivalent 'single layer'

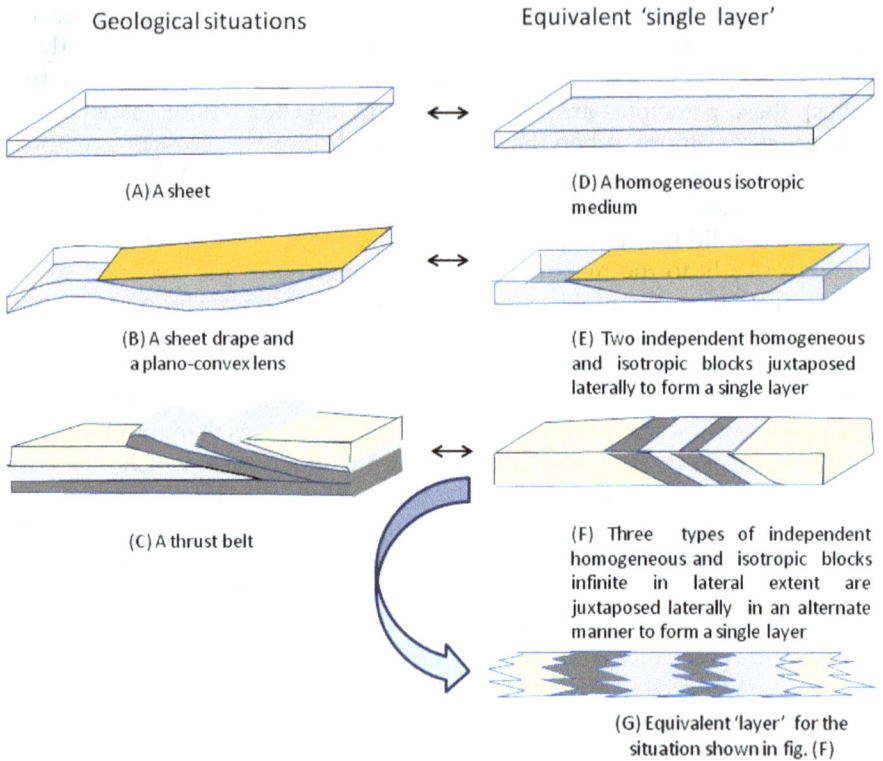

(A) A sheet

(D) A homogeneous isotropic medium

(B) A sheet drape and a plano-convex lens

(E) Two independent homogeneous and isotropic blocks juxtaposed laterally to form a single layer

(C) A thrust belt

(F) Three types of independent homogeneous and isotropic blocks infinite in lateral extent are juxtaposed laterally in an alternate manner to form a single layer

(G) Equivalent 'layer' for the situation shown in fig. (F)

FIGURE 5.1
Causes of anisotropy in the shallow subsurface.

matter simple, we can assume that for young active sedimentary basins the fluvial channels incise the older flood plains and erode the older sediments. While transporting these sediments in the downstream direction, the channels keep on depositing new sediments simultaneously, the variable grain size of which is mostly a function of the stream power and local sorting. The new channel deposits are less consolidated than the older flood plain deposits, which give us the means to choose a rational threshold velocity to divide the velocity variations into just two classes – high and low. So, we have high velocity for the bulk matrix or the older flood plain deposits, and low velocity for the channel deposits (Figure 5.1b and e). Thus, for different layers, it is possible to identify the channel belt location. Furthermore, the subsurface manifestation of the palaeochannel belts would also be governed by the nature of migration. For example, a lateral migration by sweep would produce a wide channel belt, whereas an avulsive shift would produce narrow channel belts separated by flood plains. To understand the trend of channel migration vis-a-vis the present-day location of the channel belt, we can find out the median path of the palaeochannel belt at different depths. If the rate of sediment deposition is known, the temporal change of channel belt migration can be studied with a fair degree of accuracy. Thus, lateral velocity discrimination can act as a reliable attribute to understanding the fluvial dynamics of a place of much older times than what we get from the comparison of the palaeochannels from surface evidence.

Micro-Zonation of the Basin

The specific velocity distribution within different layers can be used to undertake micro-zonation of the landscape based on tectono-morphogenic consolidation of soil properties. Accordingly, by identifying $V_{2(high)}$, $V_{2(low)}$, $V_{1(high)}$, and $V_{1(low)}$, we have four combinations, namely, (i) $V_{2(high)}$ $V_{1(high)}$, (ii) $V_{2(high)}V_{1(low)}$, (iii) $V_{2(low)}V_{1(high)}$, and then (iv) $V_{2(low)}V_{1(low)}$. If we have velocity information for three different layers, instead of four, eight type areas can be identified as follows (Figure 5.2):

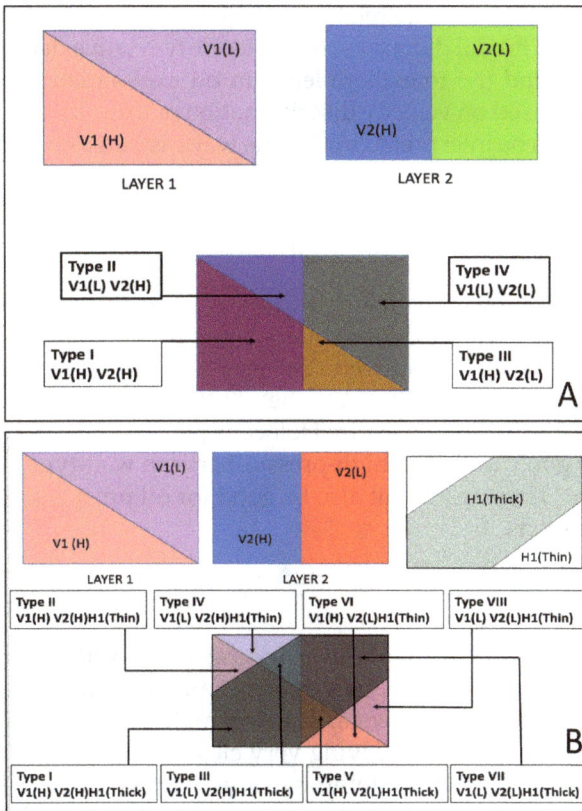

FIGURE 5.2
Mode of thematic overlapping.

   i. Type 1: $V_{3\,(high)}$ $V_{2\,(high)}$ $V_{1\,(high)}$,

   ii. Type 2: $V_{3\,(high)}$ $V_{2\,(high)}$ $V_{1\,(low)}$,

   iii. Type 3: $V_{3\,(high)}$ $V_{2\,(low)}$ $V_{1\,(high)}$,

   iv. Type 4: $V_{3\,(high)}$ $V_{2\,(low)}$ $V_{1\,(low)}$,

   v. Type 5: $V_{3\,(low)}$ $V_{2\,(high)}$ $V_{1\,(high)}$,

   vi. Type 6: $V_{3\,(low)}$ $V_{2\,(high)}$ $V_{1\,(low)}$,

   vii. Type 7: $V_{3\,(low)}$ $V_{2\,(low)}$ $V_{1\,(high)}$, and

   viii. Type 8: $V_{3\,(low)}$ $V_{2\,(low)}$ $V_{1\,(low)}$.

Significance of Tectonogeomorphic-Type Areas

For the frontal thrust belt areas of valleys with evidence of high tectonic readjustments, the tectonogeomorphic zonation can help in at least four different ways: (i) to differentiate relatively stable areas from less stable ones, (ii) to reconstruct the fluvial dynamics, (iii) to identify relative depths of groundwater prospect, and (iv) to explore leads in oil exploration. The qualitative interpretations based on velocity discrimination of three layers for a window on the surface can be interpreted in various ways as outlined in the following sections.

Type 1 Areas $\left[V_{3\,(High)}\ V_{2\,(High)}\ V_{1\,(High)}\right]$

If all three layers show high velocities, this is probably a regional high. It is either constituted of very old flood plain deposits or perhaps the older rocks thrusting upwards and weathered. The area is stable for civil construction unless there is a proven fault zone. The place was bypassed by the fluvial channels probably during the entire Holocene period. Groundwater prospect might be very good, and even high-pressure artisan well-type situations can also be present. This area might also be good for oil prospecting from structural considerations.

Type 2 Areas $\left[V_{3\,(High)}\ V_{2\,(High)}\ V_{1\,(Low)}\right]$

The low velocity of the topmost layer indicates recent flood plain deposits due to a channel lying nearby or a palaeochannel having an avulsive tendency. As both the deeper layers are having high velocities, this might represent a zone of active faults. If lying very close to the regional highs, these areas represent the slopes. Possibilities of seasonal floods make the area vulnerable to habitation. There might be oil indications for shallow structures through seepages.

Type 3 Areas $[V_3$ (High) V_2 (Low) V_1 (High)$]$

The low velocity of the middle layer indicates the presence of an older palaeochannel. As the channel has migrated elsewhere, this might represent areas with very recent uplifting tendencies, which is not so good for constructing high-rise buildings, and additional precautions should be taken for constructing large structures. Groundwater is available at moderate depth. Also, there is oil prospect.

Type 4 Areas $[V_3$ (High) V_2 (Low) V_1 (Low)$]$

Due to periodic subsidence, the channels reoccupied this zone. If palaeochannels used to flow through this area and currently come under the influence of periodic flooding, this must be a very favourable path for the channels. Vulnerable to habitation, great precautions should be taken before raising any major architectural structure. This area is structurally unstable, with less oil prospect.

Type 5 Areas $[V_3$ (Low) V_2 (High) V_1 (High)$]$

During the early to mid-part of the Holocene, there used to be a channel belt through this area, which migrated subsequently to other places. Later on, the place was never occupied by a channel. This is indicative of the fact that the area is uplifting continuously. This can be the area representing newly emerging frontier thrust belts. For general civil construction, the area is quite habitable. Good drinking water is available in deeper horizons. Good oil prospects are in adjacent areas.

Type 6 Areas $[V_3$ (Low) V_2 (high) V_1 (Low)$]$

The older channel belt has been reoccupied recently, which might be by a more powerful stream. The area is a part of the subsiding zones. If there are no active channels, it represents a relatively low-lying area where rainwater accumulates, which is good for rice cultivation, but bad for tea bushes. Good quality groundwater is available in deeper horizons at variable depths. Seasonal water logging makes the area less attractive for civil construction. It is difficult to predict the oil prospect.

Type 7 Areas $[V_3$ (Low) V_2 (Low) V_1 (High)$]$

This area used to be a regional low, perhaps the peripheral zone of the earlier thrust front through which the channels used to flow. With the advent of newer thrust fronts, it experienced a mild uplift, and the channel migrated elsewhere. Groundwater is in plenty, and the quality, in general, is good. In those areas with thinner top layers, additional precautions should be taken for civil construction. It is difficult to predict oil prospects.

Type 8 Areas $[V_{3\ (Low)}\ V_{2\ (Low)}\ V_{1\ (Low)}]$

This area is a regional low caused by a continuous tendency to generate accommodation space. These depressions are very risk-prone for civil construction. Oil prospects in this area is highly uncertain.

Multi-Parametric Attributes for Basin Analysis

As discussed in the previous section, velocity is no doubt a very useful attribute for discriminating different zones of a given area of the basin. Whether it is shallow refraction or uphole, in addition to the velocity values, we also get the lateral variation in thickness for different layers. From a simple situation of double-layered earth with the known values of V_2, V_1, and H_1 (H_2 extends to infinity), multi-layered earth models can be brought into consideration (Figure 5.3 and Table 5.1). Thus, if we use some threshold values of

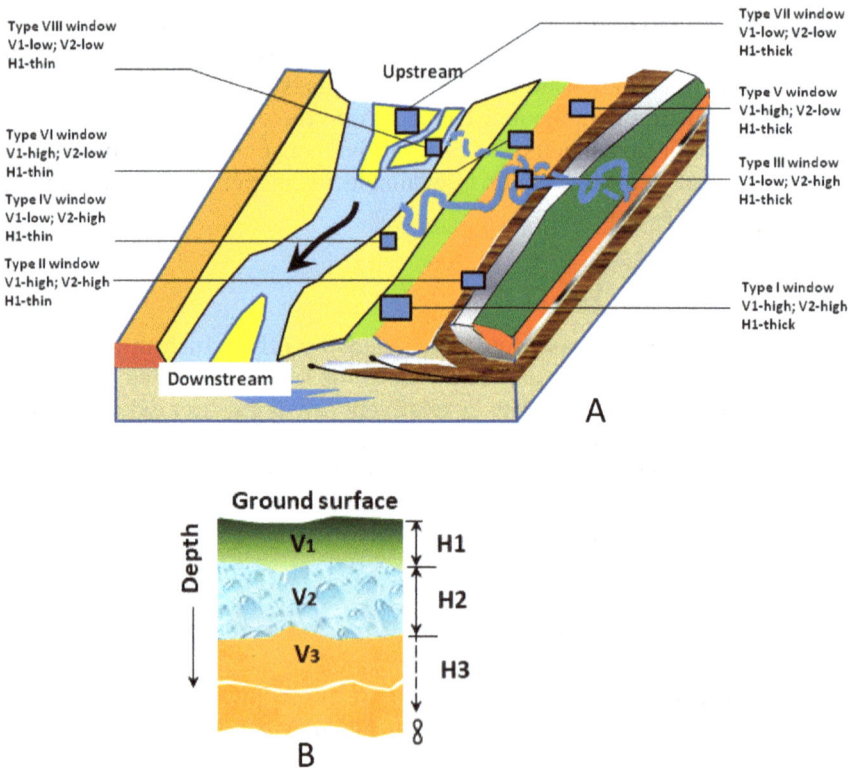

FIGURE 5.3
Multiparametric micro-zonation scheme shown as a figure.

TABLE 5.1

A Schematic Description of Tri-Parametric Micro-Zonation Model

Type	Parameters	Model Characteristics
1	V_1-high V_2-high H_1-thick	Regional high, older sedimentary rocks, bypassed by the fluvial channels, or older flood plain deposits uplifted
2	V_1-low V_2-high H_1-thick	Presence of a channel that incised the older sediments. Thicker low-velocity material is caused by lack of channel dynamism
3	V_1-high V_2-low H_1-thick	Presence of a palaeochannel of much older age, migrated long back
4	V_1-low V_2-low H_1-thick	Regional low having a continuous tendency to generate accommodation space
5	V_1-high V_2-high H_1-thin	Regional high, came into being much before the type 1, with a higher rate of erosion
6	V_1-low V_2-high H_1-thin	Recent flood plain deposits due to a channel lying nearby or a palaeochannel having avulsive tendency
7	V_1-high V_2-low H_1-thin	Presence of a palaeochannel that has migrated more recently than type 3
8	V_1-low V_2-low H_1-thin	Regional low with higher slope causes a low rate of aggradation

layer thickness to call a given layer 'thick' or 'thin', comparable types arise as follows:

On the other hand, if we have information for the parameters V_2, V_1, H_2, and H_1, this leads to 16 possibilities.

Choosing 'Threshold' Values

While dividing a layer based on velocity discrimination as 'high'/'low' and thickness discrimination as 'thick'/'thin', we are trying to emphasize more the 'degree of consolidation of the earth material and the 'accommodating ability' of a given location, rather than understanding the actual lithological composition (silt, sand, gravel, pebble, cobble, boulder, etc.).

Since channels incise the old flood plains and deposit the transported materials having different records of provenance, even under the normal rate of compaction, dehydration, etc., the velocity of seismic waves through these matrices will show considerable difference even for the highly comparable fluvial conditions. Thus, it is principally the process-based correlatable morphological signatures that we will use to isolate 'lows' from 'highs' and 'thin beds' from 'thick beds'.

For example, in the Amguri window, we have defined $V1_{200}^{400} = V1_{(low)}$ and $V1_{400+}^{650} = V1_{(high)}$. On the other hand, we have further defined $V2_{500}^{1400} = V2_{(low)}$ and $V2_{1400+}^{2300} = V2_{(high)}$. Thus, if we see from the lithology angle, there might be an overlap between $V_{1\ (high)}$ and $V_{2\ (low)}$. However, this is not going to affect process-based understanding of the fluvial dynamics and identification of the structural controls.

Study 'Windows'

Seismotectonic Atlas of India (Dasgupta et al., 2000) prepared based on a compilation of the surface geology and subsurface geophysical evidence shows two prominent blind faults running more or less parallel to the NPT belt. The fault closer to the NPT (B1B1') is about 150 km long, and another fault (B2B2') away from the NPT is about 133 km long. Four strategic windows (Figure 5.4) were used for the present study based on at least four major criteria, namely, (i) locations of producing oil fields, (ii) channels showing typical angularities suggesting strong tectonic controls, (iii) inter-channel landscape having prominent indications of palaeochannels, and (iv) prognosticated

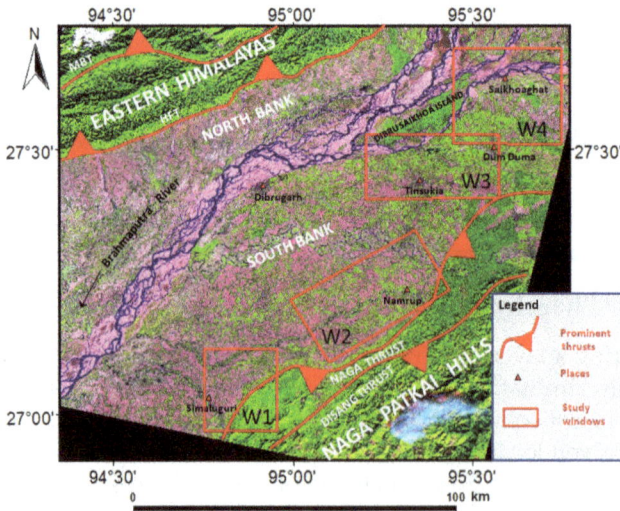

FIGURE 5.4
Locations of four study windows.

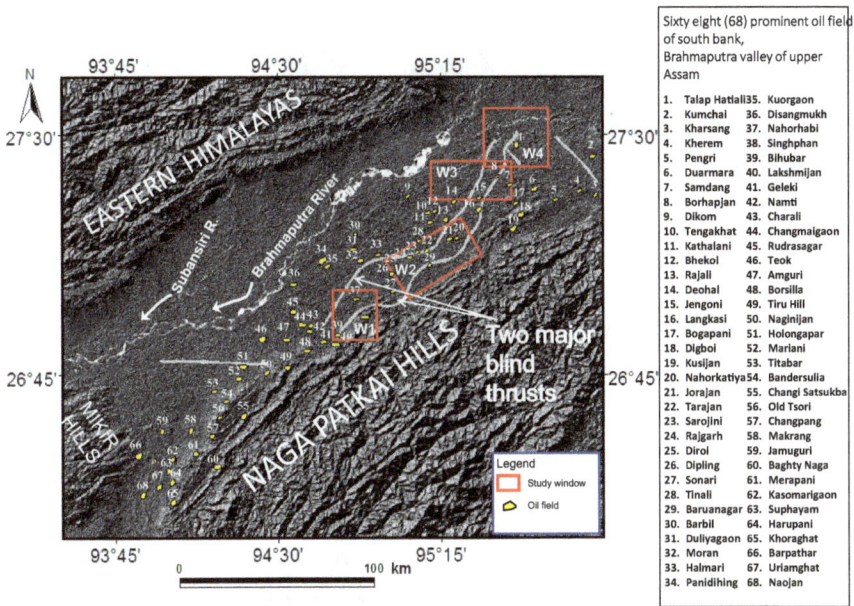

	Sixty eight (68) prominent oil field of south bank, Brahmaputra valley of upper Assam

1.	Talap Hatiali	35.	Kuorgaon
2.	Kumchai	36.	Disangmukh
3.	Kharsang	37.	Nahorhabi
4.	Kherem	38.	Singhphan
5.	Pengri	39.	Bihubar
6.	Duarmara	40.	Lakshmijan
7.	Samdang	41.	Geleki
8.	Borhapjan	42.	Namti
9.	Dikom	43.	Charali
10.	Tengakhat	44.	Changmaigaon
11.	Kathalani	45.	Rudrasagar
12.	Bhekol	46.	Teok
13.	Rajali	47.	Amguri
14.	Deohal	48.	Borsilla
15.	Jengoni	49.	Tiru Hill
16.	Langkasi	50.	Naginijan
17.	Bogapani	51.	Holongapar
18.	Digboi	52.	Mariani
19.	Kusijan	53.	Titabar
20.	Nahorkatiya	54.	Bandersulia
21.	Jorajan	55.	Changi Satsukba
22.	Tarajan	56.	Old Tsori
23.	Sarojini	57.	Changpang
24.	Rajgarh	58.	Makrang
25.	Diroi	59.	Jamuguri
26.	Dipling	60.	Baghty Naga
27.	Sonari	61.	Merapani
28.	Tinali	62.	Kasomarigaon
29.	Baruanagar	63.	Suphayam
30.	Barbil	64.	Harupani
31.	Duliyagaon	65.	Khoraghat
32.	Moran	66.	Barpathar
33.	Halmari	67.	Uriamghat
34.	Panidihing	68.	Naojan

FIGURE 5.5

Prominent oil field locations on the south bank of the Brahmaputra valley.

hydrocarbon reserves yet to be explored. The common thread among all these windows is the presence of two blind faults running continuously through all four windows.

In some of the windows, only shallow refraction data are available, and in that case, we have considered a two-layer earth model with three parameters, respectively, V_1, H_1, and V_2. In those windows where both shallow refraction and uphole data are available, we preferred the latter for the reasons discussed earlier. For uphole data, we have considered a three-layer earth model with V_1, H_1, V_2, H_2, and V_3. When the producing major oil fields of Upper Assam are plotted on a map (Figure 5.5), the relationship between the oil fields and the leading edge of the blind thrusts becomes obvious. Emerging thrust fronts are supposed to be the major causes of the neotectonic activities of the area.

Window 1: Amguri

This is a window covering ~490 km² (24.5 km × 20 km) of area (Figure 5.6) located between the latitudes 26.82°–27.04° and longitudes 94.75°–94.95°. Two tributaries of the Brahmaputra, the Disang and the Dikhau, emerging from

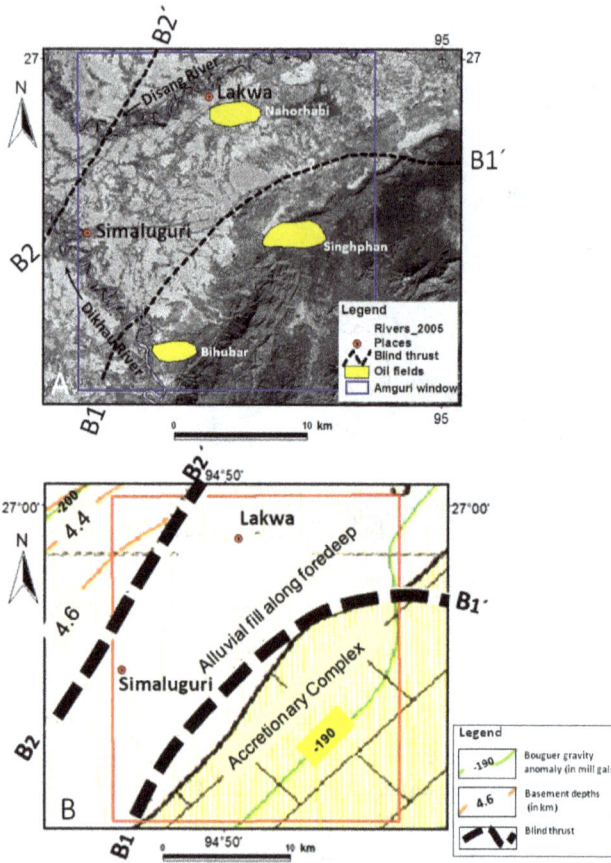

FIGURE 5.6
Window 1: Amguri and adjacent areas.

the Naga-Patkai hills, flow along the northern and southern boundaries of the window. The overall channel belt of the Disang, a highly meandering river, after coming down from the hills moves along the N-NW direction for a stretch of about 9 km and then takes almost an L-turn along an NE-SW trending 60-km-long axis; lastly, it joins the Brahmaputra River, taking another sharp bend towards the NW direction, covering a straight distance of about 23 km. As discussed in the previous chapter, the average sinuosity of the Disang is 2.06. Thus, roughly, the actual distance travelled by the river is just double the valley length it travels. The intermediate distance between the points where the two rivers – the Disang and the Dikhau – debauch in the valley is about 66 km; when they enter into the Amguri window, the distance is 25.6 km, and at the exit points through the window, it is just 7.3 km. There are at least three major oil fields in the study window, as shown in Figure 5.6a.

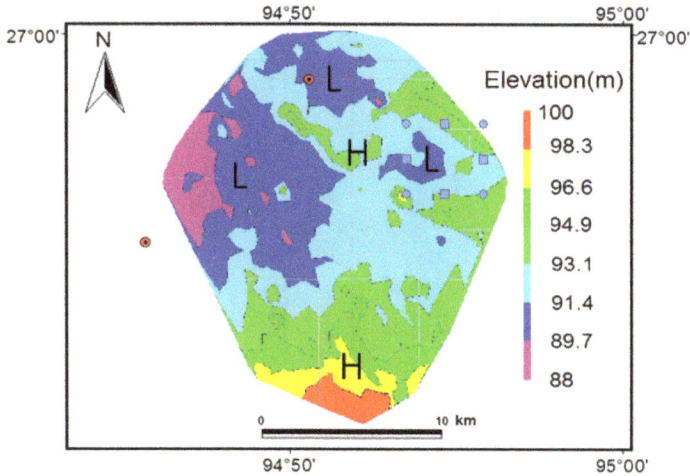

FIGURE 5.7
Elevations of different zones of the Amguri window.

Singhphan and Bihubar oil fields lie within the NPT and the subsurface blind fault B_1B_1'. The third oil field Nahorhabi is in between B_1B_1' and B_2B_2'. Figure 5.6b shows the seismo-tectonic characteristics of the area inside the window.

The Bouguer anomaly gravity values show that the area is within (–)200 and (–)190 mgal, and the basement depth shows that in the foredeep part, the sediment thickness is about 5 km. In the hilly part also, the Bouguer anomaly is too low, which indicates that the basement complex is only moving up marginally. So, the hills are mostly constituted of sedimentary rocks termed as 'accretionary complex' (Dasgupta et al., 2000), and the foredeep is, of course, constituted of the alluvial fill, orchestrated by the fluvial processes.

The elevation varies from 100 to 88 m above the mean sea level (AMSL). The elevation distribution pattern as seen in Figure 5.7 indicates the presence of well-defined 'highs' and 'lows'.

The Amguri Data

In Amguri, we have both shallow refraction and uphole data (Figure 5.8). The uphole data sets used for the present purpose were collected systematically by Geofizyka Toruń of Poland during January–February 2008 field season for Oil India Limited. Uphole surveys were conducted on 25 lines at 132 points (Figure 5.8a). Uphole data are available for 132 stations, and shallow refraction data for 512 stations (Figure 5.8b).

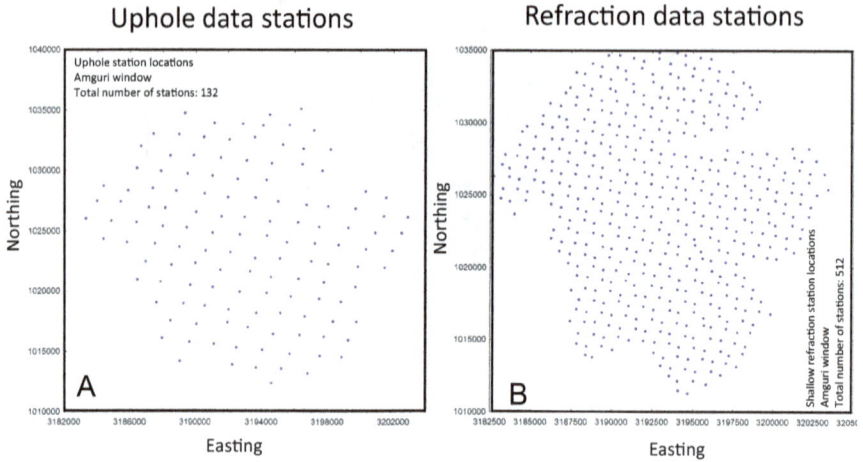

Uphole data stations

Refraction data stations

FIGURE 5.8
Locations of uphole and refraction data in the Amguri window.

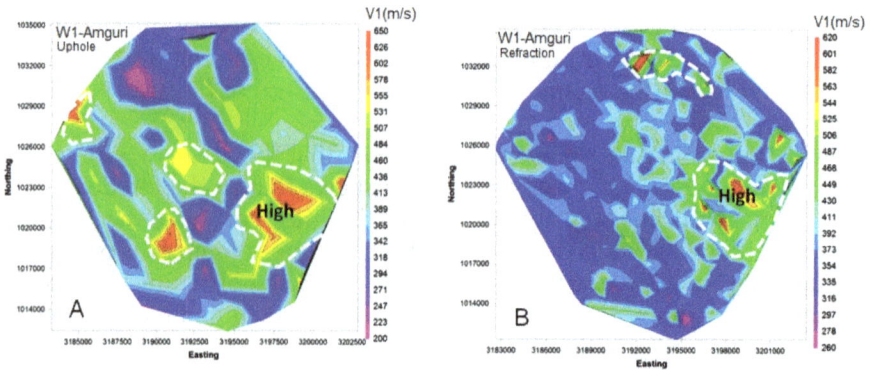

FIGURE 5.9
Velocity distribution in the weathered layer of the Amguri window.

Thus, data density is much higher for the shallow refraction than that for the uphole.

Data coverage, particularly in the northern part above the central region of the window, is much better for the refraction data than for the uphole.

However, as explained in the Introduction, despite lower data density and lesser area coverage, we have used uphole data for further processing due to its reliability and consistently higher depth range coverage. A comparative study of these two data sets shows that the velocity range for the topmost layer is almost identical (Figure 5.9a and b). For uphole data, the range is 200–650 m/s, whereas for the shallow refraction, this range is 260–620 m/s. However, the velocity range of the second layer differs significantly; the range

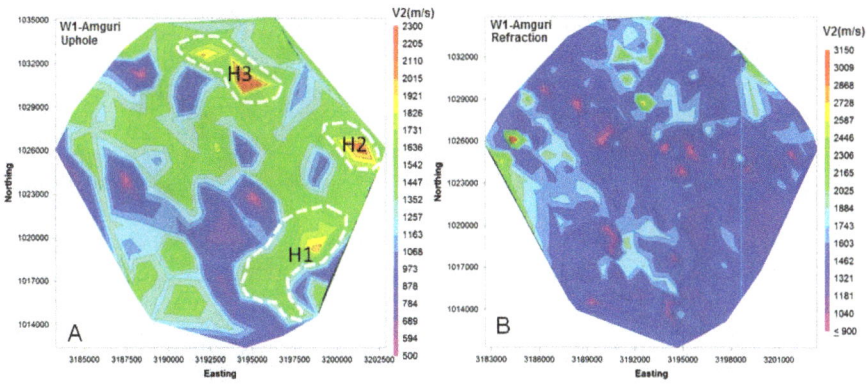

FIGURE 5.10
Velocity distribution in the sub-weathered layer of the Amguri window.

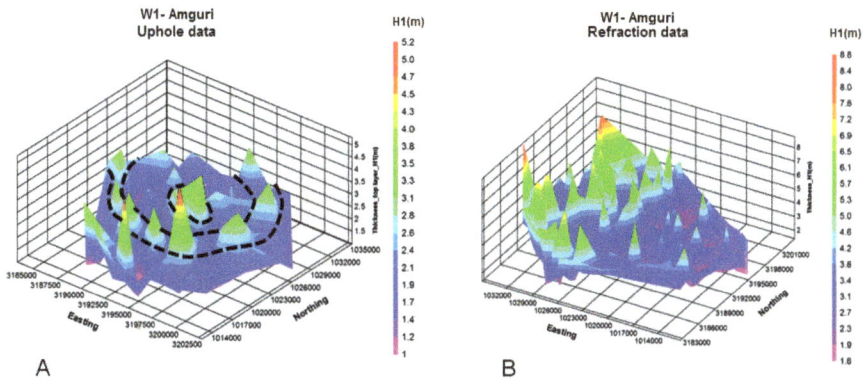

FIGURE 5.11
Three-dimensional representation of thickness distribution of the weathered layer.

is 500–2,300 m/s for uphole data, and 900–3,150 m/s for the shallow refraction data (Figure 5.10a and b). The top layer thickness variation (Figure 5.11a and b) is 1.5–5 m for the uphole data, and it is 2–8 m for the refraction data. The nature of velocity distribution shows a good correspondence in the high-velocity zones. However, we find a few more major high-velocity zones in the uphole data set. The geometry of the low-velocity zone very distinctly shows the presence of a palaeochannel in two different layers with a shifting tendency of the channel; however, the signatures are more clearly discernible for the uphole data sets.

The pattern of thickness variation of the top layer is very interesting. In the uphole data sets, it shows a central 'thick zone' surrounded by punctuated thick zones on a girdle. If this girdle is mainly constituted of low-velocity sediments, then these are, of course, the palaeochannel belt. On the other

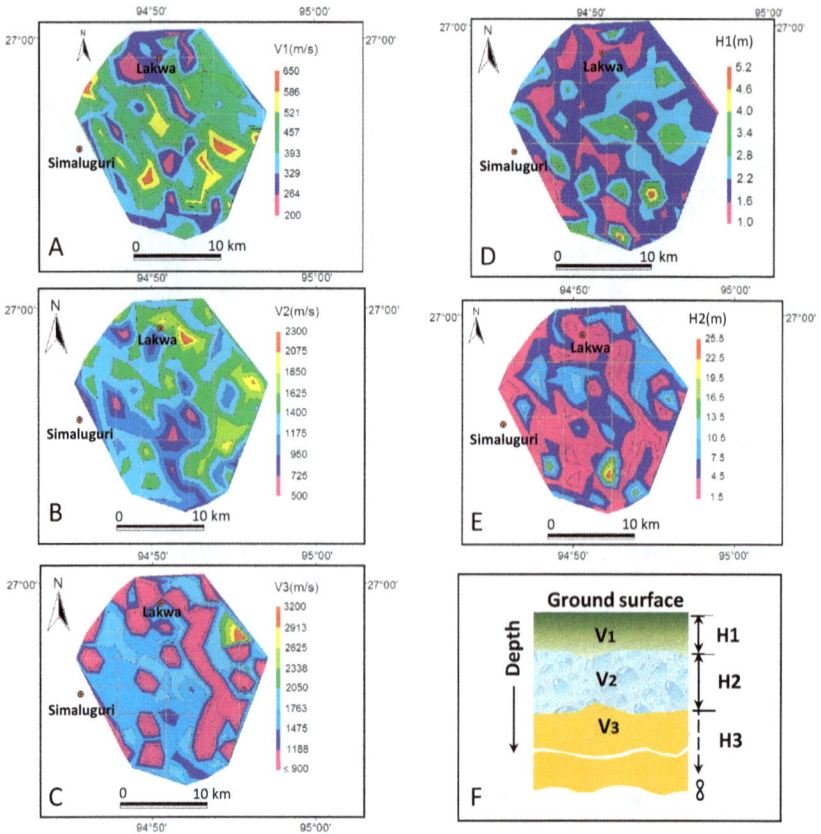

FIGURE 5.12
Velocity distribution for three layers and thickness distribution for the top two layers.

hand, if these are constituted of high-velocity materials and, the intervening space is constituted of low-velocity material, this is probably a raised platform having much older flood plain deposits, subsequently incised by channels. To resolve such situations, a process-based understanding of an area needs to be complemented by multi-parametric investigation such as the velocity of seismic waves through the material (high or low), the thickness of the layer (thick or thin), and spot elevation of the place AMSL.

Such data sets can be easily integrated into a GIS environment (Figure 5.12a–e). Of course, more the number of parameters brought into consideration, the diversity of type areas will also increase in geometric proportion. Observing the velocity distribution pattern, the low velocity range chosen for three different layers (Figure 5.12f) are 200–400 m/s [$V_{1\ (low)}$], 500–1,400 m/s [$V_{2\ (low)}$], and 900–1,475 m/s [$V_{3\ (low)}$].

The high velocity range chosen for three different layers are 400–650 m/s [$V_{1\ (high)}$], 1,400–2,300 m/s [$V_{2\ (high)}$], and 1,475–3,200 m/s [$V_{3\ (high)}$]. The thickness range of the first layer (topmost) is 1.0–5.2 m. The thickness range of the second layer is 1.0–25.5 m. For the topmost layer, a thickness range of 1.0–2.2 m has been considered 'thin' and >2.2–5.2 m 'thick'. For the second layer, 1.5–4.5 m has been considered 'thin' and all higher values 'thick'. The low-velocity zones give us a broad idea of the width of the channel belt.

Fluvial dynamics in Amguri Window

For alluvial deposits, a layer consists of a surface having a mild slope and variable thickness with a more or less planar top and a sharp erosional surface. The degree of cementation varies in a more complex way. However, generalizing broadly, a layer can be treated as a two-component system: (i) more consolidated older flood plains and (ii) less consolidated channel deposits. Commonly, seismic wave velocities through the older flood plains will be higher than those of the channel-fill zones. Typically, a threshold value of velocity is chosen to distinguish 'high-velocity' and 'low-velocity' layers by close observation of the geometrical disposition of the velocities. The challenge is to identify the locations of palaeochannels principally from the palaeo-transported materials and the deposition of suspended sediments by the channel process in different proportions at different places. These elements are to be correlated, and a continuous path and the course of the palaeochannel are to be reconstructed. The deeper the layer, the more the lithostatic pressure, and reduction in porosity causes velocities to rise in general. However, the basic logic remains the same.

Oldest Channel Belt in Layer 3

It appears that there used to be north–south-trending distinct trunks of two different rivers within the central part of the study window (Figure 5.13c); at present, there is no such modern river. There was also a third branch, but its location in the marginal part of the study window makes its characterization difficult. The eastern trunk represented a larger river than the western counterparts.

Older Channel Belt in Layer 2

The easternmost channel trunk located in layer 3 seems to have shifted (Figure 5.13b) (probably in the northern part, where the Disang River flows at present). This might have occurred due to the activation of the subsurface

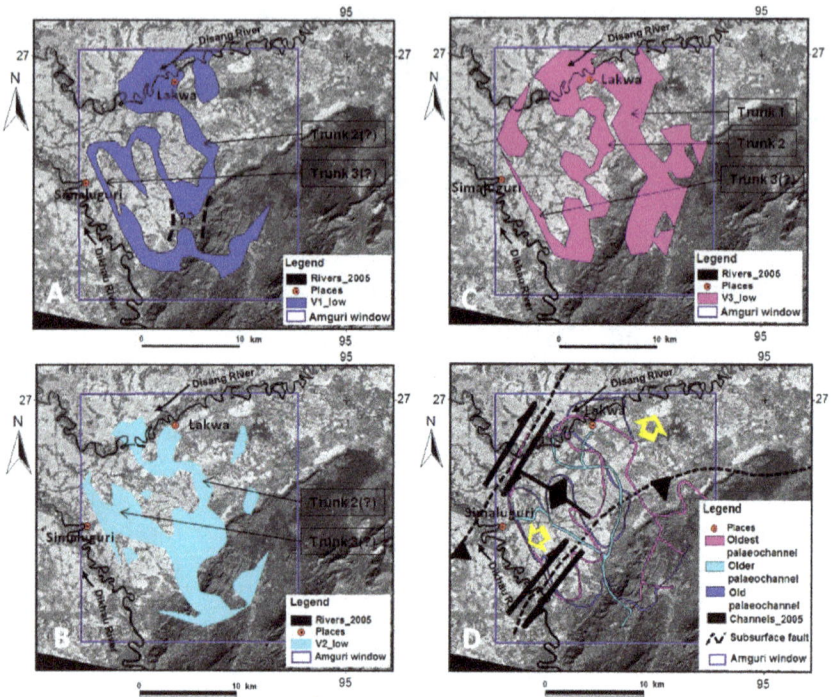

FIGURE 5.13
Identification of palaeochannel paths from the shallow subsurface data in the Amguri window.

faults representing the frontal part of the NPT belt. One remarkable thing is the shift in the channel flow direction from the south–north to SE-NW direction. Also, a bifurcation of the channel is noted. We are not sure whether the trunks of channels identified in layer 3 are identical to layer 2 or not. It needs to be substantiated by other data.

However, the continuity of the lithofacies in layers 3 and 2 strengthens the conjecture that the channel processes involved were the same.

Old Channel Belt in Layer 1

The general trend of flow through the palaeochannel remained SE-NW with dominant loops in the central region of the study area. We are not very sure about the connectivity of different segments of the channel belt (Figure 5.13a), but it seems quite obvious that the total shift of the palaeochannel to its present course took place during the period bound within the maximum sediment thickness of 5+m, which is normally termed as the 'topsoil'.

Tectonic Controls and the River Dynamics

Figure 5.13d shows the median paths of the wandering channel belts in three different layers that provide some understanding of the nature of channel dynamics during different periods and the probable mechanism. Blind faults shown in Figure 5.2b as B_1-B_1' and B_2-B_2' can be interpreted as the planes along which the leading edges of the NPT belt are reorganized. There is also a lateral movement, as shown in Figure 5.13d. This may have caused the development of a local 'high'. As a result, the palaeochannels changed their course, moving towards the peripheral zone of the study window.

Neotectonics and Oil Migration

The migration trend of the palaeochannels of the Amguri window, the descendants of which are most probably the present-day Disang and the Dikhau rivers, shows that there is indeed an immense influence of neotectonics on the landform evolution and river dynamics of the study area. In an unpublished thesis report (Gogoi, 2013), optically stimulated luminescence (OSL) dating was carried out in the Wadia Institute of Himalayan Geology, Dehradun, for samples collected from a depth of 840 cm in the Namdang area, near the NPT and very close to the present study area, and the age of the sediment was found to be 16.95 ± 2.11 ka. This shows, on average, the soil formation rate was 20 years/cm. In the present study, the maximum thickness of the top layer (layer 1) is 5.2 m. So, this involves a period of about 10^+ ka. This indicates that there was a major tectonic readjustment during the Holocene period, which however got initiated still earlier. If we ascribe the observed influence of neotectonics on the fluvial dynamics to the NPT system, then the leading edge is emerging in the westward flank of the NPT. In other words, oil migration has most probably taken place from the far eastern side to the western side of the south bank. Moreover, the oil fields like Bihubar and Singhphan located in the eastern part of the B_1-B_1' fault line are older than the Nahorhabi oil field near Lakwa.

Multi-Parametric Micro-Zonation

For shallow-focus earthquakes, lateral variation in soil characteristics (loose or tight) influences the amplitude of the advancing wavefront.

Amplitude is higher in the unconsolidated formations (lower velocity), and accordingly, the possibility of property losses is higher. There are also directional issues relating to the location of the epicentre and disposition pattern of different types of materials and anisotropy. Information relating to lateral variation in the P-wave velocity for some layers in the shallow subsurface along with the thickness variation gives us the means to undertake different types of micro-zonation schemes. For the Amguri window, we have experimented with three categories of multi-parametric micro-zonation. First is a bi-parametric micro-zonation, where we have the lateral variation in the *P*-wave velocity in the top two layers taken into consideration. We have discriminated each layer into 'high'- and 'low'-velocity zones. This gives us four type areas (Figure 5.14). Increasing the shades of discrimination will increase proportionately the number of type areas.

Secondly, we have considered the tri-parametric micro-zonation of two types. One is constituted of V_1, V_2, and H_1, and the other is V_1, V_2, and V_3. The former (Figure 5.15) subdivides the earlier bi-parametric micro-zonation based on the 'thick' or 'thin' uppermost stratum, thus resulting in eight type areas. In the second type, a three-layer earth model is used (Figure 5.16), and based on velocity discrimination alone, the study area is again divided into eight type areas.

Thirdly, we have considered the V_1-V_2-H_1-H_2 model, which represents two-layer earth where the thickness of both the layers is known, thus resulting in 16 type areas (Figure 5.17).

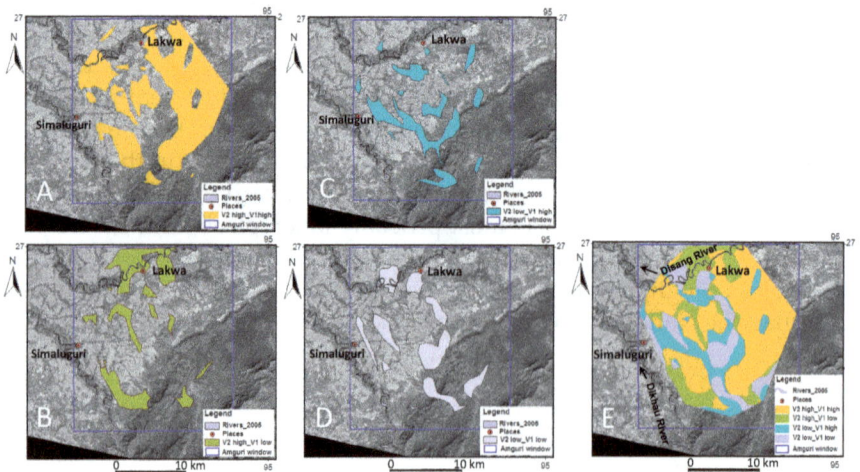

FIGURE 5.14
Bi-parametric micro-zonation of the Amguri window.

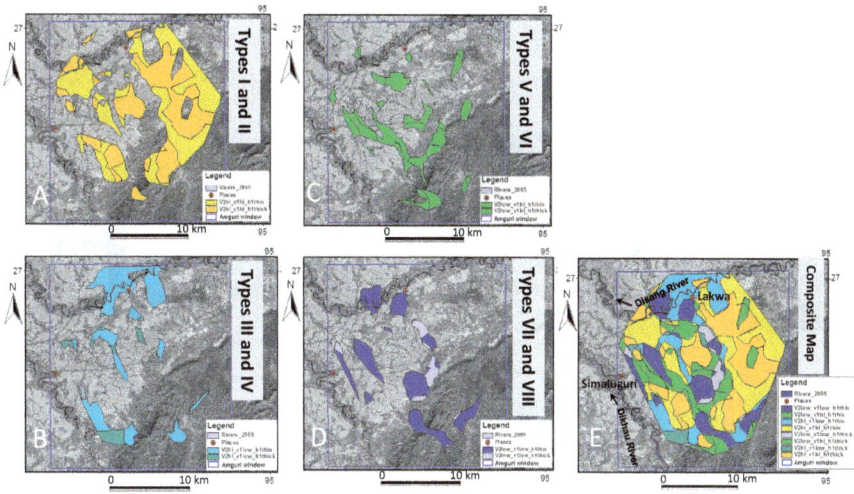

FIGURE 5.15
Tri-parametric micro-zonation of the Amguri window using velocities and thickness parameters of the weathered and sub-weathered layers.

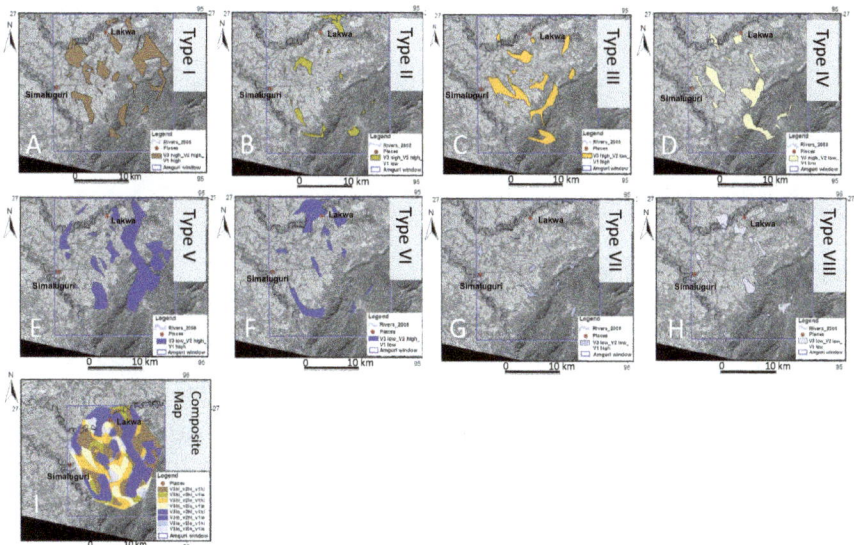

FIGURE 5.16
Tri-parametric micro-zonation using seismic velocities through three layers of the Amguri window.

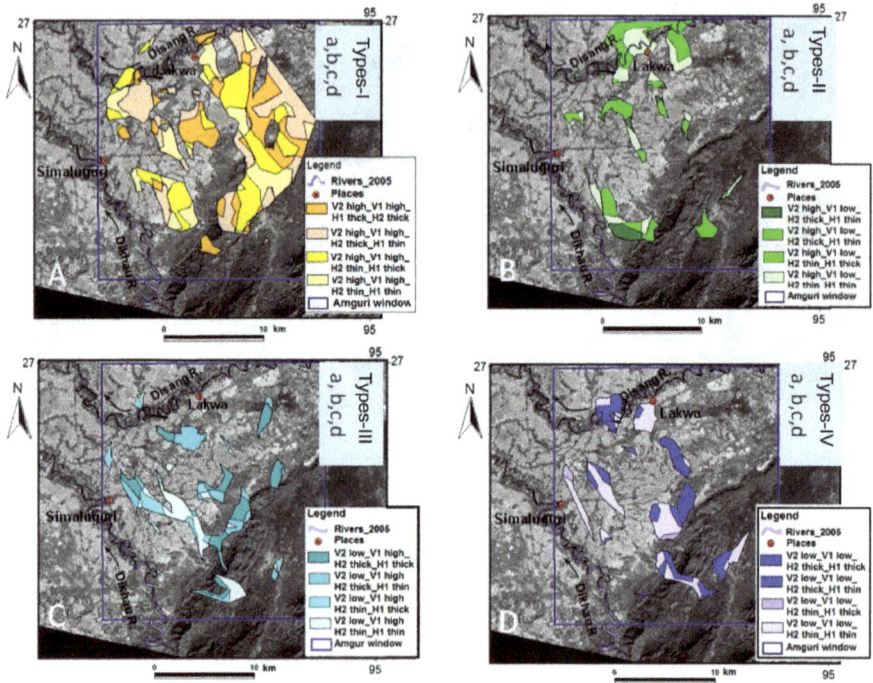

FIGURE 5.17

Tetra-parametric micro-zonation using V_1, V_2 and H_1, H_2 in the Amguri window.

Window 2: Namrup–Sapekhati

This is an NE-SW trending window with an approximately 880-km² area and located within the longitudes and latitudes 94.99°/27.17°; 95.34°/27.36°; 95.44°/27.20°; 95.10°/27.00°. This is one of the richest oil windows of the upper reach of the Brahmaputra valley (Figure 5.18a). At least seven major oil fields, namely, Nahorkatiya, Jorajan, Tarajan, Sarojini, Rajgorh, Diroi, and Dipling, seem to lie one after another in a belt, and two more oil fields – Tinali in the western part of the belt and Baruanagar is located in the eastern part of the belt. The alignment of the NPT belt, the direction of flow of the Disang River, the direction of the regional blind faults (B1-B1' and B2-B2'), and above all, the clusters of oil fields seem to be parts of an integrated whole. The seismo-tectonic map covering the window and some of the adjacent areas shows there is heterogeneity in valley fill; sediment thickness (as inferred from the basement depth variations) varies from 4.0 to 6.0⁺ km within the window. The thicker part is observed along the foredeep adjacent to the accretionary complex and the Schuppen belt belonging together to the NPT belt. Bouguer

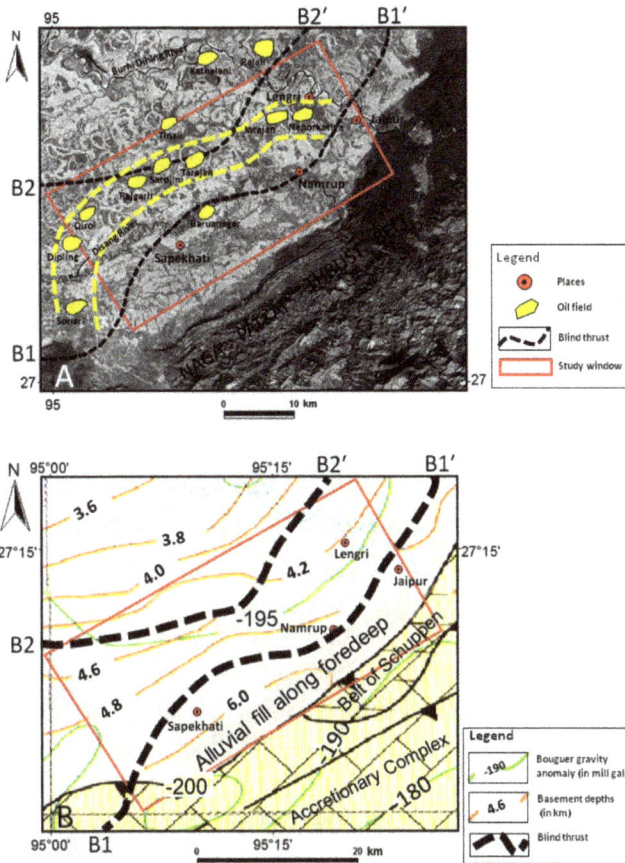

FIGURE 5.18
Location map of window 2 covering the Namrup–Sapekhati area.

gravity anomaly, in conformity with the previous situation, varies from −192 to −200 mgal. The window is very rich in palaeochannel marks and the presence of oxbow lakes. The amplitude and wavelength of the meandering palaeochannels are very much comparable with that of the present-day channels, and this gives a clear clue of river migration along a specific direction.

Namrup–Sapekhati data

Shallow refraction data coverage for the Namrup–Sapekhati window was uniform and extensive. In total, 1,228 points were covered (Figure 5.19). For a general estimate, three points were investigated for every 2-km² area.

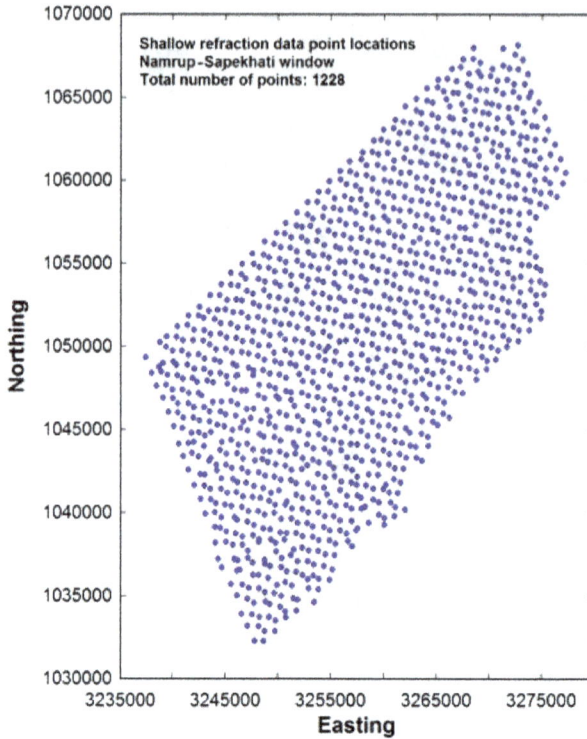

FIGURE 5.19
Shallow refraction data density in the Namrup–Sapekhati window (W-2).

Landform Variability and Quaternary Geomorphology

Spot elevations of the points of the investigation show the lowest point 102 m AMSL and the highest point 156 m AMSL (Figure 5.20a). Thus, there is quite a wide range (54 m) of variation from one place to another.

When the elevation range was divided into a few classes (seven for the present situation), it was observed that the class boundaries follow a definite trend, and the window can broadly be divided into four classes:

Class 1: 102–110 m; class 2: >110–117 m; class 3: >117–125 m; and class 4: >125 m. As we move towards the N-NE direction, the four classes show a diverging trend. Thus, in the SW corner of the window, the elevation gradient is very high. The thickness range of the topmost LVL is 1–14 m (Figure 5.20b). Following the earlier estimate of 20 years/cm rate of deposition, the period covered will be approximately 2,000–<30,000 years.

However, a thickness above 5 m is confined mostly within a narrow zone (shown by dotted lines). Thus, our observation will mostly remain confined to the late Quaternary and Holocene.

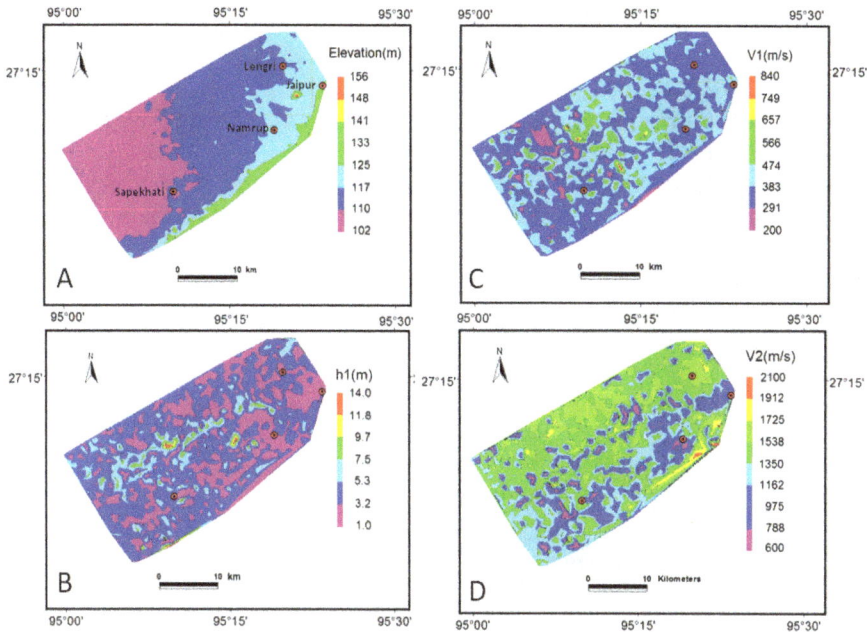

FIGURE 5.20
Lateral variability pattern of some of the parameters in the Namrup–Sapekhati window (W-2).

The 5[+]-m-thick zone is possibly due to higher stream power palaeo-Burhi Dihing River flowing through that course during the early Holocene. Moreover, there is the possibility of the absence of neotectonic activities in the foredeep part for a longer duration, and the river had its longest stay through that course.

Velocity variation in the topmost LVL is 200–840 m/s (Figure 5.20c), showing a wide range of variations. Velocity variation in the sub-weathered layer is 600–2,100 m/s (Figure 5.20d). Interestingly, for the deeper layer, there is almost a clear-cut line of demarcation, separating the low-velocity segment from the higher velocity one, indicating clearly that the preferred path of the palaeochannels was mostly along the thrust belt boundary. Subsequently, channels preferred northward migration, as evidenced by the low-velocity distribution nature of the topmost layer.

Fluvial Dynamics and Quaternary Basin Evolution

For the simplification of interpretation, we have divided the velocity distribution for the top (LVL) and the bottom (sub-weathered) layers into two-component systems, namely, 'high'- and 'low'-velocity zones. For the V_1 values, the threshold was fixed as $V_{1\ (low)} = 200–383\,m/s$ and $V_{1\ (high)} = >200–840\,m/s$. For the V_2 values, the threshold was fixed as $V_{2\ (low)} = 600–1,162\,m/s$ and $V_{2\ (high)} = >1,162–2,100\,m/s$ (Figure 5.21). The initiation of the low V_2 value coincides with the present-day entry point of the Burhi Dihing River into the valley from the NPT mountain range.

Subsequently, it shows a south-westward course and is contrary to the present-day flow direction of the river along the northwest. There is a possibility of a reduced rate of subsidence along the foredeep area of the NPT belt compared with the sediment volume influx. Accordingly, the lack of new accommodation space along the foredeep, which otherwise shows much

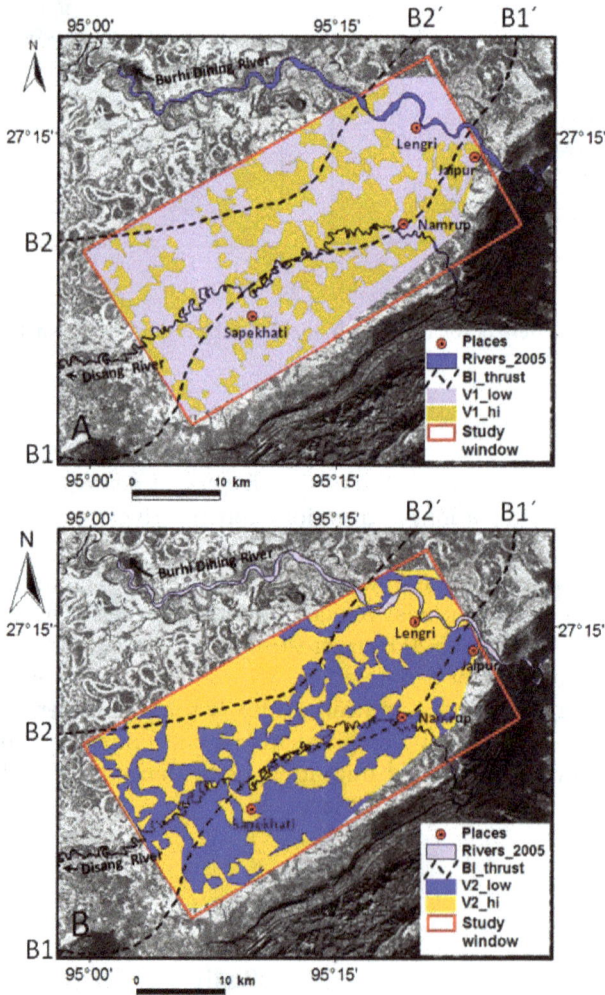

FIGURE 5.21
Velocity differentiation of layers into two-component systems.

thicker sediment accumulation (Figure 5.18b), changed the probable site of aggradation away from the foredeep. However, this would have caused the river to shift gradually, and as we described in Chapter 4, for the rivers of the north bank of the Brahmaputra valley, palaeochannel marks would be much limited. The distinctiveness of the palaeochannel marks shows the older Burhi Dihing must have avulsed suddenly. Thus, in addition to the subduction-generated tectonics, thrust belt tectonics must be playing a very significant role in basin evolution during the Holocene period in the fore-deep part of the south bank of the upper reach of the Brahmaputra valley.

Neotectonics and Valley–Range Relationship

The sharp bends of the rivers, channel meandering about straight paths for long distances along the frontal parts of the thrust belt, sag ponds, highly elevated scarps along with different segments of a flowing river, and the systematic changes in the elevation of the landforms are some of the clearest indications that the mountain range lying in the boundary of the valley is continuously influencing its evolution.

When we superimpose the velocities of two different layers (after discriminating layers as 'low'- and 'high'-velocity zones), it is observed that (Figure 5.22) the intervening space between the Burhi Dihing River and the Disang River must be experiencing uplift under the influence of multiple thrust fronts. The complexity of gross movements is giving rise to several

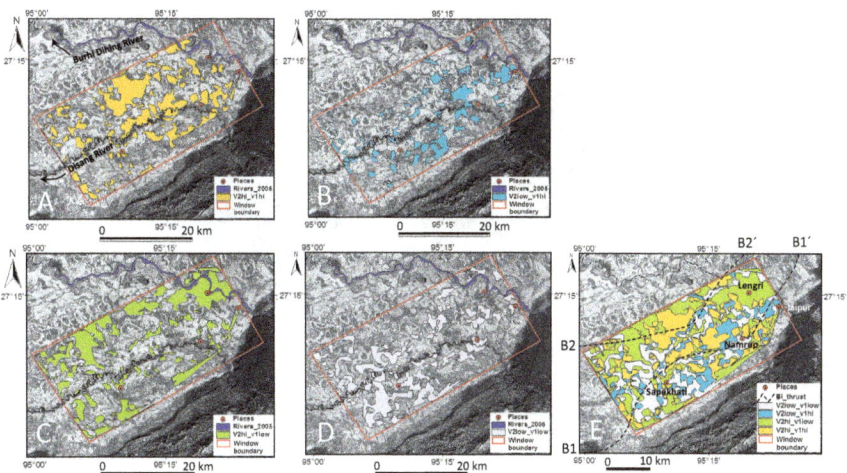

FIGURE 5.22
Bi-parametric micro-zonation of the Namrup–Sapekhati window (W-2).

local 'highs' in addition to a general trend of regional 'highs' (shown by dotted lines in Figure 5.22a).

This is the reason some of the oil field locations coincide with the semi-circular meandering loops (existing as well as palaeo-loops). Most of the regional lows having both the V_1 and V_2 values low are still showing prevalence in the adjacent part of the valley foredeep lying along the NPT belt (Figure 5.22d). Flood plain deposits due to the Disang River may not alone be responsible for the alluvial fill. Rather, the joint action of the Disang and a larger river like the Burhi Dihing might have caused it. In such a situation, the shift of the Burhi Dihing River to its present course must be much younger.

Multi-Parametric Micro-Zonation and the Oil Fields

Bi-parametric micro-zonation based on velocity characterization ('high' or 'low') has already been shown in the previous section (Figure 5.22). Adding the thickness component, we can undertake tri-parametric micro-zonation. Some of the observations relating to different type areas in the multi-parametric micro-zonation and the oil field locations are described in Figures 5.23–5.27.

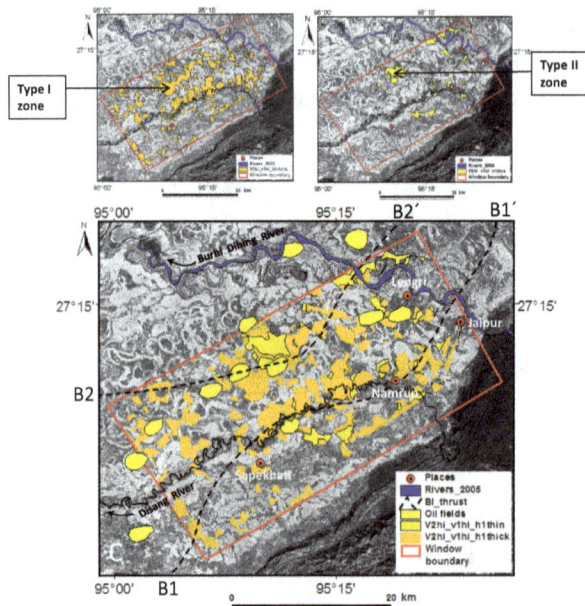

FIGURE 5.23
Tri-parametric micro-zonation based on V_1, V_2, and H_1 (types 1 and 2) in the Namrup–Sapekhati window (W-2).

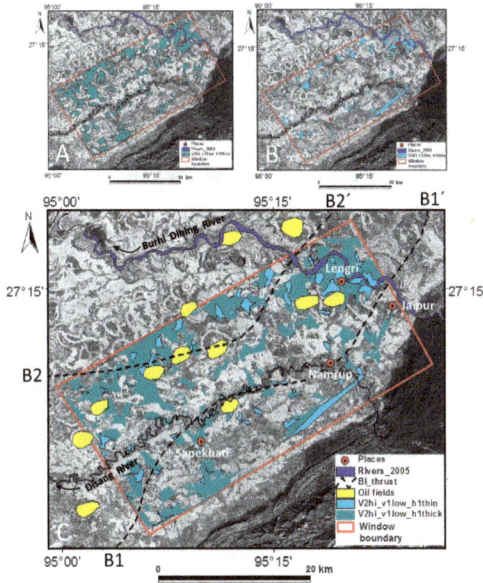

FIGURE 5.24
Tri-parametric micro-zonation based on V_1, V_2, and H_1 (types 3 and 4) in the Namrup–Sapekhati window (W-2).

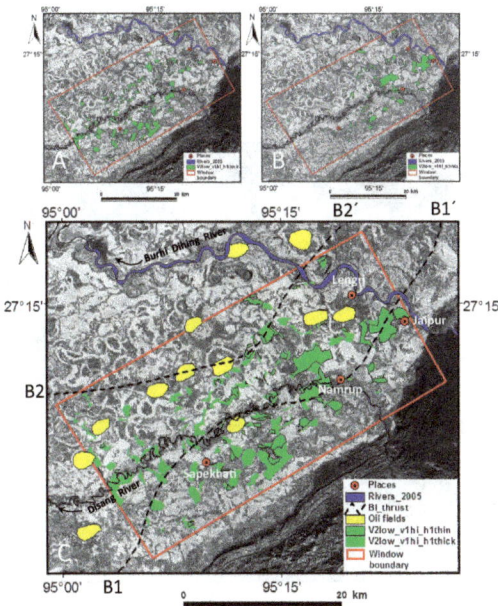

FIGURE 5.25
Tri-parametric micro-zonation based on V_1, V_2, and H_1 (types 5 and 6) in the Namrup–Sapekhati window (W-2).

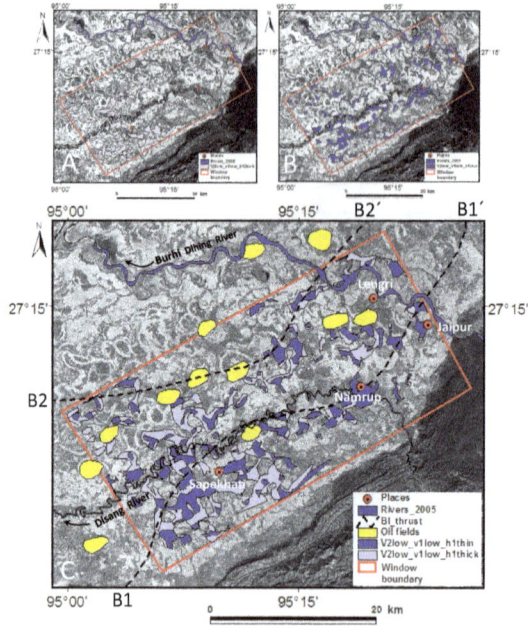

FIGURE 5.26

Tri-parametric micro-zonation based on V_1, V_2, and H_1 (types 7 and 8) in the Namrup–Sapekhati window (W-2).

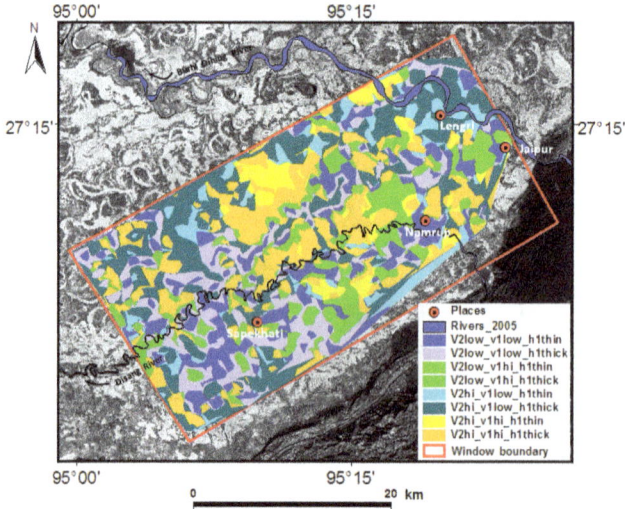

FIGURE 5.27

Composite tri-parametric micro-zonation based on V_1, V_2, and H_1 showing all the eight different types (types 2–8) in the Namrup–Sapekhati window (W-2).

Window 3: Matikhana

The Matikhana window includes important townships like Tinsukia and coal provinces like Makum and Dum Duma. Located between the latitudes 27.44°–27.62° and longitudes 95.20°–95.57°, the area covered by the window is about 697 km². In the northern end, there is a small segment belonging to the Dibru-Saikhowa forest, now turned into a Majuli-type island due to the avulsive switchover of the Lohit River in recent times. The Dibru River flows through the middle part of the window, and moving westwards, it confluences with the Lohit River. As described in Chapter 4, the confluence point of the Dibru River has systematically moved in a further upstream direction with the increasing width of the Brahmaputra River. There are at least three major oil fields, namely, Borhapjan, Jengoni, and Samdang, of which the first two are located between the two major regional blind faults, and the third one lies in the eastern part of the belt. In the immediate neighbourhood of the window, there are at least nine more major oil fields (Figure 5.28a). The seismo-tectonic map shows (Figure 5.28b) a wide range of basement depth variations (3.6–5.0 km). The sediment thickness continues increasing as we move eastwards. The nose of the Brahmaputra high (with least basement depth) terminates in the westernmost part of the window. The Bouguer gravity anomaly varies from –190 to –220 mgal.

Matikhana Data

In window 3, an uphole survey was carried out at 148 stations (Figure 5.29). Information is available in three layers. The elevation shows a wide range of variation (111–159 m) (Figure 5.30b), but practically from the eastern end to the western end, the change is quite smooth from 130 to 111 m over a distance of 36 km, showing an approximately 1-m fall in height over a horizontal distance of 2 km. In between Tinsukia and Dum Duma, there is a geomorphic high and one of the important oil fields, Borhapjan, is located on this high. The thickness of the first, second, and third layers vary, respectively, 1–9, 1.5–27, and 3–27 m (Figure 5.30c–e). Velocity variation for the first, second, and third layers are, respectively, 300–720, 500–1,700, and 800–1,950 m/s (Figure 5.30f–h). This clearly shows the wide range of diversity in the shallow subsurface stratigraphy.

Salient Features in Window 3

When we proceed from bottom to top, it is observed that for individual layers, there is a strong correlation between the lateral thickness distribution

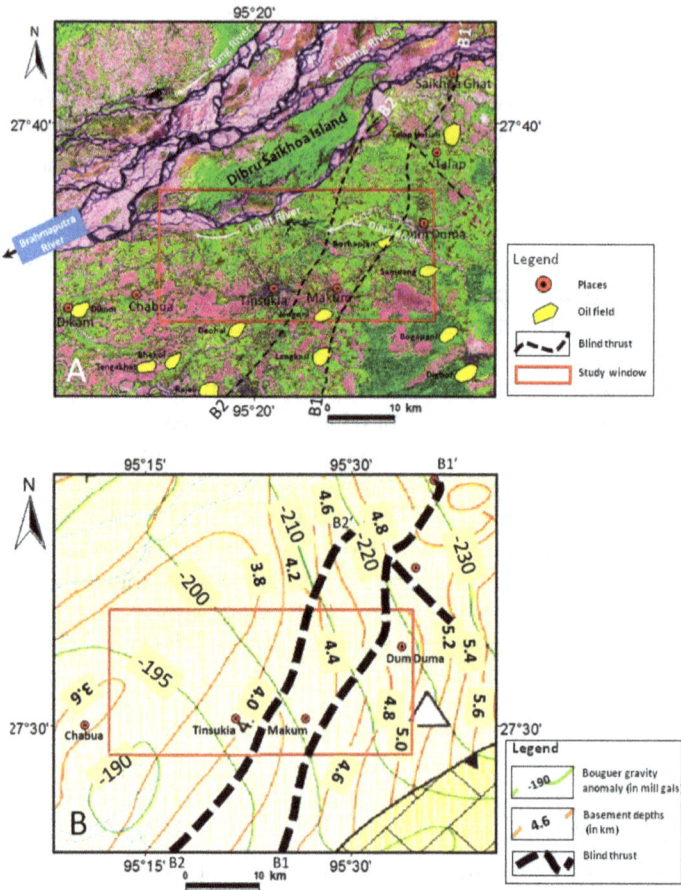

FIGURE 5.28
Location map of the Matikhana (W-3) area.

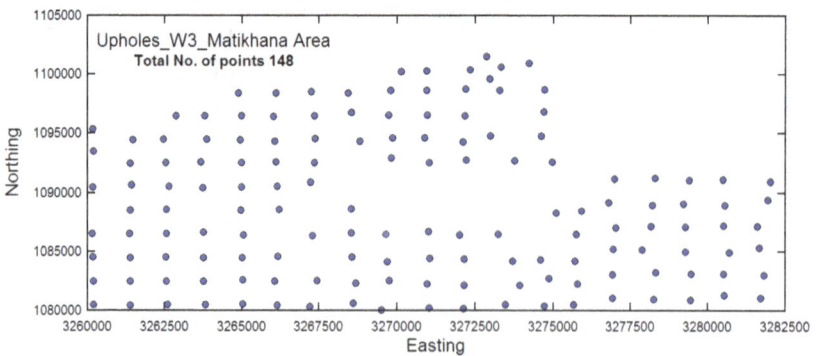

FIGURE 5.29
Data distribution in the Matikhana area (W-3).

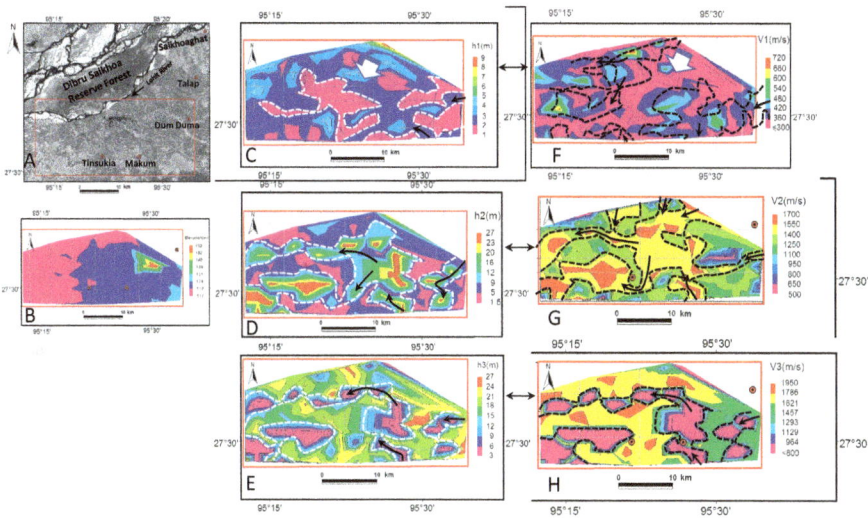

FIGURE 5.30

Lateral changes in velocities and thicknesses of three shallow subsurface layers in the Matikhana area (W-3).

and the velocity distribution. Firstly, analysing the velocity and thickness distribution in layer 3, there seems to be a meandering loop of an E-W trending channel, which was filled up with low-velocity materials (Figure 5.30e and h). Thus, the channel incising the more consolidated older flood plains deposited lesser consolidated materials. When we move upwards in layer 2, the same channel belt seems to be filled up with much thicker high-density materials. Thus, we can put forward at least three possibilities:

Firstly, the channel was captured by another river having much higher stream power, and provenance source remained the same.

Secondly, the new higher stream power channel might be having different provenance (e.g., if the earlier channel was of the NPT belt origin, the latter one was having its origin in the Mishmi Himalayas).

Thirdly, the river remained the same, but due to climatic changes (warmer climate), the rate of precipitation and the amount of sediment influx were more.

All these situations must have caused the removal of earlier sediments, and thus, we should have a local unconformity in between the boundary of layers 2 and 3.

Lately, we find a lack of correlation in the lateral distribution of thickness and velocities in layer 1, which can be attributed to the fact that the contribution of the earlier rivers in valley architecture has reduced substantially. However, flood plain deposits of the larger rivers like the Lohit have lately started to constitute a major component of the topsoil. Thus, there seems to be a major shift in the role played by different channels and the proportion

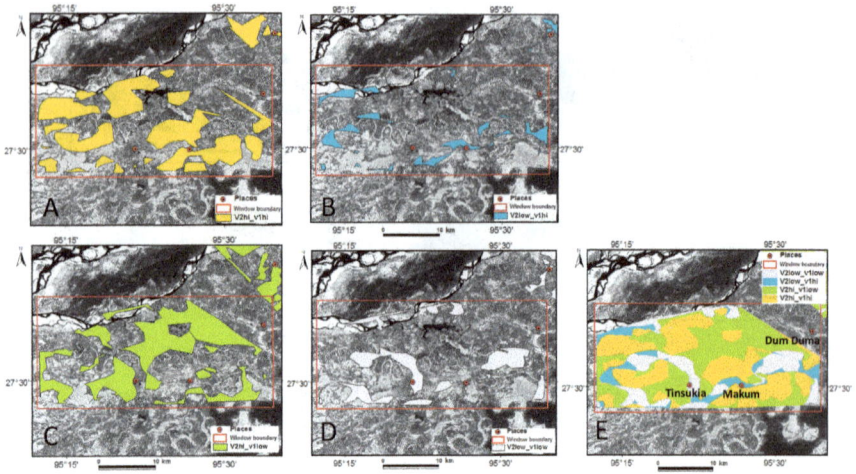

FIGURE 5.31
Bi-parametric micro-zonation based on velocity variation in the Matikhana window (W-3).

of the contributions of these fluvial systems in the late Quaternary evolution of the landforms in window 3 (Figures 5.31 and 5.32).

Multi-parametric micro-zonation of window 3 shows that the locations of the oil fields have an affinity with the zones where both V_2 and V_1 values are on the higher side. However, this cannot be interpreted as a thumb rule that wherever such conditions exist is highly prospective. Moreover, some of the palaeo-meandering loops having a 1- to 2-km radius are very often seen to bypass geomorphic highs, which we have seen for Majuli Island, and have deep-seated structural control. The overall characteristic of the window represents a 'high', and the paths along which this was cut by the channels show 'lows'.

Window 4: Mechaki

Situated between the latitudes 27.60°–27.85° and the longitudes 95.45°–95.75°, this window covers an approximately 843-km² area. Through the northern part of the area, the Lohit River runs from east to west and then bypasses Dibru-Saikhowa Island and flows through its eastern boundary, bringing the land thereby inside the channel belt of the Brahmaputra River, which caused an unprecedented width increase of the river in recent times (discussed in detail in Chapter 4). This window is also replete with palaeochannel marks (Figure 5.33a). Some of the meandering palaeo-loops have a 2-km radius. The prominent oil field of the area is Talap Hatiali, which is located between

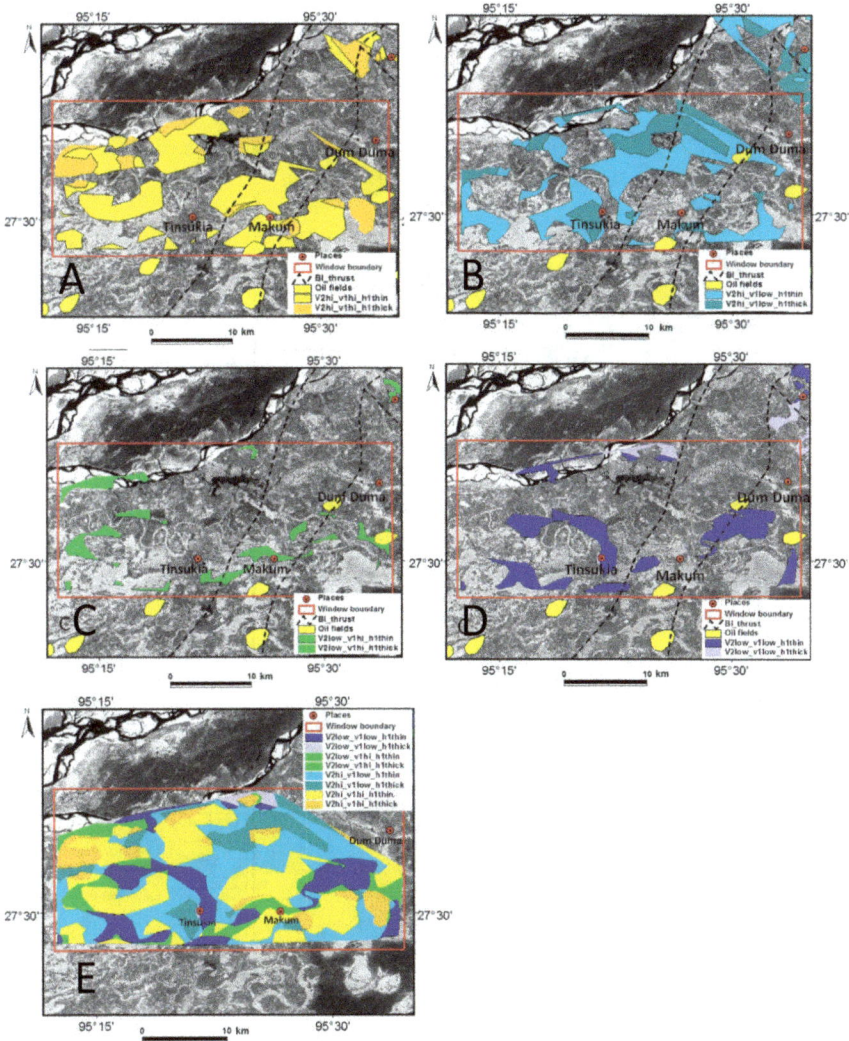

FIGURE 5.32
Tri-parametric micro-zonation based on velocity and thickness variation in the Matikhana window (W-3).

two blind faults, one is the NE-SW bound regional blind fault and the other is of localized nature running perpendicular to it and meeting it almost at a right angle. The basement depth in the window varies from 4.2 to 6.6 km (Figure 5.33b), with sediment thickness increasing drastically in the eastern fringe. The area is complex because both the frontal thrust belts of the Mishmi Himalayas and the NPT belt are competing with each other. Variable flow direction in the palaeochannels of this window is an interesting feature.

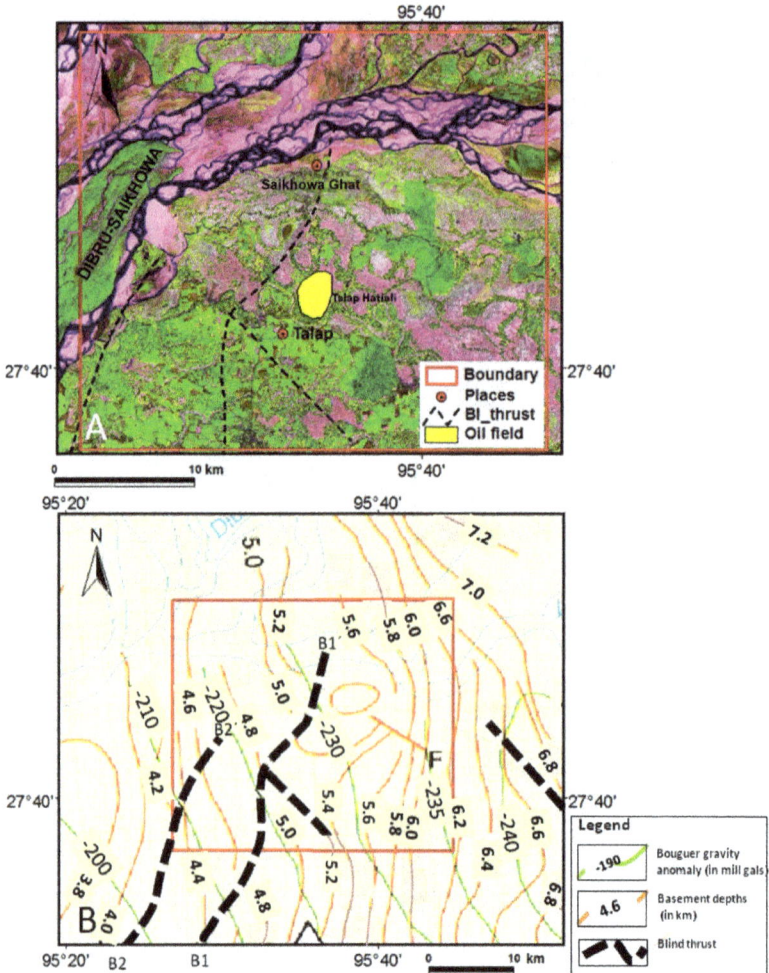

FIGURE 5.33
Location map of the Mechaki window (W-4).

Window 4 Mechaki Data

The refraction data sets (at 741 points) having NE-SW orientation have some gaps in the western fringe of the window due to some logistic problems; otherwise, the data density is quite good (Figure 5.34). Elevation varies from 120 to 141 m AMSL. From the eastern end to the western end, there is a smooth decrease in elevation. However, if we draw a profile joining the two stations Talap and the Saikhowa Ghat, we will obtain a geomorphic 'high' flanked by two 'lows' (Figure 5.35b). In the eastward direction along which elevation

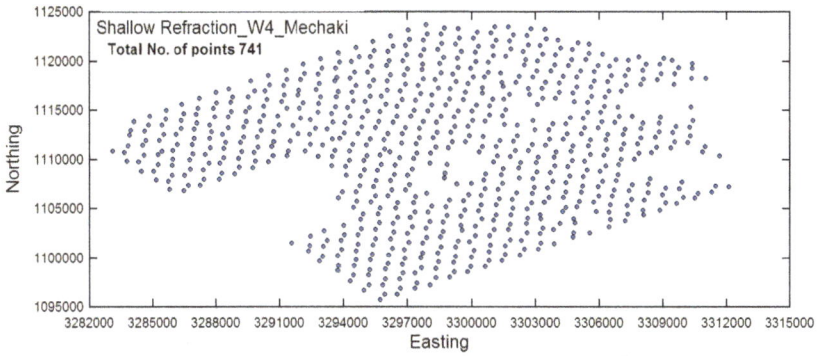

FIGURE 5.34
Data distribution in the Mechaki window (W-4).

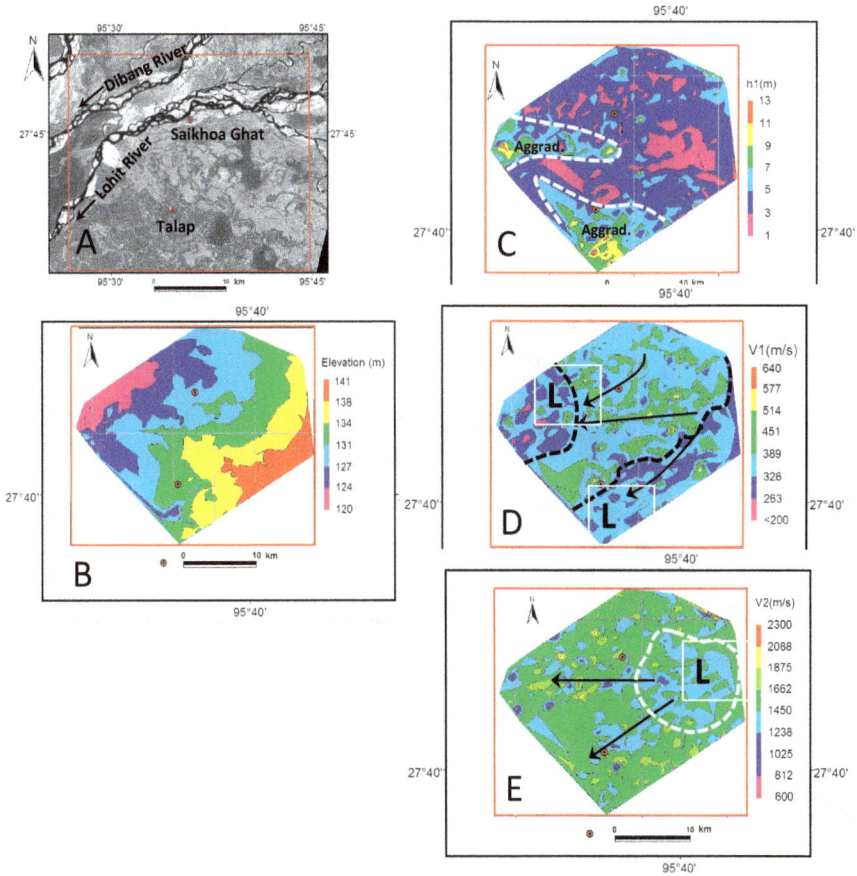

FIGURE 5.35
Palaeochannel identification based on shallow subsurface seismic velocity distribution in the Mechaki window (W-4).

increases, the basement depth also increases, which means the basement and surface elevation gradient follow a reverse trend.

Salient Features in Window 4

The lateral velocity distribution in the window shows both E-W and NE-SW trajectories of palaeochannels in conformity with the palaeochannels observed on the IRS-P6-LISS-3 imagery. Moreover, from the velocity distribution in the second layer, there appears to be a depression in the eastern half (Figure 5.35e), a lake-like situation from where channels arise. However, the location of this low seems to have shifted subsequently (Figure 5.35d). The main places of aggradation during the middle to late Holocene seem to be located in the southern and western parts of the window (Figure 5.35c). These observations are more prominently visible in the multi-parametric micro-zonation of the area (Figures 5.36 and 5.37).

Discussion

Valley filling in high-drainage density areas is complex because it is interwoven with different processes like bank-line shifting, anabranching, avulsion, channel capture, crevasse splay, meander migration, and channel cut-offs, which

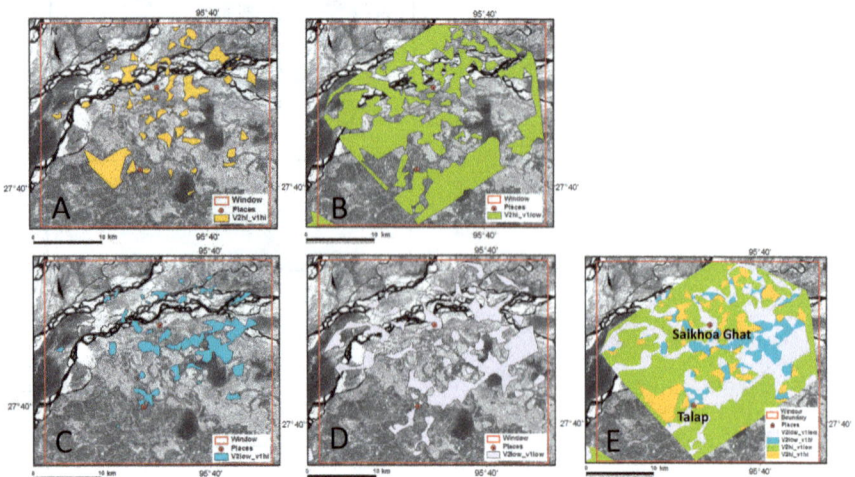

FIGURE 5.36
Bi-parametric micro-zonation based on seismic velocity distribution in the Mechaki window (W-4).

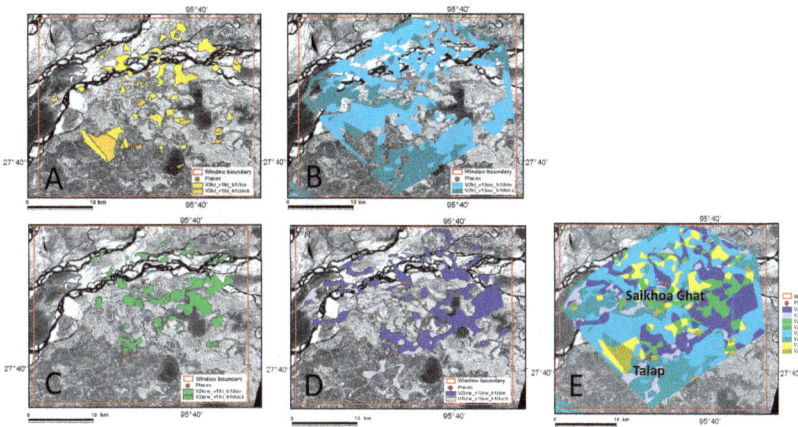

FIGURE 5.37
Tri-parametric micro-zonation based on seismic velocity and thickness of the weathered zone in the Mechaki window (W-4).

continue influencing aggradation–degradation in multifarious ways. Along with these elements, tectonic controls influence valley uplift and subsidence, thereby causing continuous changes in the definition of the accommodation space. Climatic changes, on the other hand, bring the elements of phase change like deposition at a given place. As a result, the supposed isotropic behaviour to be followed by individual strata is mostly not observed. For studying Holocene geomorphology (Aslan and Blum, 1999; Kraus and Wells, 1999; Morozova and Smith, 1999), the conventional method is to drill holes (usually <50 m), draw lithology variation by grain size analysis (sometimes, age dating is performed for frequent variations in sand/clay sequence), and then correlate the stratal boundaries. However, this scheme may not be so effective, where formation characteristics have drastic lateral variability. By studying shallow subsurface lateral variation in seismic *P*-wave velocity, we get a generalized criterion to define the 'stratal unit'. A 'stratum' is formed not only by the type of material deposited under a given condition but also by a factor of post-depositional diagenetic changes. For the present study, in a part of the south bank of the upper reach of the Brahmaputra valley, Assam, a geophysical 'stratal unit' concept was adopted to understand Holocene geomorphology, particularly in the context of fluvial dynamics. Our basic assumption is that the channel fill deposits and older flood plain deposits can be separated by velocity discrimination. Goodbred et al. (2003) in their analysis of the late Quaternary sediments in the Bengal fan have observed a huge increase in the influx of the sediments during 7–11 ky. This was a short-duration greenhouse period, and the glacial melt along with the increased rate of precipitation resulted in the surge in the sediment influx. As the Quaternary sediments in the Bengal fan were mainly transported by the Brahmaputra and the Ganges, increasing sediment influx must have acted synchronously in some of the areas of the Upper Assam basin as well as having much common provenance.

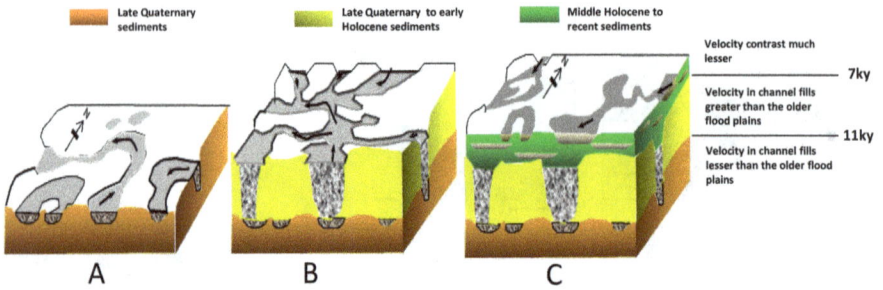

FIGURE 5.38
Diagram illustrating the nature of lateral variation in velocity during different periods of late Quaternary and construction of stratigraphy based on velocity data information.

To explain this aspect, refer to the observations described in window 3 (Figures 5.30 and 5.38).

We have defined bed boundaries for each of the observation points down the line based on sharp subsurface velocity contrasts. For each of the layers, based on the probable channel geometry (compared with the palaeochannel geometry in place or the immediate neighbourhood of the study area), first of all, channel locations were identified (Figure 5.30). The observations in different layers show that the same criterion cannot be used (like the lower velocity of channel sands than that of the older flood plain deposits) to identify the locations of the palaeochannels. Interestingly, velocity and thickness data for the respective layers have a highly logical resemblance. For example, the channel geometry for the intermediate bed shows higher positive velocity contrast with the host sediments, and when compared with the thickness variation plot (Figure 5.30d and g), it shows high average thickness and low average thickness for the identical paths lying below. This is a clear indication of the fact that the high-energy channel sands of the middle layer must have incised the earlier deposits before the high-energy material could settle there. An older scenario (as seen in the third layer), probably during the early Holocene, shows that the direction of flow of the channel was from east to west. In the later stage (as seen in the second layer), probably during the middle Holocene, it seems that channels joined from various directions at a regional low, making a lake-like situation, and from this palaeolake, two independent E-W-bound channels emerged. Although this velocity information-based seismic stratigraphy is different from lithostratigraphy, the lateral variation in thickness and velocities can provide process-based relevant information, which is not only important to understanding the nature of the sediment architecture, these can provide leads for hydrocarbon exploration in difficult terrains. We have used two sets of shallow subsurface geophysical data sets, namely, shallow refraction and uphole. Since for uphole surveys, actual boreholes are drilled, the data sets are more authentic. From the interpretation of these data sets for four different strategic windows in the south bank of the upper reach of the Brahmaputra valley, the following observations were made:

1. Shallow subsurface stratigraphy is much more complex in the valley fills having high drainage density and strong tectonic controls. However, in those windows where lateral variability in velocity and thickness is available for the three layers, the stratigraphy is observed to follow three different types of channel fills. The third layer is having less consolidated channel fills, whereas the second layer is having high-energy sediments, much thicker channel fills that must be a consequence of incision and removal of a substantial quantity of the earlier bed loads. Thus, below the mid-Holocene sediments within the channel belts, we should get unconformity surfaces under normal circumstances. Above the mid-Holocene sediments, the consolidation of both the host sediments and the channel fills are constituted of much lower velocity materials. This is perhaps due to the weaker channels bringing much lesser sediments down, whereas the adjacent large rivers bring substantial flood plain deposits, which are dumped all over including the beds of the smaller rivers. This causes uniformity in the sediment type and explains the very low lateral velocity contrast.

2. Seismic stratigraphy, making use of velocity discrimination, is more useful to identify the location of palaeochannels. This provides a key tool to identify the trends of channel migration.

3. Monitoring the trends of channel migration studied with the neighbouring structural elements can help us identify the nature of tectonic controls and their effect on the area of study.

4. A micro-zonation scheme can be worked out based on isolating different layers as 'high'-velocity (more consolidated) and 'low'-velocity (less consolidated) zones, and taking into account thickness variation ('thick' and 'thin'), then juxtaposing all these pieces of information in the GIS-based environment, certain type areas can be identified, which will open up many possibilities for exploration and seismological studies.

In the next chapter, we would like to understand certain more fundamental, basin-scale forcings influencing the fluvial dynamics.

6

Basin Evolution and Fluvial Dynamics

A basin evolves continuously due to combined actions of sediment dispersal variability (Tucker and Slingerland, 1996) regulated by syn-depositional climatic changes (Tucker and Slingerland, 1997) and the post-depositional deformations of the sediment architecture (Miall, 1991) guided by tectonic forcings (Dickinson, 1974). Thus, for a short interval of time on a historical scale, if the host material (older flood plain deposits), which is relatively static, is under continuous strain due to tectonic forcings (Gordon and Heller, 1993), channel belts witness faster changes in terms of sediment replacement as well as replenishment, a factor related to the rate of sediment influx, determined by the stream power and availability of sediments. Even a minuscule change in the surface gradient (Mitrovica et al., 1989) might be enough for a channel to migrate depending upon the geomorphic threshold of the system. The aggradation potential is higher mostly where the possibility of accommodation space generation (Miall, 1981; Posamentier and Allen, 1993; Holbrook and Schumm, 1999) is more. The continental large rivers are supposed to be more vigorous within the boundaries regulated by the lower order events of valley subsidence or upliftment (Allen and Densmore, 2000); although outwardly, river dynamics regulating the sediment architecture seems to be the fundamental cause behind the basin evolution. It is not the objective of the present study to contest whether sedimentological factors or structural controls have a larger influence on the basin evolution. We are assuming for the present study that for different orders of events, different controls will attain principality. Thus, for higher order stratigraphy, as discussed in Chapter 4, the size and dimension, stacking pattern, and density of the bed bars within the overall framework of the older flood plains are principally governed by the avulsion frequency and the residence time of channels at different places – in other words, fluvial dynamics plays the principal determining factor for stratigraphic development in the shorter time interval. However, the same logic will not hold good for the lower order regional stratigraphy, where unconformity surfaces (Coakley and Watts, 1991; Crompton and Allen, 1995) play a more significant role to act as the principal criterion for bed boundary differentiation. In this chapter, our objective is to understand the first-order stratigraphic development during the process of basin evolution and identify the role of different basin modifiers from the basement above based on the recent seismic data sets (after 2005). Subsequently, an attempt has been made to understand the controls on the pre-Quaternary basement configuration and basin filling history (Evans, 1965; Nandy, 1980, 1981, 2001; Kunte and Rao, 1989; Dasgupta and Biswas, 2000; Kent et al., 2002);

DOI: 10.1201/9781003302353-6

the nature of accommodation space generation mechanism, its unevenness, and, subsequently, the effects of changing basin configuration on the present-day river dynamics of the upper reaches of the Brahmaputra River. Moreover, we will show by integration of geophysical evidence from seismic data and the surface morphological changes that some of the relict landforms of the upper reach of the Brahmaputra valley like Majuli Island and other similar land-forms represent structural 'highs'. Also, we will explain the different stages of morphotectonic evolution of Majuli-type landforms.

Seismic Sections

There are six highly strategic seismic profiles available for the study area (Figure 6.1); some of these were shot during the same field season, and others were developed as single continuous profiles from various sources preserved in the data archival.

All these seismic sections were obtained from Oil India Limited, Duliajan, and were interpreted after identifying different reflectors based on the litho-logical information and the geophysical log data (mostly natural gamma-ray logs and the deep resistivity data).

Profile A-A'

The profile AA' (Figure 6.2a, see Figure 6.1 for location), located ~60 km upstream of the tip of the Majuli, is a merged section generated from two smaller sections.

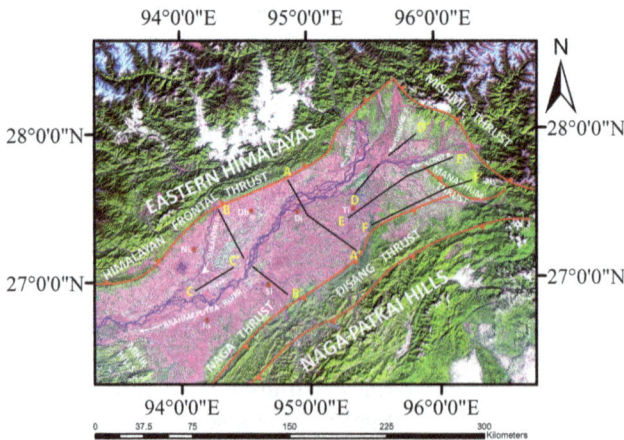

FIGURE 6.1
Locations of six 2D seismic profiles shown on the IRS-P6-LISS-3 image in the upper reach of the Brahmaputra valley of Assam.

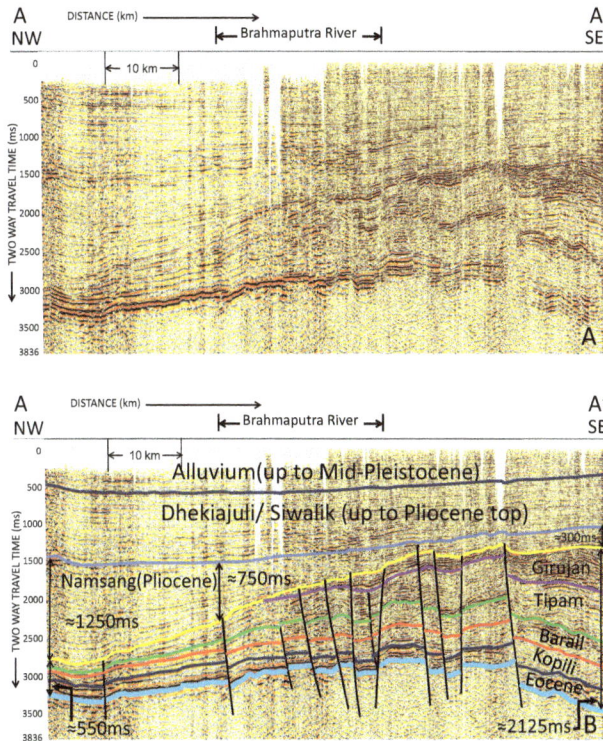

FIGURE 6.2
(a) Uninterpreted section below the profile AA'. (b) Interpreted subsurface understanding from the integrated seismic section AA/across the upper reach of the Brahmaputra valley.

Criss-crossing the entire valley, the net length of this profile is about 80 km. The NW-SE bound segment of the profile, about 25 km long, begins near the Himalayan Frontal Thrust (HFT), where the Simen River joins the valley and then ends at Dibrugarh, crossing the Brahmaputra River. The second segment of the section starts at Dibrugarh, proceeds towards the Naga-Patkai thrust (NPT) and ends near the place where the Burhi Dihing River emerges into the valley. This section, besides subsurface stratigraphy of Miocene–Pliocene sequences, shows the basin configuration and tectonic framework (Figure 6.2b, discussed in detail later). Interestingly, the Miocene top on the Eastern Himalayan side is located at about 2.8 seconds, whereas on the NPT side, this is located at about 1.3 seconds. Thus, there is a difference of about 1.5 seconds, suggesting a large variation in sedimentary thickness between the Himalayan and the NPT side.

Profile B-B′

The seismic section BB′ (Figure 6.3a, see Figure 6.1 for location) consists of two parts; there is a data gap in the Brahmaputra channel belt. The section

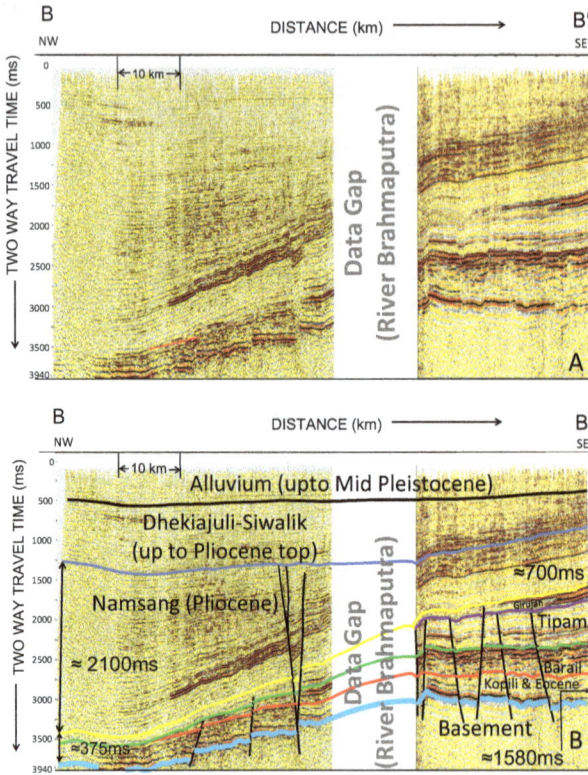

FIGURE 6.3

(a) Uninterpreted section below the profile BB'. (b) Interpreted subsurface understanding from the integrated seismic section BB/across the upper reach of the Brahmaputra valley.

starts close to the source of the Subansiri River, runs SE to the northern tip of Majuli Island, and then covers the south bank of the valley towards the east, running almost parallel to the Dikhau River. The section showing the maximum two-way travel time (TWTT) of 4.16 seconds (or 4,160 ms) covers a depth of around 6 km. The basement complex shows a general dipping trend towards the Eastern Himalayan side and several normal faults. Also, Paleocene–Miocene sediments (Figure 6.3b) of mostly marine origin thicken, as reflected by an increase in the TWTT from 375 to 1,580 ms towards the SE direction. The overlying sediments, late Miocene and younger, are mostly fluvial. Pliocene sediments are much thicker on the Himalayan side than on the topographic boundary of the NPT belt.

Moreover, as we compare AA' with BB', it is observed that the depozone (i.e., the zone of prominent deposition) in the north bank in the downstream direction of the Brahmaputra keeps on deepening (from 2,850 to 3,400 ms) and thickening (1,250 to 2,100 ms) below the topographic boundary of the Himalayan thrust belt.

Profile C-C'

The SW-NE trending seismic section along CC' (Figure 6.4a, see Figure 6.1 for location), parallel to the strike direction of Majuli Island, clearly shows the presence of a multiple-hinge anticline formed due to the contractional fault-related fold (Bally, 1997) in the basement complex itself. The hinge lines propagate in the upward direction. However, as we move up, massive sediment dumping seems to attenuate the hinge line bends. The basement shows highly faulted and fractured conditions.

Profile D-D'

This SW-NE bound profile (Figure 6.5, see Figure 6.1 for location), in the extreme upper part of the Brahmaputra valley, starts from a place called Rangagora, about 9.5 km away in the north-westward direction from Tinsukia, situated by the side of the Lohit River (about 20 years back, there used to be the Dangori River at the same place which was subsequently captured by the Lohit, one of the three main tributaries of the Brahmaputra River). This is an uplifting place where the height of the embankment above the normal level of the Lohit River is seen to increase abruptly. The profile runs for about 25 km to reach Saikhowa ghat. Then, there is a data gap of about 5 km, and crossing the Lohit River, the line continues from Morigaon onwards, more or less along the Kundil River for another 35 km. The north segment of this area is having a very dense jungle, and data acquisition is very difficult. Although data quality is not so good, the downward slope of the basement towards the Mishmi Himalayas and the continuous increase in sediment thickness is quite visible.

Profile E-E'

Profile E-E' (Figure 6.6, see Figure 6.1 for location) starts from Borhola, a place situated about 17 km away from Tinsukia towards the SW direction and sits on a gravity high, a closure having the Bouguer gravity value of −190 mgal. This, around 84-km-long profile, crosses Manabhum thrust and ends on the southern bank of the Lohit River. The seismic section clearly shows the sudden termination and upward stretch of the reflectors when these approach the thrust belt.

Profile F-F'

Almost parallel to the profile E-E' and situated at ~15 km offset and closer to the NPT belt, this profile (Figure 6.7, see Figure 6.1 for location) runs over some of the deepest depocentres of the valley, where sediment thickness reaches, at some of the places, more than 7 km. The length of this profile is about 70 km.

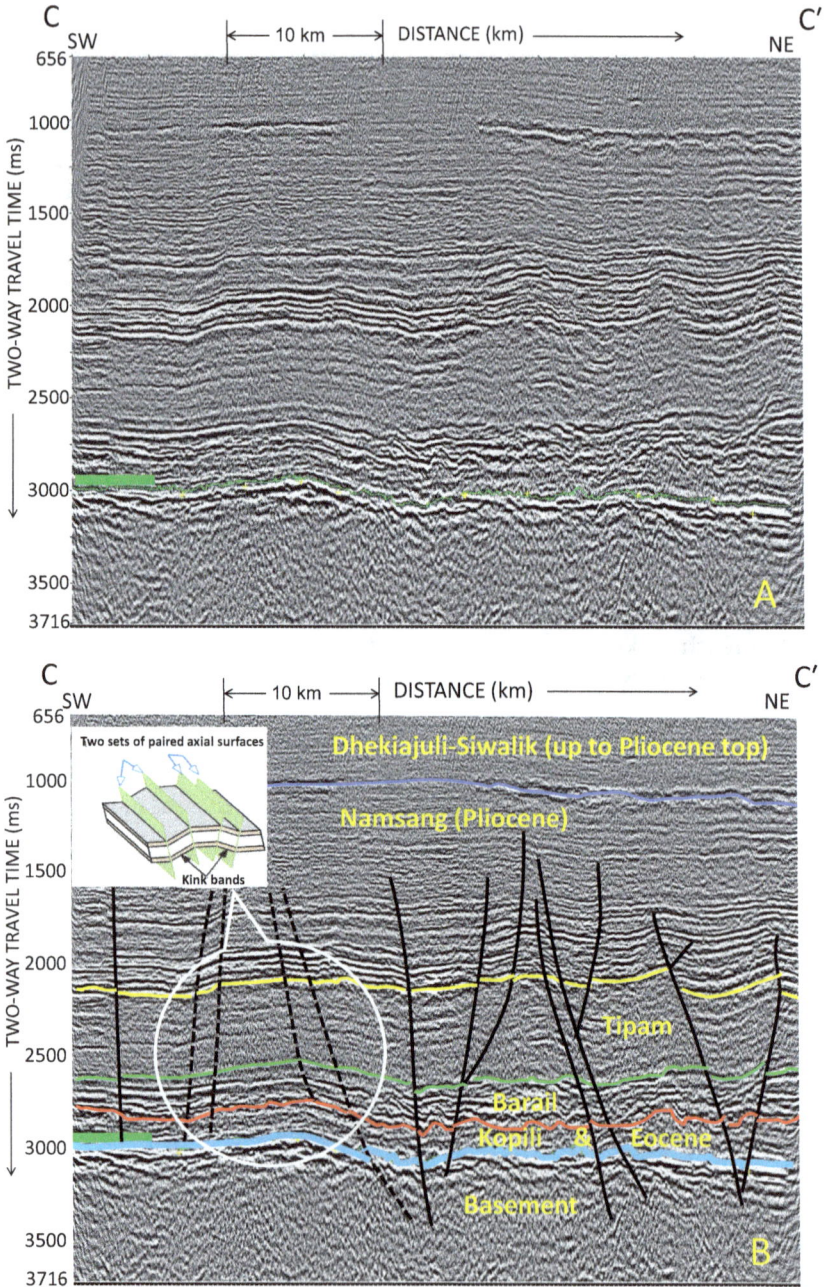

FIGURE 6.4

(a) Uninterpreted section CC′ along the strike direction of Majuli Island. (b) Seismic section CC′ below Majuli Island.

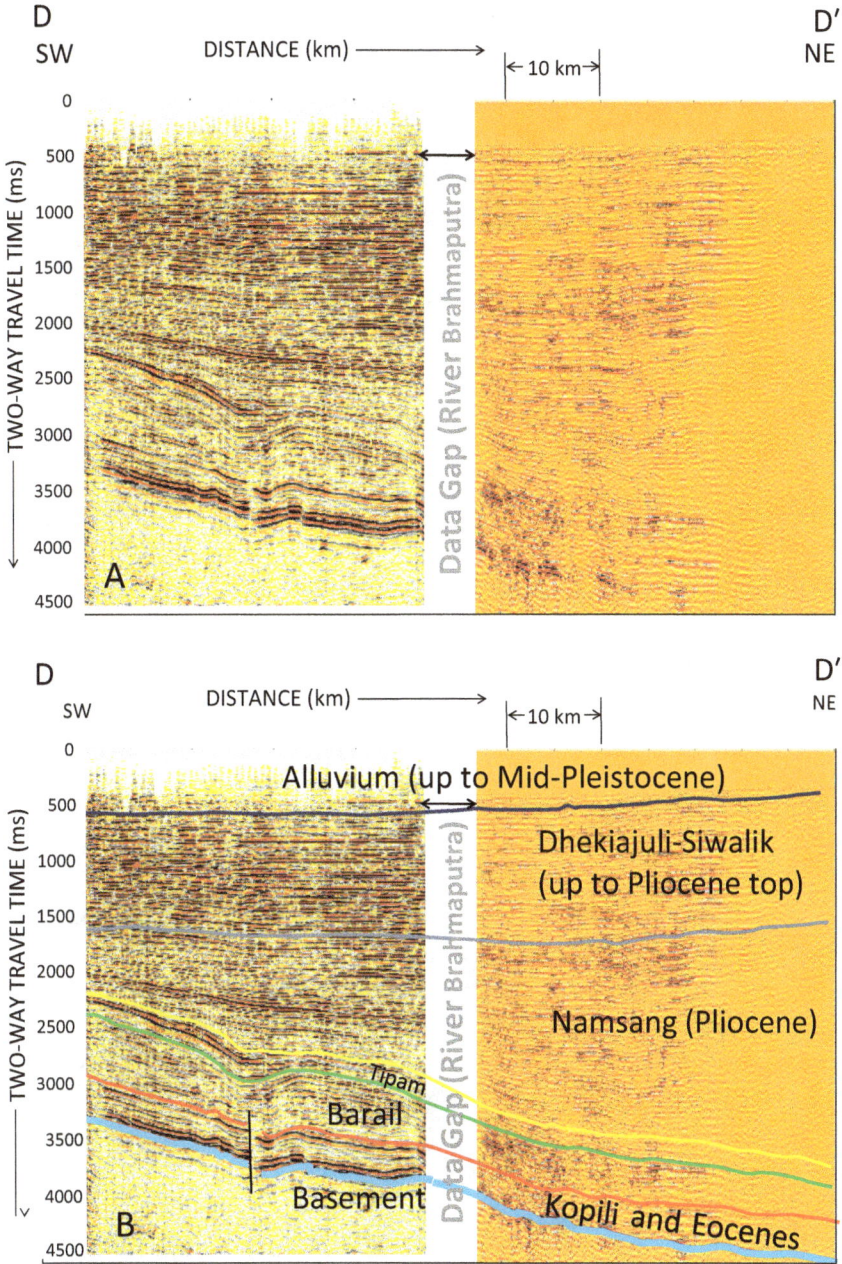

FIGURE 6.5
Interpreted seismic section below profile DD′.

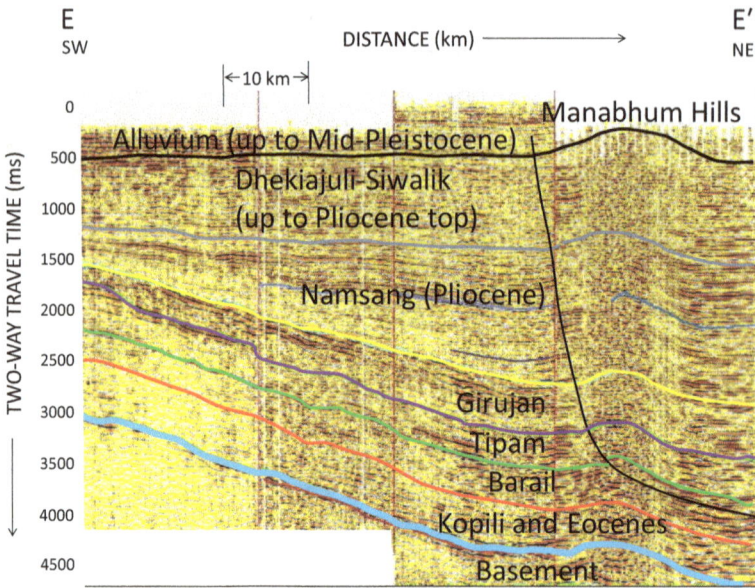

FIGURE 6.6
Interpreted seismic section below profile EE'.

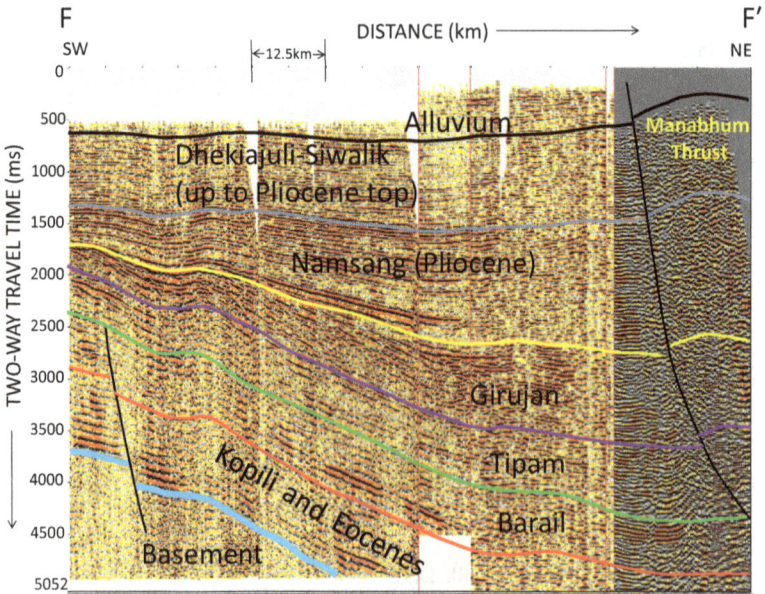

FIGURE 6.7
Interpreted seismic section below profile FF'.

Discussion

Seismic data presented above have helped understand the deeper basin fills to explain several geomorphic processes, such as spatial diversity of fluvial dynamics and genesis of Majuli-type landforms. The following sections first present a generalized stratigraphic model and basin evolution mechanism based on the seismic data and then picks up several unanswered questions on geomorphic processes discussed in the previous chapters.

Generalized Lithostratigraphy and Basin Evolution

Generalized geological successions and stratigraphy for different parts of the basin have already been discussed in Chapter 2 (see Figure 2.1 in the Appendix). Since early to mid-Miocene, the Himalayas started rising and the ancient sea regressed completely along the S and SW directions of the present study area; as a result, the basin has been filled primarily with fluvial sediments during this period, and this process continues till today. However, the lateral distribution of sediments throughout the valley was highly uneven (and still is), as can be seen in Figures 6.2 and 6.3. The seismic section presented in Figure 6.2b shows that the Miocene top (top of the marine sediments) is at about 2.8+ seconds near the HFT, and that near the NPT is 1.3 second. This time difference of 1.5 seconds being a TWTT, for an average velocity of 3,000 m/s, amounts to a depth interval of about 2.25 km. One possible explanation for such differences in basin depth is stronger subduction along the HFT and the generation of larger accommodation space (Flemings and Jordan, 1990; Beaumont et al., 1992; Gurnis, 1992; Holt and Stern, 1994; DeCelles et al., 1995) for fluvial sediment deposition. A comparison between the seismic sections BB′ and AA′ shows that the Pliocene sediments have thickened from 1,250 to 2,100 ms below the topographic front of the Himalayan thrust belt. Moreover, it is observed that the depozone in the north bank in the downstream direction of the Brahmaputra keeps on deepening (from 2,850 to 3,400 ms) and thickening (1,250–2,100 ms) below the topographic boundary of the Himalayan thrust belt.

When we generalize the stratigraphy below the profile AA′ by clubbing up the marine and non-marine sediments (Figure 6.8a) mostly of fluvial origin, we have some interesting findings.

a. Marine sediments below the fluvial sediments show a distinct trend of thickening from the HFT margin towards the present-day NPT line. On the other hand, the non-marine sediments, mostly fluvial, are the thickest about the HFT and much thinner towards the SE when compared to the location of Majuli Island (Figure 6.8a).

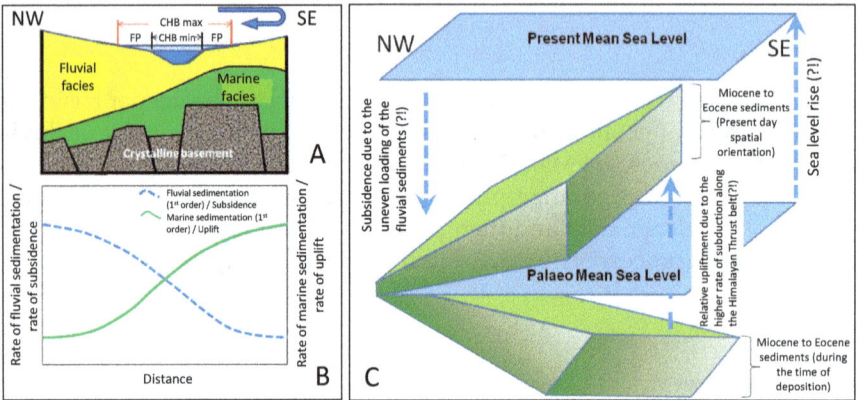

FIGURE 6.8

(a) A generalized vertical cross-sectional view of the upper reach of the Brahmaputra valley. (b) Trend of fluvial and marine sedimentation. (c) Questions associated with the present location of the marine wedge and its logical location during the time of deposition necessitating a strong tectonic intervention.

b. The timeline separating fluvial and marine sediments shows clearly that the wedge formed by mostly marine sediments (Eocene to Miocene) was subjected to post-depositional structural changes. As the marine deposition could take place only below the regional base level (mean sea level), the depth at which we get the top of the marine sediments nearby the HFT in the seismic sections must represent the palaeo-mean sea level (Figure 6.8c). The direction along which the thickness of the marine sediments increase represents logically the palaeo-deep sea side. As we move along the SE direction from the Eastern Himalayan side, the depth of the palaeo-ocean was increasing. It is observed that the Miocene top (for practical purpose top of the marine sediments) is at about 2.8+ seconds near the HFT, and the same near the NPT is 1.3 second. This time difference of 1.5 seconds being a TWTT, for an average velocity of 3,000 m/s, amounts to a depth interval of about 2.25 km. To justify this observation, we have at least three possibilities: (i) the sea level since the Miocene has risen by more than 4,000 m, (ii) the thicker side of the wedge was uplifted by at least 2,250 m, and (iii) the thinner side of the wedge experienced subsidence of not less than 4,000 m. The sea-level rise proposition is simply overruled because the stratigraphic characteristics show, in general, a transition from marine to fluvial as we move up vertically from the basement and that shows there was an overall uplift of the landmass, forcing the palaeo-sea to regress, which means a continuous fall in the sea level. In addition to this, the sea-level rise cannot justify the older marine sediment rise. The two remaining possibilities (i) and (ii) might be explained together by a subduction mechanism along the HFT, causing faster generation of accommodation space and thereby providing a more favourable site of fluvial sediment deposition.

Next, we will briefly discuss different distinct phases of the poly-history basin mostly following the scheme proposed by Kingston et al. (1983). Altogether, six major sequences have been identified above the basement complex. There is no indication of pre-Tertiary sediments.

Basin-Forming Tectonics

Usually, three elements are considered in basin-forming tectonics (Dickinson, 1974; Klemme, 1975; Kingston et al., 1983) These are (i) crust composition, continental or oceanic; (ii) plate movement characteristics, that is, convergence or divergence type; and (iii) basin position on the plate (continental interior or margin) and primary structural movements (sagging, normal faulting, or wrench faulting). From the interpretation of the seismic sections of the upper reach of the Brahmaputra valley, we observe the presence of six distinct first-order sequences (Figures 6.9 and 6.10 and Table 6.1). We will discuss these sequences/cycles from bottom to top.

Sequence 1: Margin Sag (MS-321)

From the nature of depositional events, it appears that before the formation of a fold-belt on the seaward side, this composed of continental margin sag (MS) cycles, which transformed subsequently to interior sag (IS) cycles by orogeny. MS is initiated by interior fractures in the continental graben blocks (Beck and Lehner, 1974). Grabens get filled up with non-marine sediments (stage 1). As the fractures grow, continents start diverging, margins subside,

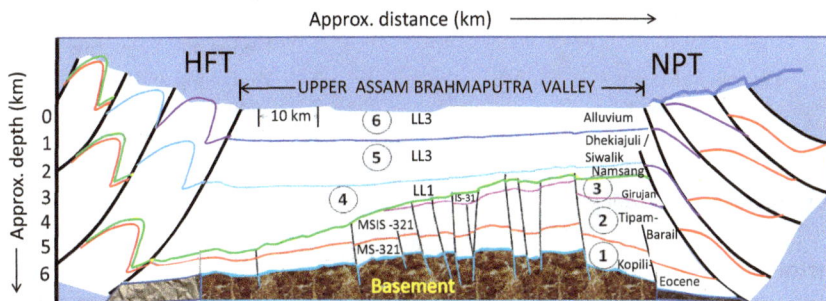

FIGURE 6.9
Interpretation for cycles of Brahmaputra basin evolution.

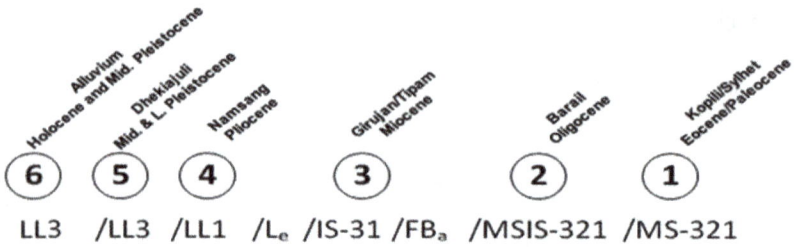

LL3 /LL3 /LL1 /L$_e$ /IS-31 /FB$_a$ /MSIS-321 /MS-321

FIGURE 6.10

Poly-history evolution formula for the upper reach of the Brahmaputra basin, Assam (method in line with Kingston et al., 1983).

TABLE 6.1

Lithostratigraphic Mode of Representing Basin Evolution History of the Upper Reach of Brahmaputra Valley, Assam

Cycle No.	Age/Equivalent	Basin/Cycle Type	Tectonic Modifier		Depositional Stage
6	Holocene to middle Pleistocene (recent alluvium)	Wrench or shear cycle (LL)	L_e FB_a	3	Wrench basin is continuously uplifted and eroded/ non-marine deposition
5	Middle & Lower Pleistocene/Dhekiajuli Formations (U. Siwalik)	Wrench or shear cycle (LL)		3	Wrench Basin is continuously uplifted and eroded/ non-marine deposition/Ends in an unconformity
4	Pliocene/Namsang Formations (M.Siwalik)	Wrench or shear cycle (LL)		1	Strong episodic wrench causing basin tilt and asymmetry/ Wrench basin due to plate convergence gets initiated/ non-marine deposition
3	Miocene/Girujan & Tipam Formations (L. Siwalik)	IS		3 1	Wedge top and base-non marine. Fold-belt formed on the side of the basin due to plate convergence
2	Oligocene/Barail Series (Murree-Dharamsala)	MSIS		3 2 1	Wedge top and base non-marine Wedge middle-marine
1	Eocene/Paleocene/ Kopili and Sylhet formations (Subathu-Kanji)	MS		3 2 1	Wedge top and base non-marine Wedge middle-marine

MS, margin sag; IS, interior sag.

marine waters invade, and marine sediments deposit (stage 2). With the onset of the formation of a fold-belt on the seaward side, it transforms into an interior basin mainly receiving non-marine sediments (stage 3). This covered the early Tertiary, Paleocene–Eocene time having glauconitic sandstones, limestones, and basaltic flows, known by Kopili and Sylhet formations.

Sequence 2: Margin Sag–Interior Sag (MSIS-321)

A phase of regression and the presence of unconformity are followed by massive sandstones having shale intercalations belonging to the Barail group. It represented gently dipping continental platform-type conditions (DeCelles et al., 1998). It represented a transitional phase of a marginal sag basin on the verge of getting converted to IS cycles. These Oligocene MSIS basins were mostly asymmetric, and many of them are found to be highly petroliferous worldwide (Kingston et al., 1983). Because of the transition stage, there was a marine intrusion in between. As a result, we get all three stages.

Sequence 3: Interior Sag (IS-31)

The presence of thick Girujan clays (reaching at some places more than 2,000 m) shows a high degree of continental basin sag with complete exclusion of marine sediments. Associated with Girujan clays, the massive Tipam sandstones are there. This was interpreted by Dasgupta and Biswas (2000) as 'an advanced stage in the sedimentation of the Indo-Myanmar gulf fillup'. Tipam sandstones are also of fluviatile origin. This early Miocene basin filling was then occurred from east to west by the rivers bringing sediments from Indo-Myanmar mountain ranges. By the middle Miocene, the Himalayas started rising, and basin filling mechanisms from the west to the east started more vigorously.

Sequence 4: Shear or Wrench Initiation (LL1)

Strike–slip, shear, or wrench basins (Freund, 1965; Harding, 1973; Crowell, 1974) are represented in this way due to the lateral (LL) nature of movement involved. Wrench basins/cycles are very often found in the areas of two or more converging plates. If the rise of the Himalayas was caused by the collision and subsequent subduction of the Indian plate below the Eurasian plate (Gansser, 1964; Molnar and Tapponnier, 1975; Bird, 1978; Ni and Barazangi, 1984), thrusting along the Indo-Myanmar belt (LeDain et al., 1984; Nandy, 2001; Khan, 2005) also caused a large-scale crustal shortening, and the upper reach of the Brahmaputra valley (our study area) started functioning as an independent basin. In the post-Miocene period, the basin was thus re-initiated by strike–slip tension. This was the Pliocene series having very thick Namsang deposits.

Sequence 5: Wrench Deformation (LL3)

A general uplift of the entire basin, as well as local uplifts due to wrench faults and simultaneous erosion, causes the sediment influx to dampen some of the block faults in the valley. Wrench structures continue to grow. This continued from the upper Pliocene to the middle Pleistocene and is known as Dhekiajuli Formation.

Sequence 6: Basin Upliftment (LL3)

This is the period from the top of the middle Pleistocene to the present. A general uplift of the entire basin continued along with the erosion. Wrench structures continue to grow. Wrench faults propagating along the foredeep areas manifest as blind faults and might be responsible for the local uplift of the basin in selective areas.

Basin-Modifying Tectonics

Basins forming during one type of tectonic movement might be subjected to other types of structural events, which are known as 'basin modifiers' (Kingston et al., 1983). These are usually of three types: episodic wrenches (L), adjacent fold-belts (FB), and complete folding of a basin area (FB3). Depending upon the degree of increasing magnitude, 'very weak to no effect' to 'very strong effect', both episodic wrenches as well as adjacent fold-belts, are expressed with subscripts 'a'–'f'. For example, it is assumed that salt/shale diapirs get triggered by a weak tectonic modifier FB_b/L_b. In the upper reach of the Brahmaputra valley, during its six cycles of history of evolution, the role of 'basin modifiers' was recognized twice.

We propose that the first basin modifier (FBa) involved the shift from cycle-2 to cycle-3, that is, from MSIS to IS; a fold-belt was formed on the side of the basin. However, the formation of the fold-belt had only a minimum structural effect on the basin itself.

The second basin modifier (Le) involved the shift of the basin from IS type during cycle-3 to wrench or shear (LL) cycle type during cycle-4, there was the influence of a strong episodic wrenching (as described earlier, in a six type scale from 'a'–'f', the subscript 'e' stands for a 'strong' effect). This stands for strong plate tectonic effects that alter significantly the basin tilt, causing marked basin asymmetry. This was the late Miocene and early Pliocene period marking the dominance of Himalayan sediments in the basin fill.

Tectonics and Sedimentation

The generalized subsurface stratigraphy and structural characteristics, based on Bouguer gravity anomaly, seismic profiles, and well logs (Murthy, 1983; Dasgupta and Biswas, 2000) across the valley are shown in Figure 6.12. For the sake of clarity, Quaternary and Neogene sediments (last 23 million years) have been clubbed together, and all pre-Neogene sediments

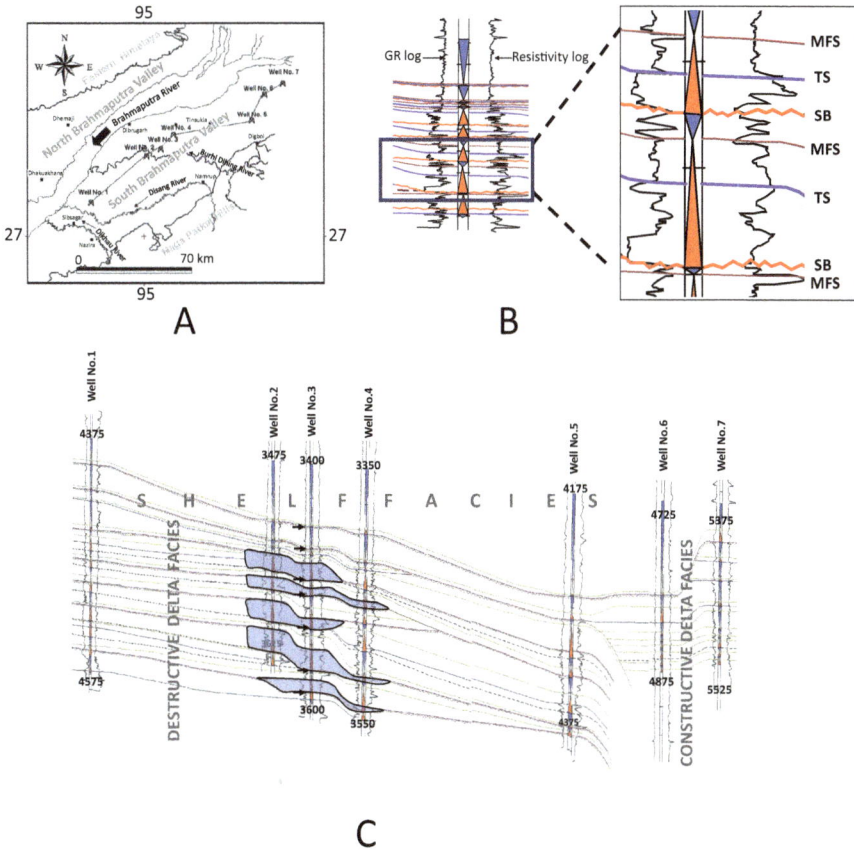

FIGURE 6.11
Sequence stratigraphic correlation of Paleogene sediments of the upper reach of the Brahmaputra valley from seven deep wells.

are grouped. The correlation of logs for the Paleogene sediments helps identify some of the higher order marine sequences and highly characteristic constructive shelf characteristics towards the northern side of the valley and destructive shelf facies towards the southern end (Figure 6.11). Now, Figure 6.12a indicates that the thickest depocentre for the low-density sediments of fluvial origin (Quaternary and Neogene) occurs on the Himalayan side, whereas the Naga-Patkai side of the valley acts as the main depocentre for the marine sediments (pre-Neogene), which pinches out towards the Himalayan side. Furthermore, the thickness of the marine sediments increases more than 10 times across the Brahmaputra valley in the NW-SE direction along the profile AA'.

This clearly shows that the Paleogene coastline was much closer to the Himalayan side (Figure 6.12c) regressing gradually in the southeast direction.

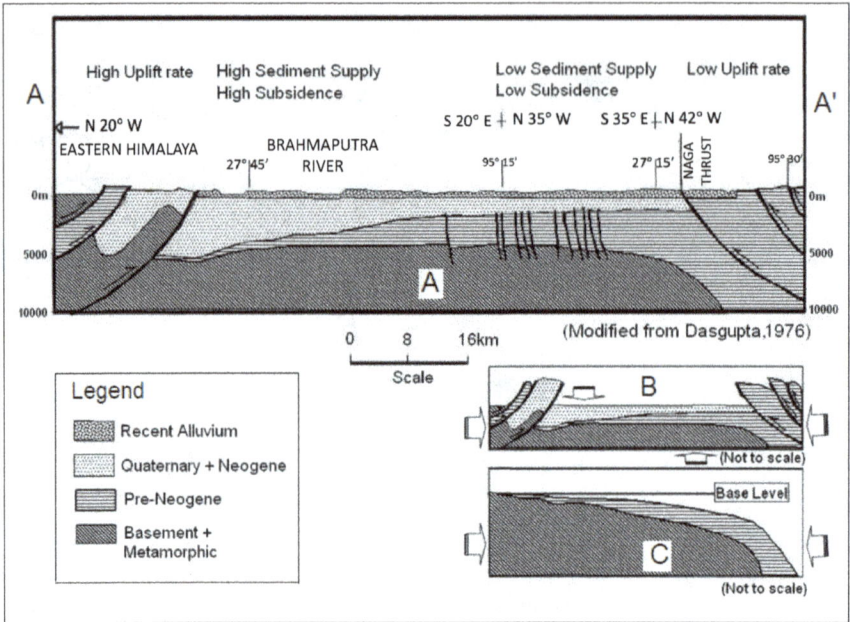

FIGURE 6.12

Sectional view along with a profile AA' passing across the upper reach of the Brahmaputra valley (modified from Dasgupta, 1976 and verified from the latest data).

It is observed in the seismic sections across the Brahmaputra valley that the elevation of the upper boundary of the marine sediments keeps on increasing as we move from the northwest towards the southeast direction.

In other words, the overburden thickness above the marine sediments is thicker near the Himalayas and thinner near the Naga-Patkai hills. There seems to be a post-depositional structural readjustment. Isostatic readjustment could have also caused the basement to move upwards. Moreover, thrusting was continuing on both the Himalayan and the Naga-Patkai sides. The Brahmaputra valley is getting arched due to the continuous subduction taking place at the Himalayan end and the Indo-Burman side, and the resulting strain is evident in the form of the fault lines, almost vertical in the middle portion of the valley as seen in the seismic sections (Murthy, 1983). This happens despite the dampening effect of the recently deposited load of fluvial sediments. Moreover, different rates of thrusting and different types of sediments present at both the ends (as discussed earlier) are manifested in different rates of external deformation.

A flat crystalline basement having poor compressibility is not supposed to create more accommodation space due to the differential sediment dispersal. A more rational explanation first supports the structural subsidence, most probably due to the overriding effects in the convergent basin margins, followed by sediment piling, and that might act as the positive feedback to thrusting in the basin margin and arching of the main basin (Figure 6.12b and c) to increase the accommodation space subsequently.

The argument given in favour of the tectonically forced rapid rate of vertical movement having a uniform rate of erosion (Burbank et al., 2003) and that given in favour of a higher rate of erosion (Wobus et al., 2005) causing isostatic compensation of the mass removed and faster rate of vertical upliftment would ultimately result in higher sediment production in high-altitude mountain fronts than the low-altitude mountain ranges. As a consequence, the sediment architecture of the intermontane basin in the upper reach of the Brahmaputra valley is bounded by high-altitude Eastern Himalayas on one side and much lesser altitude Naga-Patkai hills on the other side is expected to be uneven. We argue therefore that tectonic activities in the Himalayan front and/or faster rate of isostatic compensation are shaping the valley topography in distinctly different ways, which are manifested in variable fluvial dynamics observed for the tributaries of the Brahmaputra.

Unconformity in the Fluvial Sediments

In seismic sections, identifying unconformities is easier for the marine sediments (see Figure 3.12). However, the same thing cannot be applied for the fluvial sediments. The main distinguishing parameter is the sudden change in grain size from fine to coarse in the upward direction. If boulder beds (cobble and pebble in the downstream direction) are lying above the fine-grained materials, this is surely an indication of an unconformity surface. Thus, on seismic surfaces, unconformities of the fluvial environment will mostly appear parallel and are difficult to identify. If the velocity contrast is high and sorting is good, even the parallel unconformities can give rise to prominent amplitude standout; however, if the sorting is poor or the boulder size is bigger as well as highly angular, there might be a lot of scattering of energy and unpredictable amplitude variability. Thus, the nature of sediment transport, sorting, etc. during fluvial dynamics will determine whether unconformity surfaces will be properly discernible or not on the seismic sections. However, borehole log data in such circumstances will provide much authentic information. For the present work, we have consulted log data (mostly deep resistivity and natural gamma-ray log) for identifying unconformity surfaces in the fluvial environment.

Moreover, while identifying different cycles, we have followed the standard definition of sequence boundaries as unconformity surfaces.

Basin Evolution and Directional Shift in the Major Fluvial System

The presence of a high-energy facies above a low-energy facies, if it is having a regional extent, there are possibilities of several independent causes or perhaps juxtaposition of causes; for example, there might be changes in the course of a high-energy river, or perhaps climatic change induced higher rates of precipitation, thereby causing coarser grade of sediments to travel long distance or perhaps recent episodic uplift of the neighbouring mountains bringing the site of provenance closer to the site of deposition. It is difficult to isolate climatic factors from the tectonic changes from the origin of unconformities. However, the relative thickening/thinning of sediments between two unconformity surfaces, the presence of faults, and the faulting types in the marginal part as well as the interior part of the basin help us understand the nature of structural controls.

In the Upper Assam stratigraphy, thick Tipam sandstones are highly petroliferous, and the presence of thick Girujan clays above plays an excellent role as the cap rocks. Both Tipam and Girujan formations are of fluviatile origin. Tipam on the further south and southeast directions becomes Surma Formation, which is equivalent to the Tipam and is partly fluviatile, partly marine. Since Tipam sandstones are post-Oligocene deposits, representing the pre-Himalayan uplift period, and the NPT belts (RangaRao and Samanta, 1987) are also not present during that time, the valley width, as well as area, was much wider. That valley belonged to the Assam–Arakan basin. On the other hand, fluviatile characteristics of Tipam sandstone suggest that the Miocene continental platform was already an uplifting place, and the possibility of the main river course was along the southern extension of the Indus–Tsangpo–Suture zone (Gansser, 1980) (Figure 1.1). Myanmar-side hilly tract provided another provenance. Thus, a palaeo-Brahmaputra-like situation was there with about 200 km or more of eastward offset from the present position. The thickness of Tipam as seen in the seismic sections FF' (Figure 6.7), EE' (Figure 6.6), and DD' (Figure 6.5) keeps on decreasing as we move westwards. The great thickness of the Girujan clays is still not well explained in terms of basin evolution. Its westward extension is much more restricted than the Tipam [it becomes thinner in the seismic section EE' (Figure 6.6) than FF' (Figure 6.7) and is absent in DD' (Figure 6.5)]. Mostly lacustrine, this must be located at a much greater distance from its site of provenance. Moreover, the deposition of Tipam might have been followed by a prolonged greenhouse period with a simultaneous uplift of the basin. In the absence of the Himalayas and the absence of the monsoon precipitation (Valdiya, 2002), discharge through the earlier large rivers, responsible for the Tipam sands, became substantially weaker bringing mostly clays. But, it is not still clear why such large lakes should form at all. This can be explained by the continental sag, which could happen in the initial stage along the margins of the coalescence of three plates. Subsequently, during the post-Miocene period, due to a strong episodic wrenching, the basin got tilted towards the Himalayan side. The overriding of the Burmese plate over the Indian plate (Mukhopadhyay and Dasgupta, 1988) initiated the Indo-Myanmar thrust blocks, which isolated the Brahmaputra sub-basin from the earlier Assam–Arakan basin. Since the onset of the Pliocene period, a proto-Brahmaputra-like river regime (Robinson et al., 2014) must exist to flush out the huge volume of Himalayan sediments. As the influence of thrusting kept on increasing from the eastern end, the direction of the palaeo-Brahmaputra might have shifted from southwards to south-westwards (Uddin and Lundberg, 1999).

Accommodation Space Generation and Residence Time of Rivers

Analysis of the seismic sections shows clearly that the foredeep areas (DeCelles and Giles, 1996) of the Eastern Himalayas, and the Mishmi Himalayas (Thakur and Jain, 1975) store much thicker columns of fluvial sediments than other parts of the valley. In other words, it can be said that the residence time of the major sediment supplying rivers in the foredeep areas is comparatively longer than that in other parts of the valley. This can be explained by the fact that the rate of subsidence in the foredeep areas is higher. Syn-depositional tectonic readjustments in the foredeep zone cause regeneration of the accommodation space and subsequent reversal in tilt (Mitrovica et al., 1989) of the surface relief. This explains the unevenness of the residence time of large rivers along with the active foredeep areas of the convergent basin margin. Moreover, as the river debouches in the valley for the first time from the great heights, due to sudden loss in energy, its instantaneous sediment dumping propensity is much higher than that anywhere else, and possibilities of fan build-up are very much there in the foredeep areas (Waschbusch and Royden, 1992). A load of this sediment might act as positive feedback (Ruddiman, 2008) to the tectonic subsidence, thereby increasing the accommodation space.

Morphotectonic Evolution of Majuli Island

The SW-NE trending seismic section along CC' (Figure 6.4, see Figure 6.1 for location), parallel to the strike direction of Majuli Island, clearly shows the presence of a multiple-hinge anticline formed due to the contractional fault-related fold (Bally, 1997) in the basement complex itself. The hinge lines propagate in the upward direction. However, as we move up, massive sediment dumping seems to attenuate the hinge line bends. The isopach map surrounding Majuli Island, integrated mainly from the seismic surveys done by the ONGCL and the well data information, shows that the thickness of sediments is ~6.0 km close to the foothills of the Eastern Himalayas (Figure 4.19) and 3.5–4.4 km below Majuli Island. Sediment thickness decreases rapidly between Majuli Island and the Mikir Hills, and ultimately, the hard basement complex crops up. The basement shows highly faulted and fractured conditions.

Seismic data have also been integrated with topographic and basement configuration. The topographic map of Majuli Island (Figure 6.13a) shows a prominent high in the central part with a sharp break in slope (average slope changes from 0.17 to 0.73 m/km). This topographic break closely matches a break in the basement slope along with the same profile (Figure 6.13b). The basement depth was computed with an average velocity of 3,000 m/s from the seismic section, as shown in Figure 6.4. Figure 6.14 shows the isopach map of the region around the Majuli. There is a very prominent E-W bound fault called the 'Jorhat fault', passing through the tail end of Majuli Island. Several NE-SW bound faults are running parallel to the strike of Majuli Island.

This evidence suggests that the position of Majuli Island is structurally controlled, and such geomorphic highs are guided by basement topography.

We have also verified this hypothesis with the seismic section DD′ (Figure 6.5), EE′ (Figure 6.6), and FF′ (Figure 6.7) for the NE-SW bound profiles running near-parallel along the eastern margin of the Dibru Saikhowa reserve forest (the quality of which is not so good). The seismic sections show unmistakably a prominent basement upliftment around the 'new Majuli' area (Gogoi et al., 2022).

Based on our geophysical and geomorphological investigations, we propose a three-stage evolution of Majuli Island and similar landforms in the region (Figure 6.15).

> **Stage I** involved the development of geomorphic 'highs' guided by basement configuration. This is confirmed from the seismic sections and topographic data, and we have mapped a basement high just below the existing Majuli Island.
>
> Although the quality of seismic data around the new Majuli region in the upstream reaches of the Brahmaputra is not so good, the basement configuration and the topographic data suggest the presence of a geomorphic high at this location as well. These 'highs' can either be due to the leading edges of the blind thrust fronts in the foreland areas of the valley or due to the normal faults and 'arching' of the basin along its central part.
>
> **Stage II** involved the incorporation of the geomorphic high within the channel belt as a result of fluvial dynamics. In valleys with strong structural control, well-defined geomorphic highs generally force

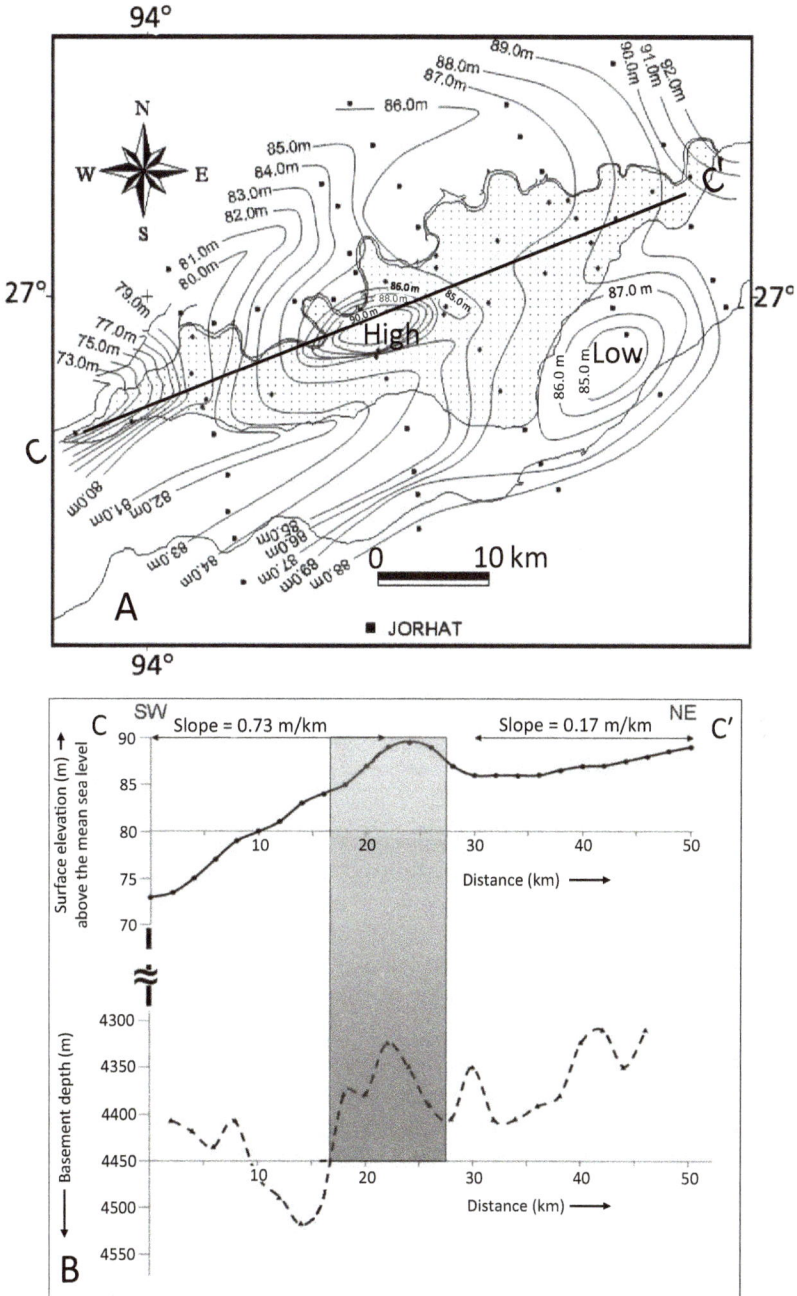

FIGURE 6.13
(a) Topographic contours around Majuli Island. (b) Surface elevation change and the basement depth variation below the seismic profile CC'.

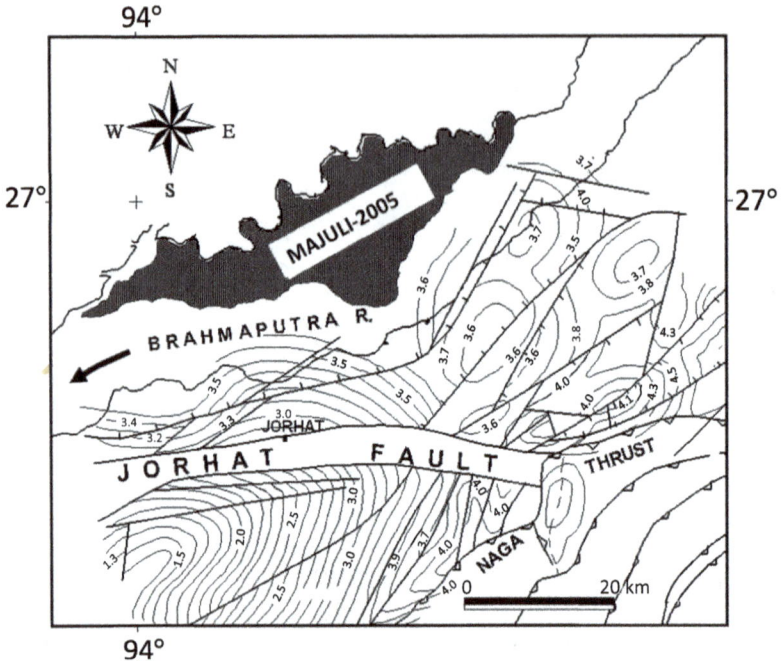

FIGURE 6.14
Isopach map (modified from Prasad and Mani, 1983) showing the basement depth or sediment thickness variation (in km) in the eastern and the south-eastern boundary of Majuli Island.

the river to bypass (Holbrook and Schumm, 1999). The development of the new Majuli Island in the upper reaches of the Brahmaputra clearly illustrates this stage where the southward migration of the Lohit River bypassed the geomorphic high, and the forested flood plain was incorporated within the channel belt at a historical time scale. Further development of the island through erosion–deposition cycles occurs due to local geomorphic processes. Our study has also shown a close relationship between the morphodynamics of the Brahmaputra River and the erosional history of Majuli Island.

Stage III involves the abandonment of Majuli Island or incorporation of the island with the adjoining flood plain. This is again affected by fluvial dynamics, as illustrated by Majuli Island configuration in 2005. The main channel of the Brahmaputra now flows south of the Majuli, and the northern branch is nearly inactive. Majuli Island is slowly getting incorporated into the northern flood plain of the Brahmaputra, while the southern and downstream edge of the island is under severe erosion.

Investigations in the basin evolution mechanism of the upper reach of the Brahmaputra valley in Assam show clearly why the foredeep areas have thicker

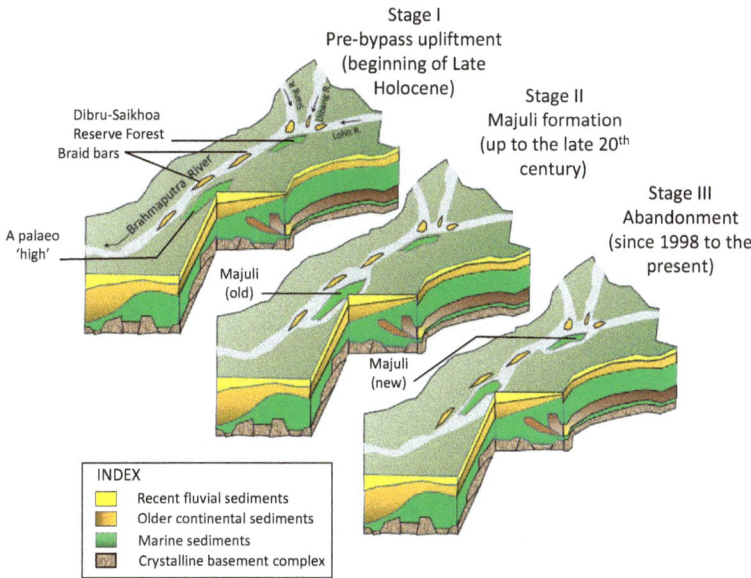

FIGURE 6.15
Evolution of Majuli in different stages.

fluvial sediments, particularly along the Eastern Himalayas and the Mishmi Himalayas. This also shows that the tectonic activities are not uniform all over the valley. Basin evolution studies show that in the post-Miocene period, the basin has gone through three cycles, and all these cycles were qualitatively identical, that is, a continuously uplifting wrench basin. The lateral asymmetry of the basin will continue to remain unless the new basin modifier becomes episodically active. Thus, the residence time of the Brahmaputra River will be longer in the north bank areas than in the south bank areas of the valley. Late Quaternary stratigraphy of the south bank of the Brahmaputra valley seems to be regulated mostly by its tributaries coming down from the Naga-Patkai mountain range. Even if for a higher precipitation period (7–11 ky), the Brahmaputra touched the eastern extremity, the residence time was much shorter.

Several well-defined 'highs' (Holbrook and Schumm, 1999) will force a river to bypass them for a valley having strong structural control. These 'highs' can either be due to the leading edges of the blind thrust fronts in the foreland areas of the valley or due to the normal faults and 'arching' of the basin along its central part. Thus, the formation of Majuli Island-like situations where older flood plains are bypassed by a large river is a very normal situation. Some episodic conditions like major earthquakes, as presently believed by the existing hypothesis based purely on the geomorphological ground (Sarma and Phukan, 2004), might trigger a situation; however, it is not at all a prerequisite condition. The role of basement configuration and tectonic setting in the evolution of such landforms, rather than a merely geomorphic process, is highly emphasized.

7

Bed–Bank Relationship and Flood Vulnerability

Large rivers show a considerable degree of reach-scale variability in the bed–bank relationship from the source to sink. Thus, rivers even during peak flood season seldom flood all the banks. Mountain-fed rivers act as conduits between the high-elevation source zones, and the sinks in seas or oceans run across reaches having variable slopes. Moreover, tributaries of different sizes join the rivers at different places. Slope, discharge, and load (both bedload and the suspended load) together play more significant components to determine the stream power and the channel patterns (Schumm, 1981) of a river at a given reach. There are reaches where slopes are the principal factor, and there are reaches where the discharge can play the main role to determine the stream power (Robert, 2003). Thus, the response of a river for equal slopes near the source zone where it is rich mainly in bedload and before approaching the sink where it is usually rich in suspended load needs different explanations. In the previous chapters, we have already seen that the Brahmaputra, having highly variable thalweg (Coleman, 1969) and the seventh-largest river in the world (Hovius, 1998), after being joined by the Ganges, the highest sediment carrier (Goswami, 1985; Latrubesse, 2008), is also one of the worst flood-causing rivers. In recent times, the Brahmaputra came to the headlines due to several reasons which were directly or indirectly related to flood disasters. We can mention a few. (i) A transboundary international issue relating to the construction of big dams in the Chinese part; (ii) the sudden influx of black clayey materials since the last quarter of the year 2017 through the Siang caused a prolonged change in the colour of the river water and drastic reduction in the fish population (the causes of which are yet to be ascertained, and there might be a connection with the unreported dam failures in the upstream side); and (iii) changing characteristics of the floods in terms of frequency, suddenness, intensity, and the magnitude of losses. The average annual flood toll of human lives increased from 41 (from 1953 to 1995) to 63 (during 2001–2017) in Assam. In 2017, the number of deaths was 158. Most of these deaths were due to embankment breaches; (iv) fourthly, the unprecedented magnitude of bank-line migration; (v) the fifth factor was related to a grand design of river-linking project of exporting surplus water from northern India to the water-starved southern provinces, and (vi) the sixth factor was related to a massive dredging project of the Brahmaputra channel belt aimed at improving the navigability of the river as a part of ambitious planning to expand the length

DOI: 10.1201/9781003302353-7

of waterways along with the roadways as a part of raising a robust infrastruc-
ture of communications. A common factor that runs under all these issues is
the hydro-sedimentary budgeting and its spatio-temporal variability annually and over different cycles.

Bed–Bank Relationship

Flood characterization essentially means spatio-temporal variability study
in the bed–bank relationship of a river system, which due to changing
hydro-sedimentary budgeting, makes different reaches of a valley susceptible to inundation due to the 'excess water' over the 'bank-full condition'.
A simple model of flooding is first to visualize an inclined trough having an inlet and outlet at the opposite ends, sediment-loaded water flowing through it. When the rate of influx (water or/and sediment) is more
than the outflow, the height of the water column keeps on increasing and a
stage comes when the spill-over happens. Now, the question is wherefrom
the spill-over begins first? For the analogy taken up in the present discussion, the obvious answer will be the downstream side. However, for a river
valley, when the flood-prone areas are mapped, a considerable degree of
arbitrariness is observed (Figure 7.1). For example, (i) both the banks of a
river may not be reaching the 'bank-full condition' at the same point of
time due to variable bank elevations; (ii) thalweg of the river might have a
considerable degree of reach-scale variability due to uneven distribution of
the sediments; (iii) different reaches of the river show different degrees of
aggradation, which might have structural implications; (iv) the channel belt
might show alternate 'nodes' and 'internodes' (Figure 7.2), where stream
power might be guided principally by the valley slope or the discharge or
perhaps a continuously variable proportion of both, which in turn deter-
mines whether the rate of lateral erosion will be more or vertical incision;
and (v) a large river like the Brahmaputra flows in certain reaches as a single
flow, or a multi-channel, or in an anabranching mode and even sometimes
showing highly anastomosing tendency. Thus, what constitutes the reach-
scale 'flow efficiency' of a channel is a complex problem of hydro-dynam-
ics, which attains greater complexity if the influence of different tributaries
is taken into account. Here, a simple geomorphological approach for flood
characterization will be explored.

Observations of river bank stratigraphy show that river bed aggradation
and the riverbank formation aided by the flood plain deposits can take
diverse forms in different reaches. Some of the possibilities are shown in
Figure 7.1.

In essence, what matters for a given reach is whether the effective accommodation space increases or decreases. An increase in the accommodation space means
for a given discharge, the possibility of flooding reduces, and if the effective

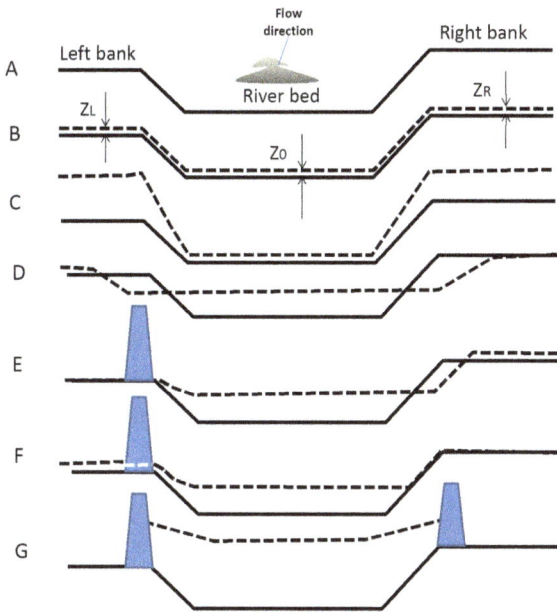

FIGURE 7.1
Bed–bank relationship shows some of the many possibilities.

accommodation space decreases, flood vulnerability for that reach increases. Flood plain construction might follow symmetrical characteristics, and in other circumstances, it can be grossly asymmetrical. Raising embankments arbitrarily along the river banks introduces a higher degree of reach-scale asymmetry, and the river bed aggradation rate becomes faster. As a result, the accommodation space decreases at a much faster rate, and even for normal or reduced precipitation than the average of the past decades, flood vulnerability increases.

Assumptions for Flood Characterization

The Brahmaputra shows alternate widening and narrowing patterns in the planform. We have made the following assumptions:

Assumption 1: Narrower the channel belt, deeper it is.

Assumption 2: For the multi-channel reaches of the braided rivers, wider channels have a deeper thalweg. Accordingly, for a braided channel belt, the location of the deepest thalweg can be assumed to be the median path of the widest channel (see Figure 7.2).

These two hypotheses oppose each other. However, it can be explained by the fact that the stream power is proportional to the

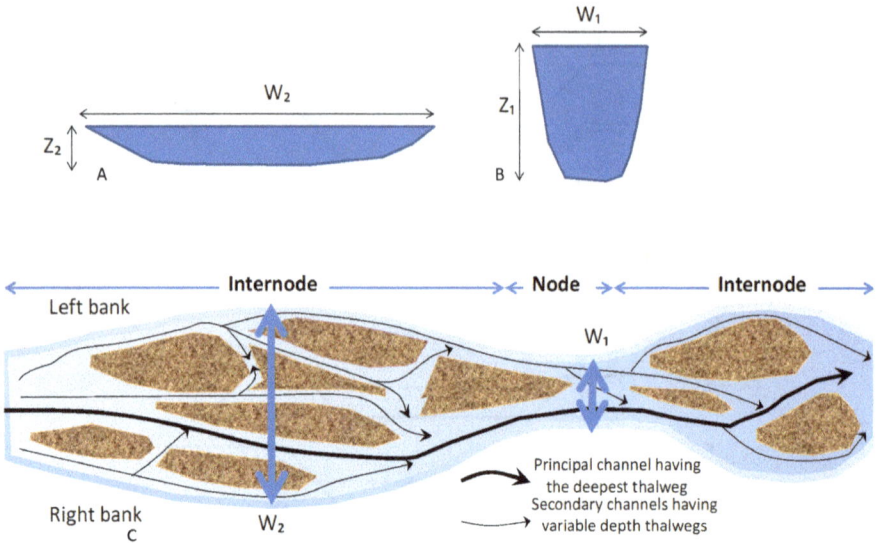

FIGURE 7.2
Illustrations showing assumptions on channel morphology.

product of the slope and the discharge, and the constant of proportionality can be a factor in bank material consolidation.

For valleys having strong structural controls, slopes show considerable lateral variability ('kinks' in the elevation variability profile). High-slope segments witness increases in the stream power, which helps greater bed incision and lesser aggradation potentiality and hence increased depth and narrower widths (nodes). On the other hand, zones of subsidence, showed enhanced aggradational tendency, more numbers of sandbars, and multiple channels. Stream power is determined mainly by the discharge volume; accordingly, the wider the channel (among different channels of the multi-channel segment of the river), the higher the stream power and the higher the rate of bed incision (for 'internodes').

Thus, for 'nodes', slopes play the decisive role, whereas for 'internodes', relative discharge volume plays a more important role to determine the depth of the thalweg. Moreover, the average stream power at 'nodes' is more than at the 'internodes'.

Assumption 3: We have further assumed that the continuous flow of the river from its upstream to downstream end can be replaced by 'cascade flow' (discrete), which essentially means that when the bank-full condition of a given reach is surpassed and inundation of the banks take place, the turn of the next downstream reach comes immediately afterwards as if 'flooding in succession'. However, the delay time is so small that the process becomes closer to the continuous flow condition. The advantage of the assumption is that the bed–bank relationship of individual reaches can be treated as principally responsible for the inundation of that reach. The impact of inundation in the upstream reaches on the downstream reaches is neglected.

Data and Interpretation

To introduce comparability in the morphodynamics of the Brahmaputra River in terms of the reach-scale variability in the depths (during 2015) and the widths (for 1915, 1975, and 2015), two indices were used, which are given as follows. This facilitates normalization. The difference in the absolute width variability and the normalized width variability can be compared in Figures 7.3 and 7.4.

$$\text{Depth Index} = \frac{\text{Reach depth} - \text{Average depth}}{\text{Maximum depth} - \text{Minimum depth}} \tag{7.1}$$

And,

$$\text{Width Index} = \frac{\text{Reach width} - \text{Average width}}{\text{Maximum width} - \text{Minimum width}} \tag{7.2}$$

'+'ve values of both the indices mean higher than the average.

Our general expectation is, as discussed earlier, reaches showing a widening tendency will also undergo a shallowing tendency principally due to the

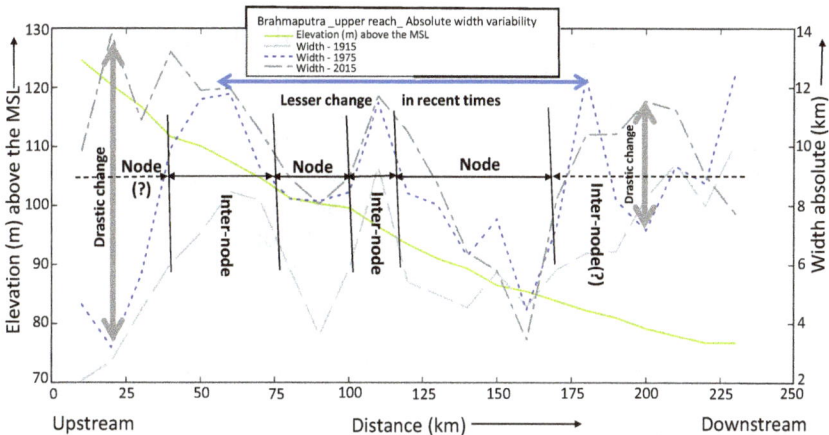

FIGURE 7.3
Absolute width variability of the upper reach of the Brahmaputra River during 1915–2015.

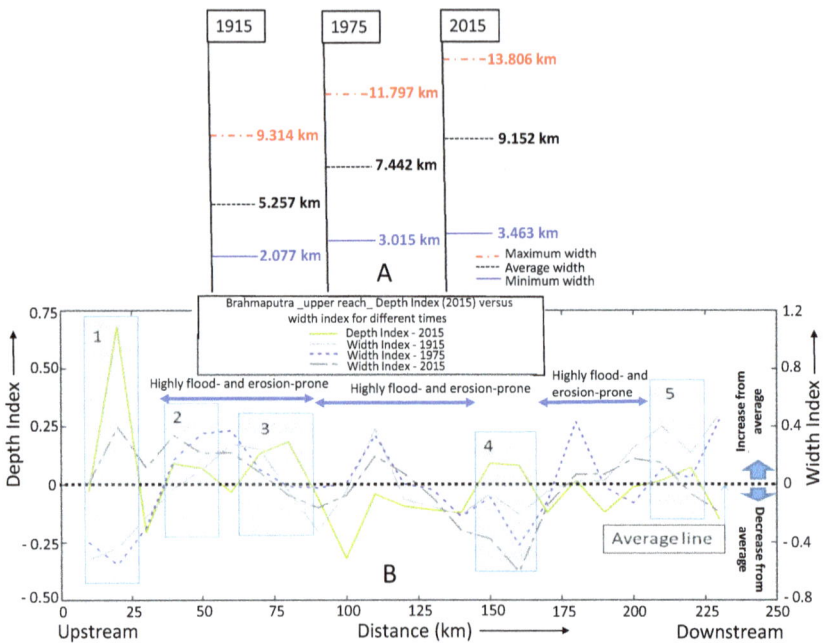

FIGURE 7.4
Variability in the depth–width indices.

increased rate of aggradation. We are interested to investigate the temporal changes in width associated with those reaches, which are deeper than the average depth. We have identified five zones (shown by boxes in Figure 7.4), where we find the channel is deeper than the average depth value.

In **box-1** (Figure 7.4b), width indices during 1915 and 1975 vis-à-vis the depth index were showing the opposite trend (fitting well with the assumptions we made). However, both the depth and the width indices during 2015 show a similar trend.

> **Explanation:** While considering the depth index, the channel belt included a newly evolved relict island the Dibru-Saikhowa. This was essentially due to the course correction of the Lohit River, which helped increase the width of the Brahmaputra channel belt. Thus, the reach-scale width index is showing a much higher value.

In **box-2**, deeper reaches are also associated with an increasing width index.

> **Explanation:** This is probably due to the massive erosion of the banks constituted of loose bank materials. Some of the shallow subsurface loose sand units help rapid 'toe-cutting'. Massive

lateral erosion helped to redistribute the sediments downstream but did not cause an effective shallowing tendency at the place of erosion itself.

Box-3 represents a transition zone, where some of the deeper depth reaches (on the upstream side) are wider and some are narrower (on the downstream side).

> **Explanation:** The river beyond these reaches is going back to its normal logic of aggradation.

For **box-4,** the channel belt of the river at different points of time shows similar characteristics (lower width index than the average), whereas the depth index shows a higher value than the average.

> **Explanation:** Deeper the reach narrower it is.

Box-5 represents the downstream end of Majuli Island. Compared to the oldest channel width index, the recent width indices show a narrowing trend, and the depth index is a little deeper than the average depth.

> **Explanation:** This is probably due to the anthropogenic intervention of various kinds of river management schemes operational in this area for a long time. A series of embankments were raised along both the banks of the river, which restricts changes in the widths of the channel belt, although the average width of the river has increased remarkably over the years (see Figure 7.4a).

The increasing values of the sandbar/channel area ratio in the planform for a given reach over time is an indication of aggradation. As can be seen in Figure 7.5, during 1915, the downstream side was showing a higher rate of aggradation. In 1975, that is, for the post-1950 earthquake phase, the upstream side was showing, in general, a higher rate of aggradation than the downstream end.

In 2015, the pattern of aggradation was more conformable with 1915, which suggests a state of stabilization has started coming into force; the downstream side has started showing a higher rate of aggradation again.

The anomalous situations arising due to a high degree of variability in the sandbar/channel ratio can influence the bed–bank relation, which is shown in Figure 7.6 by a second-order approximation of depth variation as a function of variation in width during the year 2015 (to explain the observations, we have used the logical arguments extended for the assumptions 1 and 2). Now, we can measure the latest place elevations. How to know whether a place had gone through a certain elevation change or not over the last 100 years? By knowing the old elevations of the river bed and the bank only, an estimation of temporal changes in the bed–bank elevation differences can be made. But there are hardly any direct means available to measure the

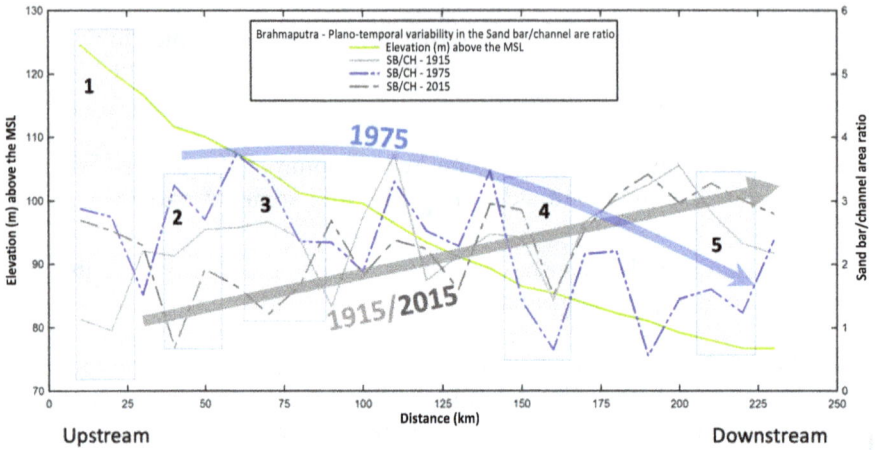

FIGURE 7.5
The trend of aggradation during 1915–2015 projected in the variability of the sandbar/channel area ratio.

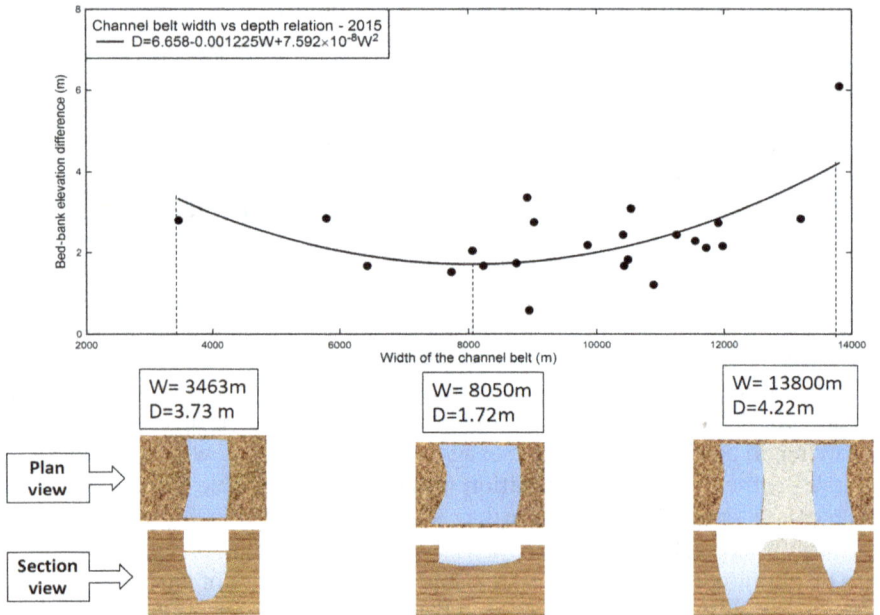

FIGURE 7.6
A second-order approximation of polynomial fit showing an empirical equation for bed–bank relation in 2015.

older place elevations. To have a reasonable estimate of the older bed–bank elevation differences, a further assumption was made regarding the correlatability of bed–bank elevation differences with the width of the channel

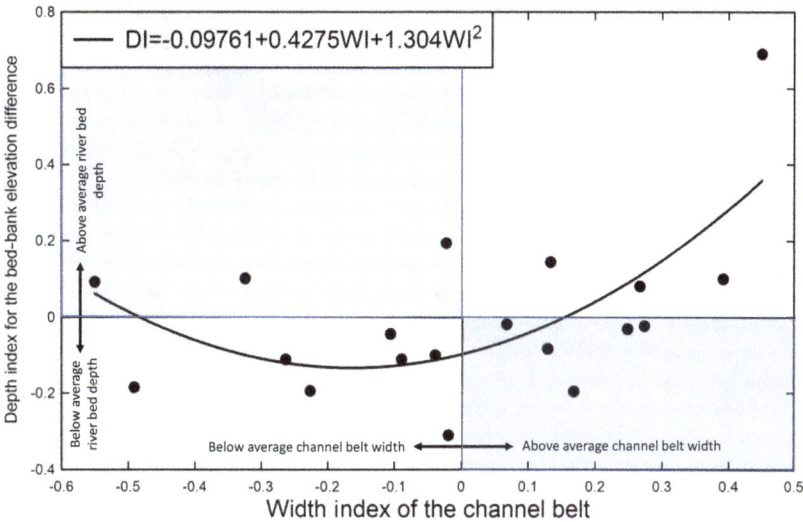

FIGURE 7.7
A second-order approximation of polynomial fit showing an empirical equation for bed–bank relation in 2015.

belt. Accordingly, a second-order approximation of polynomial fit (similar to that shown in Figure 7.6) was executed for the variability of the depth index (Eq. 7.1) concerning the width index (Eq. 7.2), as shown in Figure 7.7.

A comparative study of the absolute bed–bank elevation differences computed for the older channel belt (1915 and 1975) from the empirical equation obtained by second-order approximation of polynomial fit and the bed–bank depth indices (Figure 7.8) gives interesting results.

It was observed earlier (See Figure 7.3) that the width of the middle segment of the Brahmaputra channel belt has not changed substantially in recent times (1975–2015). The trend of aggradation was increasing (See Figure 7.5). However, only from the observations of the trend of temporal variation in bed–bank depth index (see Figure 7.8), does it become clear that the relative elevation difference between the river bed and the river bank has drastically reduced in recent times, which suggests faster incubation of flood disaster in the middle segment of the upper reach of the Brahmaputra River.

Flood Characterization

The flood characterization carried out is mainly based on the areal extent of inundation due to flooding concerning a unit increase in the head in the water level in bank-full condition. Each of the reaches selected every 10-km

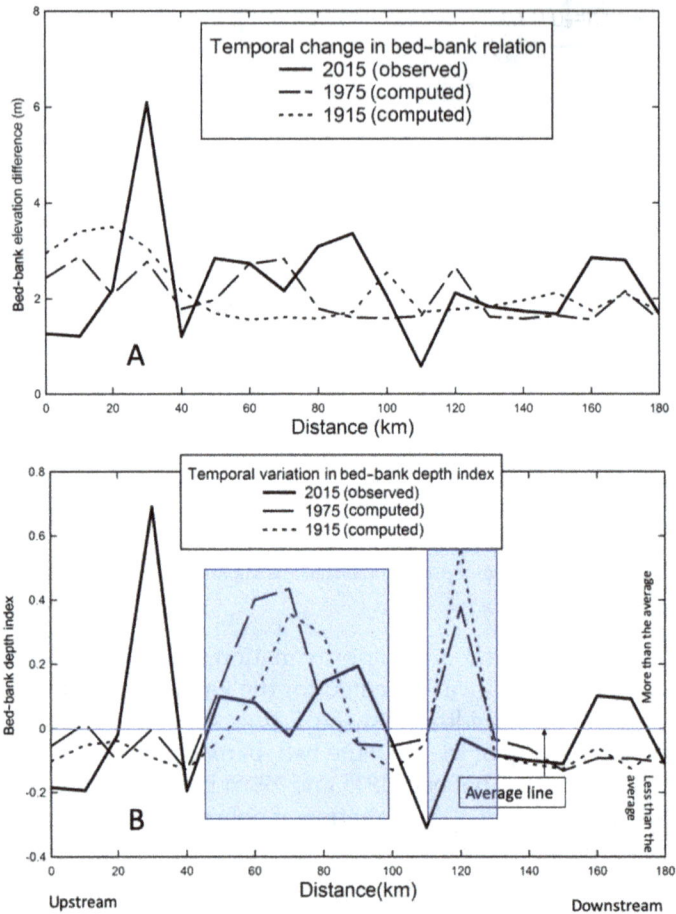

FIGURE 7.8
Comparative advantage of the normalized depth index over absolute depth.

interval was further subdivided into five sub-reaches, every incremental rise was identified in the DEM of the Google Earth, and the final equal-interval contours of probable inundation were plotted in the ArcGIS platform.

Using the DEM data, areal extent of flooding is delineated (Figure 7.9), and simple volumetric calculations are performed. To keep the calculation simple, the presence of ridge and swale geomorphological features of the flood plain are not considered.

The areal extent of the inundation areas had been calculated, as shown in Table 7.1. From the data thus obtained and calculated assuming no difference in elevation between both the banks, the following observations are made:

- A merely 0.5-m increase in the water level head can cause a great extent of flooding of 4,000+ km² area.

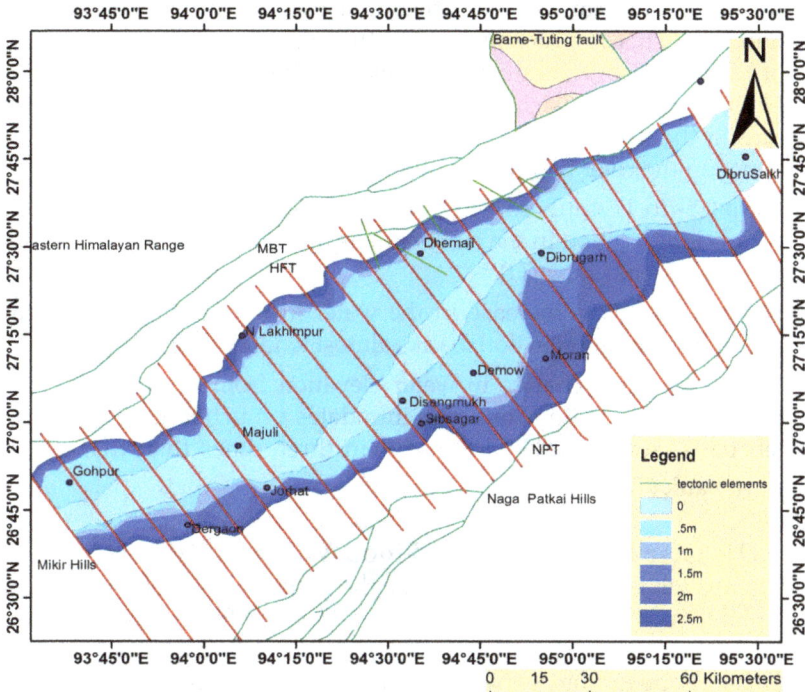

FIGURE 7.9

A map showing the areal extent of flooding with incremental shallowing of the river bed.

TABLE 7.1

Net Areas of Inundation Calculated for an Incremental Increase in Water Level

Increment in Water Level (m)	Right Bank (km²)	Left Bank (km²)	Left +Right bank (km²)
0.5	2,919.3	1,309.5	4,228.9
1	3,114.2	1,604.8	4,719.0
1.5	3,312.5	2,260.0	5,572.5
2	3,666.8	3,404.9	7,071.8
2.5	3,945.8	3,935.7	7,881.6

- In the upstream part, the left bank is more prone to flooding, and its areal extent is much larger than that of the corresponding right banks (Figure 7.9). This can be attributed to the impact of Himalayan orogeny near the frontal thrust belt in terms of upliftment as well as a higher rate of sediment influx, raising the effective elevation of the relief.

- However, in the middle reaches of the study area, both the right and the left banks are equally vulnerable to flooding.

- In the downstream reaches, the right bank is more vulnerable to flooding. This is probably due to the sediment influx from some of the major south bank tributaries like the Burhi-Dihing, Disang, Dikhau, and Dhansiri, which compensate for the sediment influx due to the north bank rivers. Moreover, shifting aggradation tendency in the river bed reduces the bed–bank elevation difference.
- For a 0.5- to 1.5-m increase in the water-level head, the areal extent of flood inundation in the right bank is much higher than that in the left bank areas (Table 7.1).
- However, with an increase in the water level over 1.5 m, the areal extent of inundation in both banks almost tends to be equal (Table 7.1).
- The incremental change in water elevation from 1.5 to 2.0 m compared to the bank-full condition inundates a much larger area of the left bank (south bank), by more than three times than that of the right bank (Table 7.2).

Flood Disaster Incubation and Flood Vulnerability

Flood incubation is a term used to define the growth of potentiality of a flood in a region. Many anthropogenic, as well as natural factors, are responsible for intensifying the flood vulnerability of a region. Embankment construction, a major issue for the people of Assam, is one of the most important attributes that increase the potentiality of flood vulnerability. Flood occurrence due to embankment breaching is a common cause of flood in Assam.

Embankments that are built for the mitigation of floods decrease the bed–bank elevation difference (or, depth of the river bed) rapidly, which in turn reduces the accommodation space and greatly influences the flood vulnerability of the region (see Figure 7.1). For a more or less similar annual pattern of river runoff, the river water overtops in the subsequent years and thus starts flowing above the bank. This overtopping water entering the pore spaces of the embankments decreases the stability of the embankments and hence acts as an incubator of the flood.

TABLE 7.2

Incremental Increase in Areal Extent Due to Every 0.5-m-Increase in Water Level than in the Bank-Full Condition

Increment in Water Level (m)	Right Bank (Increment in Areal Extent) (km²)	Left Bank (Increment in Areal Extent) (km²)	For Both Banks (Increment in Areal Extent) (km²)
0.5			
1	194.9	295.3	490.1
1.5	198.3	655.2	853.5
2	354.3	1,145.0	1,499.3
2.5	279.0	530.8	810.0

By plotting the average elevation of the bed with an increase in unit amount of sediment from the present-day elevation of the river bed versus distance and the present-day reach-scale elevation of both the banks (Figure 7.10), the following observations were made:

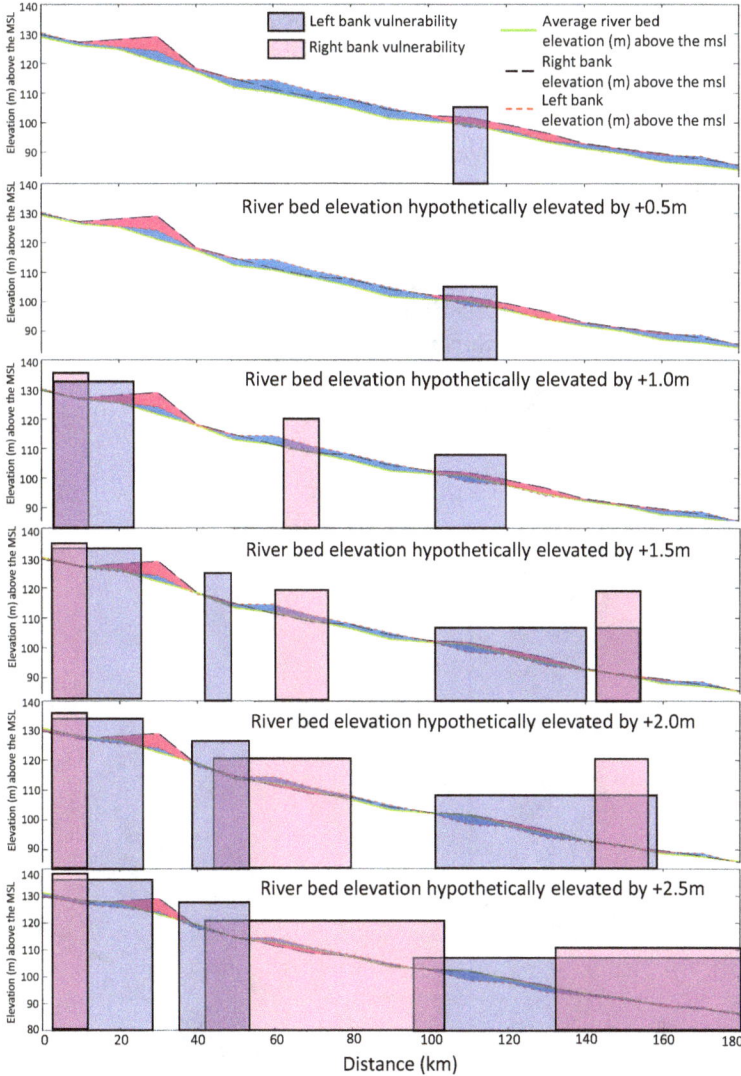

FIGURE 7.10

Graph showing the relationship between the unit increase of sediment deposition in the river bed with flood disaster incubation and the associated vulnerability of the respective zones.

- For a 0.5- to 1-m increase in the river bed due to sediment deposition, the water-level overtops the left bank much more than the right bank.
- With further increase in the sedimentation of the river bed, it can be seen that much of the bank line in both right and left become susceptible to flooding, or the reach-scale flood vulnerability increases with an increase in sediment deposition.
- From the present position of the river bed elevation, if the rapid rate of aggradation due to the construction of embankments causes a 2.5-m rise, the threshold is reached, which makes both the banks equally vulnerable to flash floods due to embankment breaching.

There is ample evidence of palaeo-megaflood events in the higher Himalayas (Figure 7.11a), which are usually called the glacial lake outburst floods and landslide lake outburst floods. These were mostly climate-induced or large-scale tectonic readjustment-induced events. However, ever since huge dams are constructed at different levels mostly to tap hydro-electric potential, possibilities of flood triggering due to dam outbursts have started attaining similar and in certain situations much bigger proportions.

In nutshell, for flood characterization, sediment budgeting, and spatio-temporal changes in the pattern of aggradation plays a crucial role.

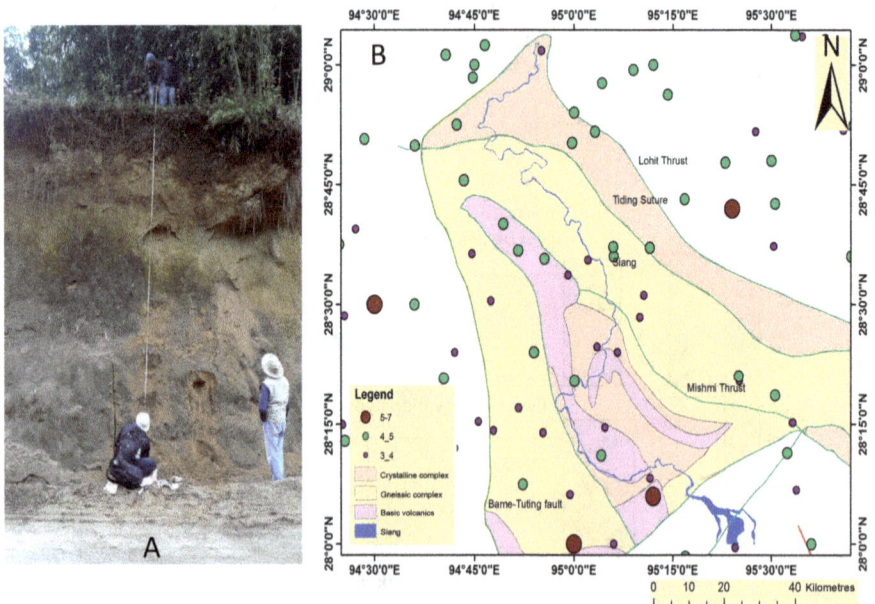

FIGURE 7.11
Evidence of palaeo-mega-flood deposits near Geku in the Siang Himalayas.

The reach-scale study of the river bed and the river bank relationship can give important clues to assess the trend of flood vulnerability. From 1915 to 1975, the principal site of aggradation was concentrated mostly in the upstream part of the Brahmaputra valley, which is evident from the reach-scale sandbar/ channel area ratios computed for the length of 230 km. However, from 1975 to 2015, the site of aggradation shifted further downstream in and around Majuli. A comparative study of width indices for different times (1915, 1975, and 2015) with the depth index computed during 2017 helps understand the stream power variability of the river flow and its consequences on the bank erosion and bed incision (Lahiri et al., 2021).

Raising embankments help incubate flood vulnerability; five-stage increments in river bed thickness show the spread of vulnerability along both the right and left banks of the Brahmaputra River. The areal extent of multi-stage flood inundation was generated in case of flood triggering due to single or multiple dam failures or embankment breaching. For flood head increases from +0.5 to +1.5 m, more of the north bank areas are vulnerable; however, for a +2.5-m increase in the flood water head, vulnerability of both the banks reaches uniformity.

8

The Golden Corridor

The natural and human resources of Assam were exhaustively studied and written for the first time in the middle part of the 19th century (1841) by William Robinson in his book *A Descriptive Account of Assam*. He was an inspector of schools. His sense of wonder, joy, and the spirit of exploration rings well in the following words. Introducing the intermontane Brahmaputra valley to the external world, he wrote,

> Its unexplored mineral resources, among which gold and iron are abundant; its animal and vegetable productions; the descent, customs, and languages of its numerous mountain tribes, present subjects of inquiry which deserve, and if vigorously prosecuted, will abundantly repay, the researches of the lover of nature and the observer of mankind.

(p.1–2)

Another smaller volume in the book form came earlier with the title 'Topography of Assam' in 1837. It was written by John M'Cosh. He was a professional medical practitioner. The most exhaustive district-wise coverage of Assam having a highly detailed and quantitative account was being compiled by William Wilson Hunter in his 'A Statistical Account of Assam', published in 1879. He was the director-general of statistics to the government of British India. From all these accounts and our studies, keeping some flavour of the historical context, a summarized overview of the natural resources of the Brahmaputra River and the valley is presented in the following text.

The upper reach of the Brahmaputra valley, extending from the confluence point of three large rivers – the Siang, the Dibang, and the Lohit to the Mikir Hills along the NE-SW direction, with an average width of 100⁺ km and length of 200⁺ km – is a golden corridor with a multi-layered abundance of natural resources, and it belongs to those rare places on earth where every square kilometre of the land surface is prospective but associated with 'plays' that can surprise highly knowledgeable geoscientists even when equipped with the latest technology and artificial intelligence. To have a generalized understanding of the explorable natural resources, we can use a simple criterion of two types of remote sensing: satellite images to monitor the surface topography and geophysical investigations for the subsurface. A set of geophysical tools are only effective for shallow investigations (up to the depth range of 100 m). Since resource exploration without the prospect of exploitation is not pragmatic, there is little utility to going beyond the basement depth beyond, which is very difficult to drill. For the upper reach of the

DOI: 10.1201/9781003302353-8

Brahmaputra valley, the maximum depth of investigation is less than 10 km. Accordingly, it was shown earlier in Figure 1.3 that prospecting for natural resources can be carried out in three segments: *surface, shallow subsurface*, and *deep subsurface*. Resource exploration and exploitations have their baggage of hazards and knowledge gaps. In this section, a limited segment of prospects will be covered, which are usually not discussed.

Surface Resources

Gold on the Surface

> Gold dust is found in the bed of the Subansiri, and from time immemorial it has been the practice of a class of men called Hanuáls to wash the streams for gold. Nearly every one of the rivers in the Subdivision produces gold, and there can be scarcely a doubt that the adjacent mountains are the source of the supply.
>
> (p. 413)

W.W. Hunter in his book 'A *Statistical Account of Assam*' mentioned earlier, by the reference to the subdivision above meant the present boundary of North Lakhimpur district and the adjacent areas. In our present work, the first-order tectonogeomorphic zonation (See Figure 4.32) shows this area within the Subansiri Depression. Hunter observed further,

> .... Gold washing used to be extensively carried out in former years, and it is stated that the right to wash for gold was farmed out by the old Assamese kings for Rs. 27,000 per annum. This branch of industry has now nearly died out, owing perhaps to the great increase in wages in late years since tea cultivation.
>
> (p.299)

From the aforementioned excerpts, it is clear that with the colonization of the Brahmaputra valley, tea replaced gold in at least one of the bigger districts of Assam, the North Lakhimpur. With the death of the gold extraction industry, a particular clan was jobless. To grow the tea industry, plantation workers from different corners of the Indian sub-continent were imported; in a similar way, sugarcane workers were exported to Mauritius. In the present third decade of the 21st century, one-third of the Assam population is known by the tea tribe which is, in fact, a conglomeration of different tribes, castes, languages, and cultures, forcibly uprooted and subsequently integrated with tea plantation. Pre-colonial profession-based identity (mentioned previously by Hunter as 'Hanuáls', which is a phonetic expression of a title now used as 'Sonowal') was made redundant and replaced by a commodity-based identity.

Hunter had given a more detailed account of the quantitative measure of gold extraction and a description of traditional knowledge utilized for this purpose in the following words:

> .... When a party of from eighty to a hundred Hanuáls commenced washing for gold dust in the Subansiri...two to two and a half pounds of gold were obtained by the party during the season, but the yield would doubtless be much greater with proper appliances. The process adopted for washing out the gold dust is exceedingly simple

> *(p.414)*

This account goes back more than 150 years (exactly to the year 1870). Considering the present gold rate (on an average Rs. 50,000 for 10 g of 24 karat gold in 2022), this amounts to about 2 crores (pure gold) of rupees in a season (for a party), which is not a great deal of income for 100 families but could provide a living with dignity. New advanced appliances were not developed. The samples of traditional appliances were kept at the state capital museum in Guwahati. A state of the art that was being continued from time immemorial was so neglected that practically it was forgotten by the planners and the policymakers and also from the general memory of the people, except for a few villages located very near the places where gold extractions were being carried out for generations.

Hydropower

The Brahmaputra and some of its tributary rivers originating in the great heights of the Himalayas experience a sudden fall (Figure 8.1; see Sarma, 2005; Wasson et al., 2022). This causes a huge potential for hydroelectricity generation.

The estimated magnitude is 66,500 MW. A blueprint for tapping this resource has already been prepared; work in some of these projects is in progress and some are to be initiated very soon (Figure 8.2. See https://in.boell.org/en/2019/05/08/infrastructure-development-northeast-hydropower-natural-resources-legal-and-institutional).

There is enough ground to believe that many investors are interested to invest money into these projects. There is every chance to convert this corridor into an energy hub and export electricity by laying cross-country high-voltage power grids. However, many new research works suggest that large river dams, an essential component of producing hydroelectricity, are

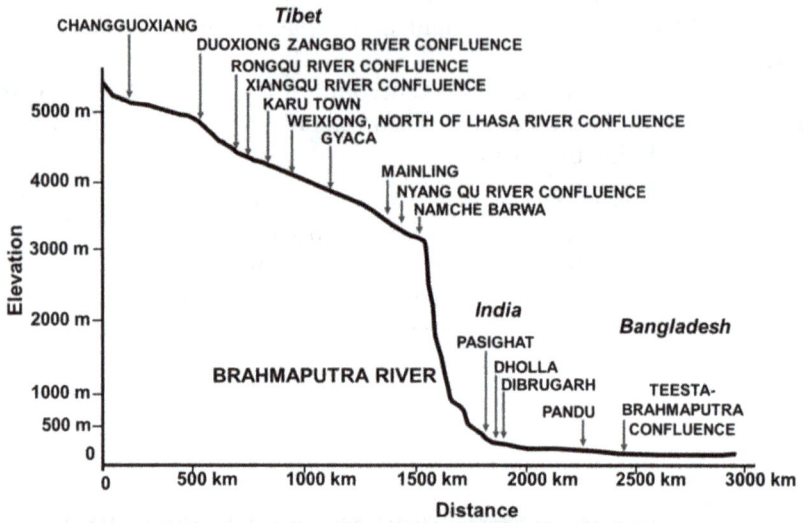

FIGURE 8.1
Elevation differences at different locations of the Brahmaputra River.

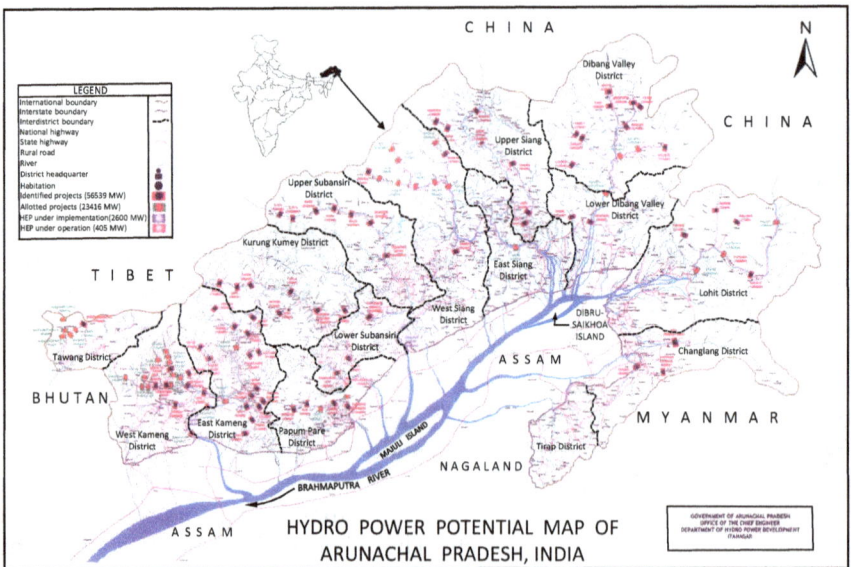

FIGURE 8.2
Blueprint of proposed dams on the Brahmaputra and its tributary rivers.

comparable to that hydro-bombs accentuating future disasters of unpredictable proportions. Moreover, representing one of the richest biodiversity pools in the world, many aquatic creatures are soon going to be extinct. No economic theory can estimate the amount of monetary loss of a species. The construction of large dams also alters sediment budgeting of the watersheds as well as other riparian areas located in the valleys. Since the majority of the valley population (more than 70%) sustains by agricultural practices, drastic changes in the sediment load characteristics are going to influence the flood plain formation and that, in turn, will increase the elements of uncertainty in agriculture manyfold. An alternative measure could be the installation of small hydroelectricity units in large numbers. As the mountains adjacent to valleys have a diverse population, constituted of different tribes, big dams displace them. Smaller hydroelectricity projects, on the other hand, do not displace the tribal population from their age-old habitats. This helps check massive internal migration. Moreover, simple technical training could help the villagers to manage these small projects. Thus, the advantages are twofold, bringing light to the remote villages and empowering the villagers.

Alternative Energy Resources like Using Solar Power and Wind Power

Compared to some other parts of India, annual sunshine days, computed from the averages of historical weather data collected from 1961 to 1990, in the position of the Brahmaputra valley is (Source:https://www.currentresults.com/Weather/India/annual-sunshine.php) within the lower bracket (with 2,277 hours/year) having the highest in Ahmedabad, Gujarat: 3,020 hours/year and the lowest in Kolkata, West Bengal: 2,108 hours/year. However, most of the European countries and upper regions of both North and South America are poorer in annual sunshine days (Figure 8.3) (Source: Jong et al., 2017) than the Brahmaputra valley. Tapping solar power is most effective for most of the places in Arunachal Pradesh, where the Brahmaputra and some of its major tributaries originate. These places have a much lower population density (17 persons/km^2) than the average population density of India (382 persons/km^2, according to the census report 2011). Thus, despite the higher per unit cost for solar energy, the net cost will be almost insignificant compared to the investment cost of hydroenergy.

Wind energy, in general, may not be a good alternative (Figures 8.4 and 8.5) due to inconsistency in the flow direction. But, technological improvements, particularly the installation of sensors capable to monitor the changes in the direction of the wind blowing and automatic adjustment of the face of the fan, can help. Since the average width of the Brahmaputra River is about 10%

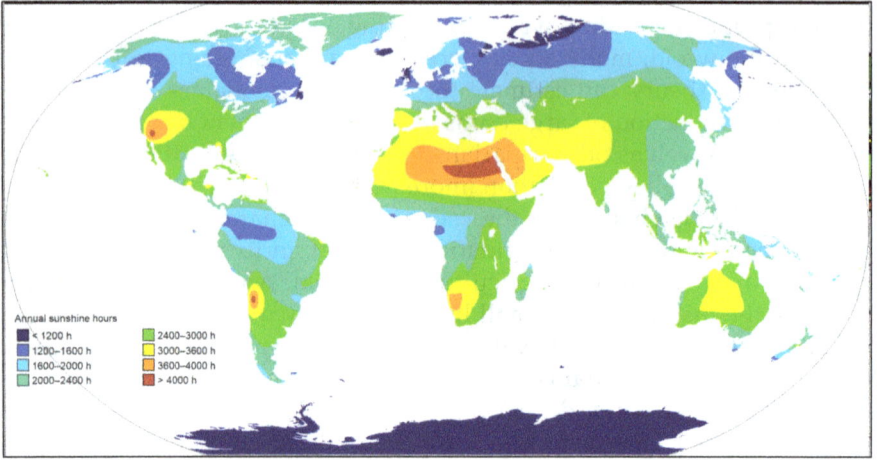

FIGURE 8.3
Annual sunshine hour map of the world.

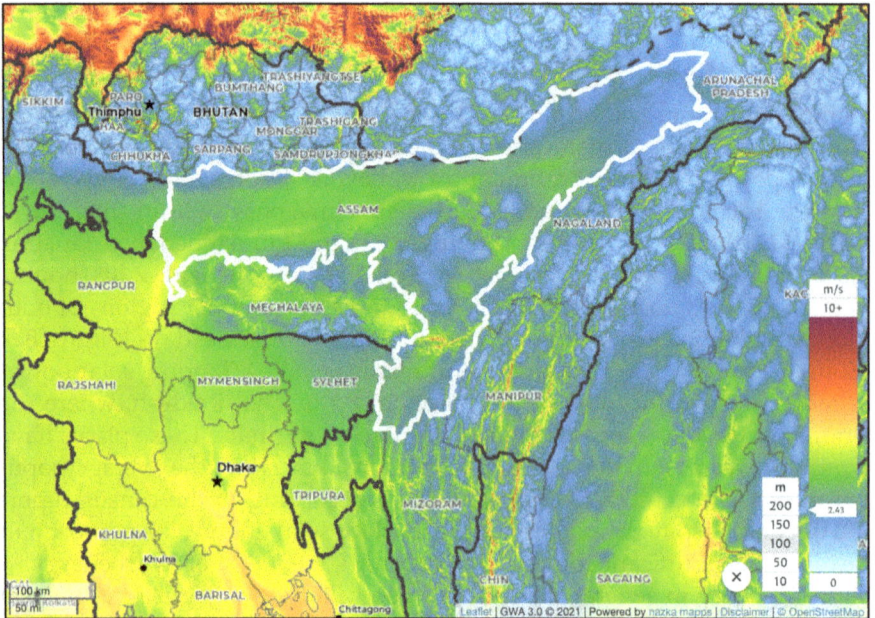

FIGURE 8.4
Average wind speed map of Assam.

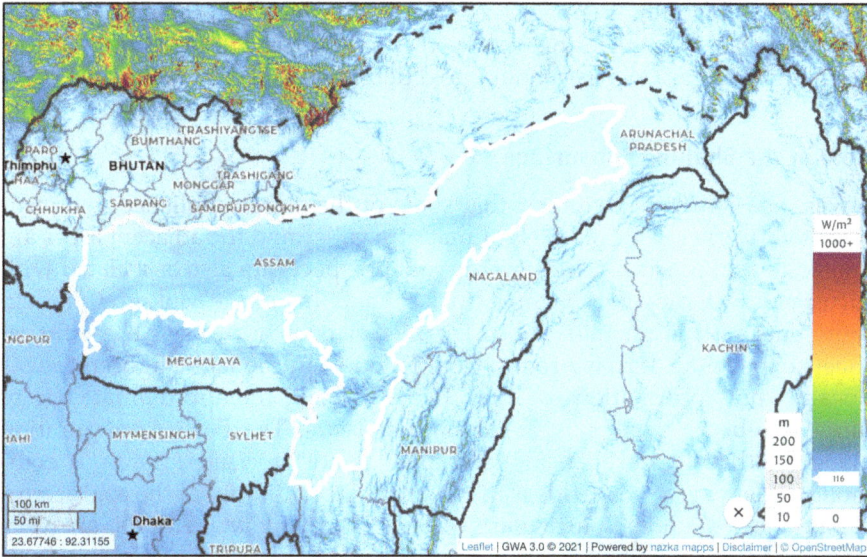

FIGURE 8.5
Average wind power map of Assam.

of the valley area, it also acts as a secondary agency in controlling the wind, which is evident in the average wind speed map (Figure 8.4) and the average wind power map (Figure 8.5).

Lately, the Himalayan stretch of the Brahmaputra was seen to cut through shales, which are supposed to be rich in hydrocarbon. Although shale oil is an attractive option for alternative energy sources, the associated environmental degradation calls forth a very cautious approach.

Shallow Subsurface Resources

Gold in the Shallow Subsurface

This is now common geoscientific knowledge that economic mining of gold is carried out from the placer deposits. Subansiri is the largest tributary of the Brahmaputra River (contributing 7.92% of the total flow of the Brahmaputra), and as could be seen in Figure 4.31a, it had avulsed during the post-1950 earthquake period. The Subansiri shows a partly braided and partly meandering characteristic. In the foothills of the eastern Himalayas where gold mining was historically being carried out, there are braid bars around where secondary gold deposits are usually found. We have shown further (Figure 4.32) that along the northern bank of the Brahmaputra River, there is a long stretch showing a frequent trend of subsidence, which we have named as 'Subansiri

Depression'. Mapping the palaeo-courses of the Subansiri River in the shallow subsurface by conducting refraction or uphole surveys can help locate the sandbars, around which secondary gold deposits are highly probable.

Coal in the Shallow Subsurface

It was mentioned in Chapter 2 that some of the oldest published geoscientific literature available on the Brahmaputra valley and the adjacent areas are related to coal prospects. Most of the coal prospects are also associated with thrust belt tectonics. For example, surface mapping of the Makum Coalfield represents a thrust sheet bounded between two major thrusts. These thrust sheets expose the Tertiary rocks. Coal is found in the late Oligocene rocks (locally known formation is 'Tikak Parbat'). Thus, the approximate age of these coal-bearing units is 23–28 million years. The basin configuration along the thrust controlled foreland areas is not precisely known. A common problem in the coalfields is the frequent discontinuity of the coal seams. Geophysical methods, particularly electrical resistivity methods, are very effective in mapping coal units in highly faulted areas. Now, Oligocene was the time during which the Tethyan sea was receding gradually due to the pre-collisional convergence of the India and Eurasia plates. Thus, the possibility of the presence of coal beds should have a longer regional continuity. In the places where these older rock units were uplifted, we get coal in the shallow subsurface or simply as outcropping units. For example, we get coal of similar characteristics in the state of Meghalaya, where the Shillong massif was supposed to experience 'pop-up'. Extending the identical logic, we should expect coal on a comparable scale along the Eastern Himalayas as well. It has already been proved from the reports of various scientific labs that the Tertiary coal of Assam and the adjacent areas is of excellent grade for 'coal to liquid fuels'. A change in policy from converting coal-to-coke to coal to liquid fuels could help add great value and increase the amount of revenue collection.

Coal Bed Methane (CBM)

CBM prospect in Assam is very high due to the simple common ground of thrust-induced faults and folds, bringing the deeper rocks near the surface and causing entrapment of oil. Some oil experts believe that in those coal units below which oil pools are also present, thrust-induced fractures cause both the liquid and gaseous hydrocarbons to leak, move above, and get trapped inside the coal units in the form of the CBM. Fluvial morphology and planform dynamics in recent times help identify the exact locations of thrust-induced blocks in the foreland areas of the valley.

Deep Subsurface Resources

Gold in the Deep Subsurface

As could be seen in the seismic sections (Figures 6.2 and 6.3) across the upper reach of the Brahmaputra valley, the foreland adjacent to the Eastern Himalayas have fluvial sediments of more than 2 km in thickness. Since the 'Subansiri Depression' shows periodic subsidence due to tectonic readjustments, we can expect to have shifting 'sandbar bowls' for different vertical cross-sections cut across the 'Subansiri Depression' of different sizes (depending upon the extent of the residence time of the river at a given place) both laterally and vertically down. Gold mining at greater depths is difficult. However, with the development of high precision control over directional drilling, effective strategies can be framed to undertake gold mining at greater depths.

Deeper Oil Prospects

In Figure 5.5, locations of 68 oil fields are shown along the Naga-Patkai thrust belt. Also, two blind faults are shown. These are mapped by using seismic data and geomorphological evidence. We do not have a clear understanding of how thrust belt tectonics is operating in the subsurface of the valley. Also, we are not very sure about the volume of oil-bearing reservoir rocks present within the compressed thrust belt. Based on multiple pieces of evidence, there is a fair degree of certainty about new prospects if proper modelling for the thrust belt and advancing frontier of thrusting in the foreland areas are given priority. These are the challenges in the immediate future.

In the complex areas, locating the basement (constituting the crystalline metamorphic rock bottom of the basin) using geophysical methods (seismic, gravity, and magnetic) is not always accurate due to the assumptions based on which these methods work. There are instances where a deep drilling programme brings up sediments of much older age than anticipated from below the basement depth. Thus, sub-basement prospects can open up new possibilities for oil in the golden corridor. We have already got oil from the fractured basement.

9

Epilogue

On 22 March 2021, a giant hanging glacier collapsed and dropped from an elevation of about 6,600 to 2,700 m to the Yarlung Tsangpo valley bottom in Tibet. The coordinates of the centre of the landslide were latitude 29.815° and longitude 94.932°. A very large mass displacement of about 100 million tonnes was estimated (Source: https://www.planet.com/). The event was reported first time by Göran Ekstrom of Columbia University and confirmed by Dan Shugar from Planet Labs at the University of Calgary. After a few days, reports in small headlines started appearing in local newspapers about the increased muddiness of the Brahmaputra water. On the same date, the honourable home minister of the country, Mr Amit Shah, during his election campaigns said to make Assam flood-free in the coming 5 years the way in the past 5 years Assam has become militancy-free and corruption-free. While elaborating on the line of action, he said,

> Now our aim is to make Assam flood-free. We will create large water reservoirs and water channels to divert the floodwaters of Brahmaputra and use it for irrigation and to give relief to the people of the state from the flood.
>
> *(Source: https://www.time8.in/*
> *will-create-reservoirs-to-solve-flood-problem-in-assam-amit-shah-in-majuli/)*

As mentioned earlier in Chapter 2, the average discharge of the Brahmaputra River at its mouth is 19,830 m³/s (Mahanta et al., 2014). To have a reliable estimate of this discharge volume, let us consider a lake having dimensions of a length of 2 km, a width of 1 km, and a depth of 20 m. A simple calculation suggests that on average, 43 such lakes of the size mentioned will be filled up by the discharge volume of a single day. These examples are quoted to show the huge gap existing between the magnitude of the morphological changes that affect the flow regime of the Brahmaputra, and not only the lack of understanding among the highest office-bearers of the peoples' representatives but also a false understanding, blatant reductionism, and disregard for scientific studies help incubate disasters. On the other hand, if not adequate, many scientists are engaged seriously to understand the complexities of river dynamics and valley formation.

A large number of studies, particularly in the lower reaches of the river in Bangladesh, bear testimony to the international attention this river has received (see Coleman, 1969; Bristow, 1987; Thorne et al, 1993; Curray, 1994; Goodbred and Kuehl, 1998, 2000; Uddin and Lundberg, 1999; Richardson

DOI: 10.1201/9781003302353-9

and Thorne, 2001; Wasson, 2003; Goodbred et al., 2003; Garzanti et al, 2004; Darby et al., 2015; Reitz et al., 2015; Whitehead et al., 2015; Grall et al., 2018; Pickering et al., 2019). By contrast, the upper reaches of the river in Assam have received inadequate attention so far (Goswami, 1985; Valdiya, 1999; Sarma and Phukan, 2004, 2006; Kotoky et al., 2005; Sarma, 2005; Singh, 2006; Lahiri and Sinha, 2012, 2014, 2015; Barman and Bhattacharjya, 2015). The Brahmaputra River divides the upper Assam valley into two distinct geographic zones: the north and south banks. In addition, this reach of the river hosts one of the largest alluvial islands in the world, the Majuli, which is known for its unique cultural heritage (Sarma and Phukan, 2004) apart from its geomorphic significance. The Brahmaputra River shows significant geomorphic diversity in this region, which is strongly manifested in the morphodynamics of the river at a historical timescale. Moreover, the tributaries joining the main river from these two banks are quite different in terms of their morphometric characteristics and temporal dynamics. The wide-ranging curiosity among the geomorphologists about understanding a large river like the Brahmaputra was possibly having five broader aspects. (i) The Brahmaputra and its tributaries show a very high degree of morphological variability. The Brahmaputra, considered one of the top 10 large anabranching mega-rivers of the world (Latrubesse, 2008), is the seventh-largest tropical river (Hovius, 1998; Latrubesse et al., 2005; Tandon and Sinha, 2007) in terms of mean annual discharge (~20,000 m^3/s in Bangladesh) which passes through three most populous countries, China, India, and Bangladesh. (ii) The sediment load transported by the river is subjected to aggradation–degradation in multifarious proportions at different places on its way before it ultimately meets the ocean; the sediment budgeting for different transacts is thus influenced by different constraints. (iii) The Himalayan orogeny and the orogeny of the Indo-Myanmar part of the valley (Figure 9.1) influence the sediment influx through the Brahmaputra River system. Being an antecedent river, from the subsurface fluviatile sedimentological archive, it is possible to know about the competing characteristics of different orogeny cycles. (iv) The Himalayan orogeny has vastly influenced the climatic regime of not only the Indo-Gangetic plain but also the entire South Asia. Climatic changes, in turn, influence the sediment flux continuously and therefore influence basin filling and fluvial dynamics.

(v) Finally, the study of fluvial dynamics of the Brahmaputra River system and long-term basin filling episodes has significant implications for oil exploration. The incised valley deposits are considered the best reservoir rocks, provided proper sealing and capping devices were present. As the ancient sediments at depths show evidence of the regression characteristics of the marine environment (with the increasing convergence of India and Eurasia plates and regression of the Tethyan sea) all over the basin, source rock potentiality was high. For a regressive sea, the length of shelf incision would keep on increasing with time along the direction of regression. As the basin at its different stages of development was tectonically

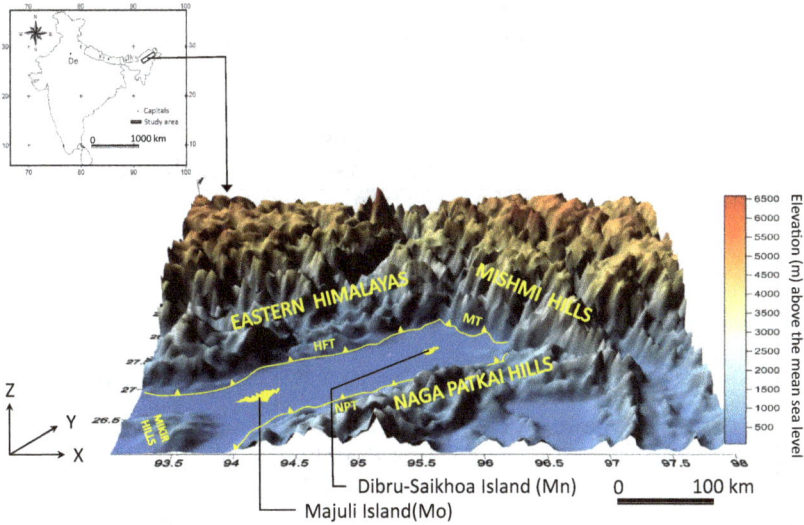

FIGURE 9.1

Upper reach of the Brahmaputra valley in Assam is shown as a simplified relief with the intermontane valley.

modified, the migration of oil must have gone through many complex processes. Thus, there are wide-ranging prospects that can be better understood if the palaeo-fluvial dynamics (Uddin and Lundberg, 1999) could be comprehended.

The present study was principally focussed on fluvial dynamics of the upper reach of the Brahmaputra valley. The study attempted to transgress different scales of fluvial dynamics. It starts with the surface observation of fluvial dynamics for the Brahmaputra and its tributaries in the last 100 years based on the comparative study of the old maps and the recent IRS-P6-LISS-3 images in a GIS environment after following usual procedures for georeferencing and mapping. The understanding of fluvial dynamics was extended further to understand the nature of ongoing tectonic controls in the study area. After that, fluvial dynamics in the late Quaternary are reconstructed in the shallow subsurface (<100 m) in selected windows based on uphole and shallow refraction data sets. These data sets were used for the first time to understand palaeo-fluvial dynamics and neotectonics in the foredeep areas of a tectonically active valley. The studies on neotectonics from the shallow subsurface geophysical data were carried out by developing a new concept of multi-layered velocity discrimination to map palaeochannels and then identify the trends of movements of the palaeochannels. Subsequently, deep subsurface (<10 km) studies were conducted based on the recent 2D seismic data, which were processed for long profiles from various data sets by Oil India Limited, Duliajan, to understand different cycles of the poly-history basin following the line of interpretation proposed by Kingston et al. (1983).

The basic purpose of the lower order basin evolution study was to understand the cause of differential residence time of the Brahmaputra River during its lateral excursion through the valley and the variability observed in the accommodation space in different parts of the valley. A brief study was conducted to understand the changing bed–bank relationship of the Brahmaputra River to aid vulnerability assessment of the flood disasters. Summarized accounts of the findings from these studies are as follows:

Morphodynamics of the Brahmaputra River and Tributaries

Reconstruction of planform changes over a period of 90 years in the upper reaches of the Assam valley shows that the 240-km-long channel belt is widening all along its course in the region. From the average width of 9.74 km in 1915, the channel belt has widened to the average width of 14.03 km in 2005 (44% widening), and in certain reaches, the average widening is as high as 250%. The mean shift in the north bank for the period 1915–1975 was 1.45 km, whereas a shift of about 0.7 km (approximately half) was recorded for the south bank during the same period. However, the mean shift in the north bank during the period 1975–2005 was just 0.06 km, while the mean shift in the south bank was 2.05 km (~34 times) during the same period.

In three different units of the upper reach of the Brahmaputra River, the nature and extent of shifts were quite different. However, the south bank in the upstream unit shows maximum migration. It is in this unit, in the last one and half decade, a new island was formed that earlier used to be the Dibru Saikhowa reserve forest and a part of the older flood plains of the south bank of Brahmaputra valley.

Furthermore, morphometric characterization and temporal variability in several planform parameters were documented in 37 reaches of the Brahmaputra River, each 4.5–9 km long, in the upstream (unit 1), midstream (unit 2), and downstream (unit 3) units. Unit 1, with an average channel slope of 0.35 m/km and mean elevation of 114 m, shows a continuous increase in the channel belt width (W) and channel area (CH) (CH/CHB= −34%) and a drastic increase in the sandbar area (BB/CH= +76%) during 1915–2005. On the other hand, for the same period, unit 3 with an average channel slope of 0.21 m/km and a mean elevation of 77 m shows only a minimal increase in the CH (CH/CHB=+15%) and a proportionate decrease in the sandbar area (BB/CH=−18%) Unit 1 is characterized as a major zone of aggradation, unit 2 also shows aggradation but at a much slower rate, and unit 3 shows an overall degradation.

Seismic, gravity, and deep well-log data were integrated with the trends of fluvial dynamics in the last 100 years. A tectono-geomorphic zonation (Lahiri and Sinha, 2012) of the upper reach of the Brahmaputra valley into

(i) the central uplift, (ii) the slopes, and (iii) the depressions has been proposed, which helped understand the first-order aggradation–degradation patterns in the valley. Concerning the 1950 earthquake, the co-seismic subsidence and inter-seismic uplift have influenced the dynamics of the Brahmaputra channel belt and bank-line migration significantly and also resulted in the formation of a large river island (~300 km^2 area), which was earlier a part of the Dibru-Saikhowa reserve forest.

The Brahmaputra River receives tributaries along the north bank originating in the Eastern Himalayan watershed as well as along the south bank draining through the watershed of the Indo-Burman Range. Significant differences in fluvial dynamics of the northern and southern tributaries were noted. The general trend of avulsion of the north bank tributaries of the Brahmaputra River is towards the southwest, and definite structural control of very recent origin is inferred. The palaeochannels of the southern tributaries, on the other hand, are much older and suggest stability at shorter timescales. Such difference in fluvial dynamics of the north and south bank tributaries is attributed to the influence of different tectonic regimes. Tectonic forcings related to Himalayan orogeny are much stronger than those of the Naga-Patkai thrust (NPT) belt belonging to the Indo-Burman orogeny. As a result, the bank-line migrations, as well as channel avulsion of the north bank tributaries, are more frequent than the south bank tributaries.

Erosional and Evolutionary History of Majuli Island

Between 1915 and 1975, the surface area of Majuli Island reduced from 787.87 to 640.5 km^2 (18.7% reduction) and then to 508.2 km^2 by 2005 (35.5% reduction as compared to 1915). The average rate of erosion in the last 30 years or more has increased considerably from 2.46 km^2/year (1915–1975) to 4.40 km^2/year (1975–2005). We have investigated in detail the possibilities of influences of geomorphic parameters like channel belt area (CHB), channel belt width (W), braid bar area (BB), channel area (CH), and bank-line migrations on the trend of erosion of Majuli Island and making use of statistical tools found that wherever braid bar deposits increased, the actual CHs decreased. Moreover, the increase in the CHB was principally due to the increasing area of braid bars. The BB/CH ratio of the Brahmaputra in the old Majuli bearing unit reduces drastically from 2.34 in 1915 to 1.24 in 1975, and then again it rises to 2.49 in 2005. When we compare the nature of erosion in different parts of Majuli Island from 1975 to 2005, it is observed that the rate of erosion in the lower Majuli (~32.3%) was just double the upper Majuli (~16.3%). Furthermore, the thalweg of the Brahmaputra River migrated away from Majuli Island in its upper part during this period. On the other hand, in the entire lower part, the thalweg migrated towards Majuli

Island, and this might have contributed to a higher erosion of the lower Majuli. In addition, the Subansiri River has also been migrating towards the south since 1950 and also accelerated the rate of erosion of the lower Majuli.

Several seismic sections were studied close to Majuli Island, and one seismic section on Majuli Island lying along the SW-NE trending strike direction of the island clearly shows the presence of a multiple-hinge anticline formed due to the contractional fault-related fold (Bally, 1997) in the basement complex itself. The hinge lines propagate in the upward direction. However, as we move up, massive sediment dumping seems to attenuate the hinge line bends.

Integration of seismic data with topographic and basement configuration helped us understand the evolutionary history of Majuli-type landforms. A major topographic high in the central part of the Majuli with a sharp break in slope closely matches a break in the basement slope along with the same profile. In addition, a prominent E-W running Jorhat fault at the tail end of the Majuli and several other NE-SW bound faults parallel to the strike of the Majuli are also noted. These pieces of evidence suggest that Majuli-type geomorphic highs are guided by basement topography and structures and not just a product of aggradation and fluvial dynamics, as suggested by earlier workers.

Multi-Parametric Micro-Zonation in the Shallow Subsurface: Implications for Palaeo-Fluvial Dynamics

Four strategic windows in the south bank of the upper reach of the Brahmaputra valley were used for the present study based on at least four major criteria: (i) locations of producing oil fields, (ii) channels showing typical angularities observed in the valleys having strong tectonic controls, (iii) inter-channel landscape having prominent indications of palaeochannels, and (iv) prognosticated hydrocarbon reserves yet to be explored. The common thread among the areas selected is the presence of two blind faults (retrieved from the seismotectonic map prepared by Dasgupta et al., 2000) running across all four windows. Uphole data are available for two windows and shallow seismic refraction data for the remaining windows. Discriminating 'low' velocities from 'high' velocities, based on some suitable threshold criteria, palaeochannel belts could be identified. A comparison of the palaeochannel belts with the present-day locations in the GIS environment helped identify accurately the avulsive trend. It was also possible to assign relative ages of the palaeochannels based on stratigraphic relationships.

Cyclic Evolution of the Poly-History Basin

Using six seismic sections, two across the valley and four along the NE-SW direction, the basin evolution study was carried out following the line of interpretation proposed by Kingston et al. (1983) where basin evolution is mostly guided by 'basin-forming tectonics' and 'basin-modifying tectonics'. Sites of preferential fluvial aggradation were identified, and the unevenness observed in the residence time of the Brahmaputra River (or the Brahmaputra-like palaeo-river) during lateral migration was investigated. It is observed in the seismic sections across the Brahmaputra valley that the elevation of the upper boundary of the marine sediments keeps on increasing as we move from the northwest towards the southeast direction. In other words, the over-burden thickness above the marine sediments is thicker near the Himalayas and thinner near the Naga-Patkai hills. Analysis of the seismic sections clearly shows that the foredeep areas of the Eastern Himalayas as well as the Mishmi Himalayas store much thicker columns of fluvial sediments than other parts of the valley. The poly-history basin of the upper reach of the Brahmaputra valley experienced six first-order cycles. During the late Miocene and early Pliocene periods, the basin experienced a strong episodic wrenching effect due to the plate tectonics resulting from the interactions between the India and Eurasia plates, as well as the India and the Burma plates. This caused basin asymmetry, tilting more towards the Himalayan side. This is the reason that explains the higher rate of accommodation space generation in the Himalayan foredeep areas of the basin.

Characterization of Changing Bed–Bank Relations

For a high sediment-loaded river like the Brahmaputra, if the bedload is prin-cipally responsible for the construction of the river bed as a continuous pro-cess round the year, suspended load constructs the bank during the floods episodically. Embankments raised to check floods practically cause cessation of bank construction and increase the pace of river bed construction. The Brahmaputra, in the upper reach, flows through a tectonically controlled val-ley, where the relief shows a high degree of slope variability. Thus, the aggra-dation–degradation potentiality for different smaller reaches varies a lot. As a result, the difference in the bed–bank elevations varies too. The question is whether there is any relation between the widening and deepening (or, shallowing) of the Brahmaputra channel belt or not. How can the chang-ing morphological status at different points of time be compared? Using the 'method of least squares', an empirical relation was 'fitted' for the present-day channel belt width and the bed–bank elevation differences. Subsequently, assuming the validity of the empirical law in the past, bed–bank elevation differences were computed for the measured variability in the W. A simple normalizing technique was proposed and applied in the present work by introducing width and depth indices to have a planform comparison of different reach-scale variabilities of the bed–bank relationship during three different years – 1915, 1975, and 2015. This helps identify flood vulnerability of different reaches.

Natural Resources of Geological Significance

One of the traditional practices in the upper reach of the Brahmaputra valley was gold extraction from the sands of some of the tributary rivers of the Brahmaputra. The Subansiri River, contributing 7.92% of the Brahmaputra total flow, got its name derived from the Sanskrit word *'swarn'* meaning gold. An argument was put forward for calling the place a golden corridor, not metaphorically, but strictly based on historical evidence as well as great future possibilities of reviving gold mining at different depths. There are wide-ranging possibilities for utilizing solar and wind energy. Further exploration of two large energy resources, coal and oil, needs a precise understanding of thrust belt tectonics in the foreland areas. A combination of geomorphological dynamics and an understanding of deeper subsurface controls interpreted from the geophysical data could open up new possibilities for resource exploration in the Brahmaputra valley.

Major Conclusions

1. The Brahmaputra valley in upper Assam is strongly influenced by the Himalayan and Naga-Patkai belt orogeny and can be broadly subdivided into three types of first-order tectono-geomorphic units, namely, central *uplift*, *slopes*, and *depressions*. The spatial variability in morphodynamics of the Brahmaputra River is strongly linked to these units.

2. In a 240-km-long stretch of the Brahmaputra, the longitudinal profile shows several 'highs' and 'lows' that have led to uneven sediment dispersal, resulting in reach-scale aggradation and degradation.

3. The general trend of avulsion of the north bank tributaries of the Brahmaputra River is towards the southwest, and definite structural control of very recent origin is inferred. The palaeochannels of the southern tributaries, on the other hand, are much older and suggest stability at shorter timescales. These differences in dynamics of the northern and southern tributaries are in response to different tectonic regimes of the Himalayan thrust and NPT, respectively.

4. Co-seismic and inter-seismic forcings influenced the morphodynamics of the valley. The 1950 earthquake (M 8.7) caused subsidence in the Subansiri and Lohit depressions, which helped generate more accommodation space. Post-1950 earthquake, aggradation in the Lohit depression decreased the bedload/suspended load ratio and increased lateral erosion tremendously on the most upstream side, where very recently, a new relict island has emerged. The site of effective aggradation shows a switchover from the extreme upstream unit to the extreme downstream unit, hosting the older Majuli Island, bypassing the intermediate unit. Subsidence in the Subansiri depression caused many channels of the north bank to avulse in the southwest direction. The south bank tributaries remained mostly unaffected.

5. Majuli Island represents a geomorphic high sitting on a 'high basement' topography. This is confirmed by seismic sections around the Majuli and the correlation of geophysical and topographic data. A three-stage evolution of Majuli and similar landforms – (i) *pre-bypass upliftment*, (ii) *Majuli formation*, and (iii) *abandonment*, involving the development of a geomorphic high, and fluvial dynamics of the main channel has been proposed. The role of basement configuration and tectonic setting in the evolution of such landforms was emphasized, rather than a merely geomorphic process.

6. Shallow subsurface studies show the confirmation of a major depositional episode during the late Quaternary. A velocity discrimination and thickness variability-based zonation of shallow subsurface layers and their superposition in the GIS environment is a novel approach followed in this work. This was subsequently used for a multi-parametric micro-zonation for selective windows on the south bank of the Brahmaputra valley.

7. The inversion observed in the stratigraphic sequence of the Brahmaputra basin means the direction of thickening of the marine wedge is opposite to that of the fluvial sediments, which can be explained by a post-Miocene strong episodic wrench causing basin tilt and asymmetry and then onwards formation of a wrench basin due to continuous plate convergence. This also explains the higher residence time of the Brahmaputra River closer to the Himalayan foredeep belt than that with the foredeep of the NPT belt.

8. The empirical relation between the W index and the bed–bank elevation difference index (for the year 2015) based on the 'least mean square method', when applied to the older river morphology (for the years 1975 and 1915), gives a much realistic means of trend analysis compared to the absolute width and depth variability. This helps identify flood-vulnerable banks of the Brahmaputra River.

9. New ideas and technologies should be brought into use to revive gold mining on the surface as well as at different depths in the subsurface.

10. Both the coal and oil sectors need an infusion of new ideas, particularly by developing precise modelling of thrust belt tectonics in the foreland areas of the valley by using a multi-disciplinary approach of combining remote sensing data collected from satellite imagery to monitor the changing riverscape and other landforms and geophysical data to have high-resolution deeper subsurface analysis to identify the role of structural controls in greater detail.

Scope for Further Work

Some of the very recent works (Rao et al, 2020) suggest that the estimates of sediment budgeting conducted so far (Wasson, 2003; Wasson et al., 2022) between the *sink* (the Bengal fan and its south-eastward continuation has an average thickness of 16.5-km-deep sediment stack) and the probable *sources* have a considerable mismatch. We are not very sure whether in recent times, there is a resurgence in large-scale landslides or not. A very recent incidence of a giant landslide was mentioned at the beginning of this chapter. Another incidence was reported on 20 December 2017. The Siang River became almost black, which normally looks bluish (source: https://www.ndtv.com/india-news/brahmaputra-river-now-muddy-and-black-could-be-poisonous-says-iit-guwahati-study-1789898). A study conducted in the IIT Guwahati at that time revealed that the particle pollutant level shot up to 1,249 nephelometric turbidity unit, which under safer limit should not exceed 5 nephelometric turbidity unit. In 2002, there was a massive flood which substantially changed the morphology of the Siang River, which is locally known as the 'China flood'. It is difficult to say very conclusively whether the anthropogenic intervention in the form of mega-dam construction and the associated deforestation is the main reason or not behind the increasing sediment influx in the last few decades. However, Quaternary stratigraphy studied in the lowermost part of the Brahmaputra valley where it becomes the Ganges–Brahmaputra–Meghna system, there is evidence of glacial lake outburst flood-driven discharge (Pickering et al., 2019). The present work was focussed on a very short timescale of around 100⁺ years to monitor major geomorphic changes and fluvial dynamics on the upper reach of the Brahmaputra valley relief and was extended subsequently to relate with stratigraphic and structural features in the shallow subsurface and the deep subsurface. Some of the additional work needed to provide further insights into the complex geomorphic, stratigraphic, and tectonic setting of this basin may include the following:

1. Age data for strategic core samples of the late Quaternary sediments to compute sedimentation rates in different parts of the valley and to relate this to differential tectonic regimes.
2. Preparation of shallow subsurface seismic facies based on the velocity model is no doubt useful, but to explain finer variations in the lithofacies, high-density vertical electrical sounding is needed.
3. For mapping the subsurface reliably in the foredeep zones of the thrust belt area by geophysical methods like vertical electrical sounding, the effectiveness of the electrical methods should be tested frequently as a part of dynamic simulation studies in the laboratory for various models of thrusting.
4. Open-hole wireline log-based sequence stratigraphy for the deep wells should be conducted in as detail as possible for superior subsurface control.
5. Basement configuration in more detail needs to be studied by combining different types of geophysical data like high-resolution satellite-borne magnetic and gravity data with 3D seismic.

Appendix 1: Field Visit

The present study involved many rounds of field visits, which were meant principally to meet the following four broad objectives:

1. To have a feel of some of the prominent geomorphologic features.
2. First-hand observation of the nature of bank material erosion going on in some of the highly vulnerable areas.
3. Observation of some of the geological elements which have strong structural implications.
4. Understanding the nature of geological hazards in Majuli, the largest river island in Asia that has a high density of human habitation.

In this section, the experience of the first field visit conducted from 7 to 13 July 2008 is described.

The northern part, as well as NW side, of the study area, is constituted of the Eastern Himalayan foothills. Eastern Himalayan syntaxis constitutes the NE corner. NE to further SE is flanked by the Naga-Patkai foothills. The Mikir Hills stand like an inselberg on the SW side. The Brahmaputra River acts as a great valley divider (Figure A1.1). In the following account, some of the site-specific brief descriptions are presented.

Site 1: Kahai Spur, about 8 km away from Dibrugarh in the NE direction (N27°30′15.4″- E94°56′11.6″):

This was one of the oldest spurs raised around 1956 and still stands strong (Figure A1.2). The spur was raised to divert the flow away from the bank line and check the toe-cutting and arrest bank collapse to save the Dibrugarh town. The objective was achieved. The southernmost branch of the Brahmaputra River passes through it. The net width of the river is about 13 km, and the branch width is about 400 m. Regular maintenance is very essential for these types of spurs. During the site visit, some concrete porcupine structures were seen lying in a very casual manner.

Site 2: Oakland Ghat (N27°31′56.8″ E95°02′37.6″):

From this place onwards, a stretch of around 12 km length in the upstream direction, up to Balijan is highly erosion-prone. This whole area is known as 'Rohmoria' which could draw wider media attention due to the voluntary local efforts to check bank-line erosion, strong protracted movements of the local people in the form of long-duration oil blockade and visits of big politicians, including the Prime Minister. Timbers cut illegally in the upstream side forests were floated and then collected at

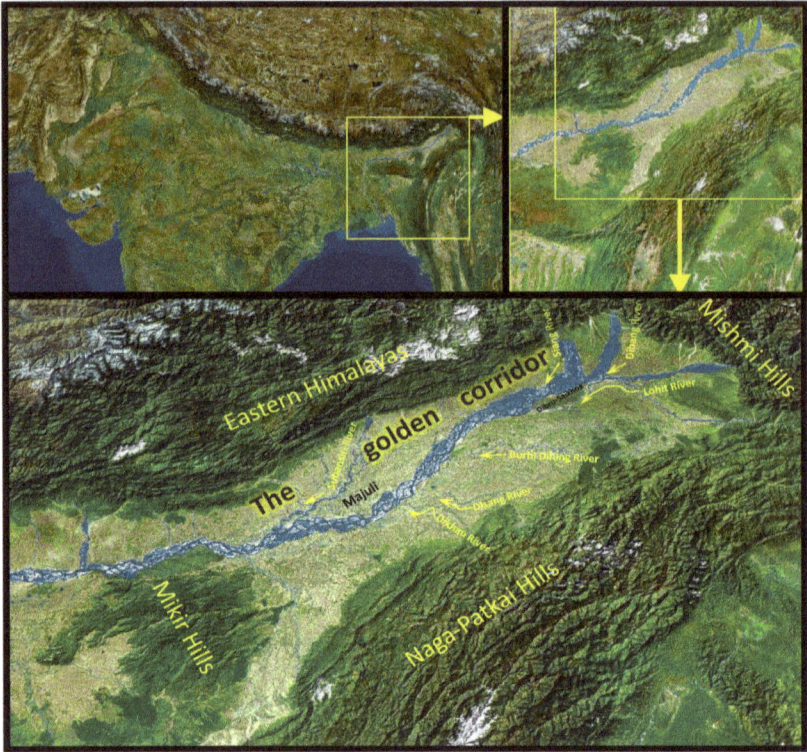

FIGURE A1.1
Study area covering the field visits.

different sites downstream (Figure A1.3). Earlier, there used to be a man-made canal (Ananta Nala) meant for irrigation and illegal trafficking of timbers. That old canal was subsequently occupied by the main channel of the Brahmaputra River. Long back, Oil India Limited (OIL) diverted the Burhi Dihing River to the Noa Dihing, thereby changing the water budget. As a consequence, erosional problems decreased in the Burhi Dihing but increased in the Noa Dihing.

As a part of erosion management, OIL raised 34 spurs (2004–2007) with the help of scrapped iron pipes (Figure A1.4). These seemed to be working well as long as regular maintenance work was carried out properly. A boat ride along the bank in the upstream direction from Oakland to Balijan revealed variable water depths (from 10–12 to 40–50 ft). Recent flood plain deposits and older ones were distinctly visible (Figure A1.5). The bank material was quite muddy with minor gullies with thin layers of sands (Figure A1.6). This indicated that the river had occupied the position very recently and had been eroding the older flood plain, possibly the flood plain of the Dibru River, captured lately by the Brahmaputra River. Very old tea bushes having very deep-seated roots were seen to collapse indiscriminately

FIGURE A1.2
Kahai Spur, one of the oldest raised in 1956 near Dibrugarh.

FIGURE A1.3
Oakland ghat with stacked piles of illegally cut timbers.

FIGURE A1.4
OIL raised spurs to protect against bank erosion.

FIGURE A1.5
Recent flood plain deposits.

FIGURE A1.6
Collapsing riverbank.

FIGURE A1.7
Collapsing tea gardens.

(Figure A1.7) without offering much resistance. Most large-scale erosion seemed to be due to the vertical collapse of muddy layers. At places, very high-velocity flow and strong eddies along the flow were observed. Piping was very common. Bank collapse seemed to be the main mechanism of lateral erosion. At places, ~20-cm-thick sand layer capped the muddy bank, which was light brown with mottles. That layer was supporting agriculture. There was also a thin iron-rich red mottling zone separating the light brown mud and the grey sand layer.

Site 3: Guijan (Rongagora) – About 12 km away from Tinsukia town in the westward direction (N27°34′54.1″- E95°19′57.8″):

This was a 3- to 4-m-high scarp (Figure A1.8) along the Brahmaputra River. This particular spot had an additional attraction for spotting river dolphins, an endangered species. Interestingly, even during the high-flow time, the bottom of the scarp was at a higher elevation than the river water level. Some of the workers claimed that the Rongagora scarp was a neotectonic feature. Most probably, the scarp existed much before the Brahmaputra River had underwent rapid bank-line migration. As a result, the scarp line had been modified by the Brahmaputra River.

FIGURE A1.8
A 3- to 4-m-high scarp in the south bank of the Brahmaputra River near Guijan, Tinsukia.

Site 4: Digboi Peak [N27°23'16.7"-E95°37'56.8"] Distance from Dibrugarh: 72 km

This was the oil field of vast historical importance, where the first producing well in Asia is located (Figure A1.9). There are very few oil fields in the world that are situated amidst such types of hilly areas having thick vegetative cover. From the Digboi Peak, the surrounding area having mountain and valley terraces was very clearly visible.

Site 5: Borjan (Naga thrust exposure) [N27°21'31.3" E95°32'13.5"] Distance from Dibrugarh: 63 km

This represented the ridge–valley border (Figure A1.10). An about 2-m-high scarp was followed by a ridge behind along the road section. The top surface of the scarp was cemented sand (Tipam sand), which was highly weathered followed by brown oxidized sand and mud (Figure A1.11). Local people informed that rock exposures were present along Nala (manifestation of Naga thrust). The area was highly vegetated and extremist-infested, which prohibited detailed field mapping.

Site 6: Margherita (Burhi Dihing Bridge) [N27°17'7.6" E95°39'42.5"] Distance from Dibrugarh: 77 km

This is a very sharp cliff line about 25–30 m high along the left bank of the river composed of boulders and gravel and had been interpreted as an

FIGURE A1.9
First oil-producing well of Asia in Digboi.

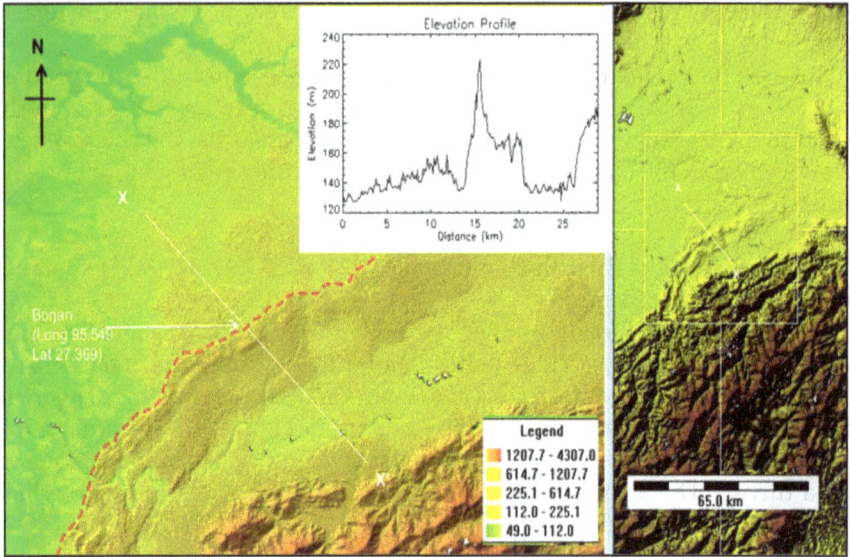

FIGURE A1.10
Location of the Naga thrust.

FIGURE A1.11
Field exposure of the Naga thrust.

FIGURE A1.12
River terraces near the Naga Thrust belt.

old fan deposit. The river followed the thrust. The cliff line was probably a tectonic feature. The right bank of the river was relatively flat, around 4–5 m above the bank line. At least three levels of terraces (Figure A1.12) could be seen along the right bank cliff. There might be successive incisions by the river in response to the tectonic uplift. Terrace sediments were constituted mainly of pebbles and poorly sorted (Figure A1.13). However, towards the base, large boulders were found. Also, faint laminations could be seen at some levels, principally constituted of variegated clays.

Site 7: Lekhapani [N27°18′47.7″- E95°48′2.9″] Distance from Dibrugarh: 89.6 km

The Tirap River, a tributary of the Burhi Dihing, flows by this site. Bedrock (Dihing formation) was found exposed. Except for the exposure, the water depth was quite high (8–10 m), and high-velocity flow was observed. Most probably, Khumsai thrust was passing about the site. Dihing formation looked pebbly, quite similar to the terraces observed at site 6 near Margherita but was much more indurated.

Site 8: Tipong Coal mine [N27°18′14.3″-E95°51′54.8″]

From the studies conducted by previous workers, it was known that the coal found in this area was of the Tertiary age; most probably, those were

FIGURE A1.13
The lithology of the uplifted river terrace.

Oligocene coal formed in the deltaic environment. The host rock was usually shale, siltstone, and subordinate sandstone. This area was having quite a large number of rock unit exposures that belonged to the Tertiary age. We could observe (Figure A1.14) quartz boulders embedded in soft sandstone beds (Dihing formation) and then salt and pepper-type typical massive sandstone (Tipam) of Miocene age. The location mentioned here was the place where massive sandstone changed to shaly type (Barail top?!)

Majuli Visit

Large motorboats are available to ferry passengers from Nimati ghat near Jorhat to Kamalabari ghat at Majuli Island (Figure A1.15). The bank near Kamalabari was highly erosion-prone. The exposed bank section was seen to have more cohesive silty material overlain by loose sandy material of about 1.5 m thickness. During high flows and recedes, upper sand got washed away first, followed by the lower layer which got submerged and dissolved. The rates of bank failure for those two layers were different, and the cliffs were not vertical. This was in contrast to the Rohmoria area near Dibrugarh, where bank failure of the embankment was common and the cliff was seen to be near vertical.

FIGURE A1.14
Outcrops showing rocks of Tertiary age in the Tipong Coal mine.

FIGURE A1.15
Ferry service from the Nimati ghat near Jorhat to Kamalabari ghat at Majuli Island.

Ten specific locations (Figure A1.16), severely affected by erosion were visited.

Site 1: Karatipar [N26°55′03.7″ E94°00′39.6″]

There was a small branching off the Brahmaputra connecting the Subansiri River. It used to be a much larger channel 3–4 years ago (i.e., during 2004–2005). On the upstream side near Kordoichuk, some porcupine structures (Figure A1.17) were raised, which were mainly responsible for the reduction of flow. That measure, even if temporary, could save some of the villages, like Baghgaon and Chamaguri, from extinction.

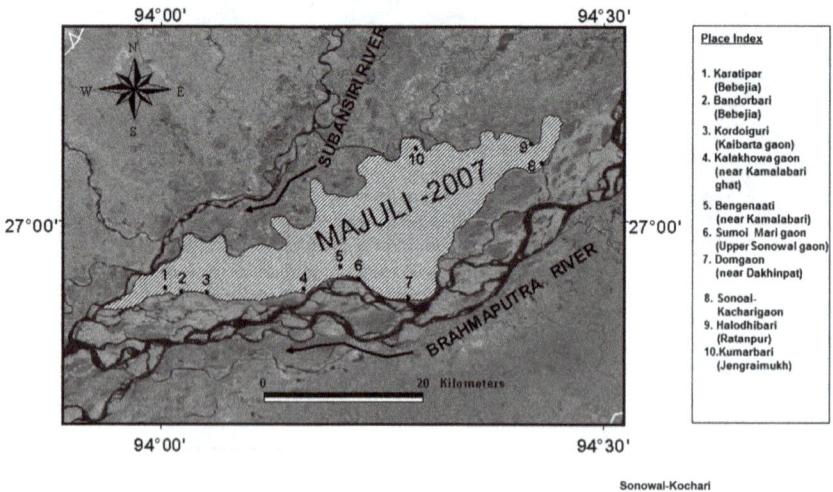

FIGURE A1.16
Ten locations visited in Majuli where massive erosion was going on.

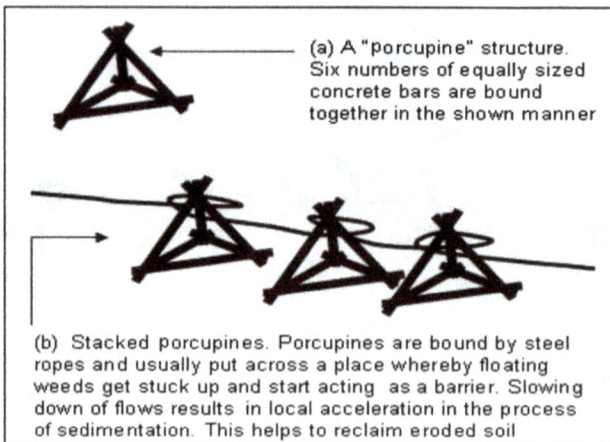

FIGURE A1.17
Porcupine structures used to reduce erosion.

Site 2: Bandorbari [N26°55′20.2″ E94°01′48.7″]

Compared to 2006 satellite imagery, it was observed that a very small channel had become a major channel in about 1 year. The width of the channel had increased by more than 400 m. Top 1- to 1.5-m sandy soil tended to cave in easily. That was underlain by muddy sediments (older flood plain). Very rapid erosion was going on. Banks started collapsing once the level of river water started falling. During our visit to the site, the water level was quite higher. The riverbank was very close to habitation. The inland bank surface having large trees and grass vegetation was quite older and stabilized (Figure A1.18). The new bank line was very close to the stabilized part. The preventive measures to check bank erosion were not only temporary but highly inadequate. Some 'bamboo porcupines', which were supposed to be put as dampeners, were lying idle by the riverside.

Site 5: In between Bhogpur Satra and Bengenaati Satra [N26°55′20.6″ E94°10′00.4″] Distance from Dibrugarh: 91.5 km

After 1975, the river shifted inwards by more than 5–6 km. At the site of observation, the river made a convex bend. Several prominent villages, like Upper Sonowal Gaon, Kaniajan, and Sukunamukh (etymologically the name came from a dried-up palaeochannel which was the Tuni River that used to divert excess water during the high-flood situation earlier), were completely submerged. Some bamboo porcupines (Figure A1.19) were raised. This helped divert the main current, resulting in a braid bar formation ahead in the downstream direction. The sandbar observed near this place has emerged in 2008 itself (Figure A1.20).

FIGURE A1.18
Banks having large trees could stabilize erosion.

FIGURE A1.19
Success to resist bank erosion by using porcupine structures near Kamalabari ferry ghat.

FIGURE A1.20
Bamboo porcupines for resisting bank erosion.

Site 3: Kordoiguri [N26°55′01.9″- E94°03′02.9″]

This was an area of bar growth and siltation. In 2007, some concrete porcupines were erected, and the erosion was checked. However, that enhanced the severity of erosion near Bandorbari. During dry periods, water used to recede completely, leaving the area completely dry. Earlier, the village was located at least 300 m inside the water body location compared to the channel position in July 2008. Repeated erosion had caused the village to shift behind the embankment. The area behind the embankment was heavily vegetated with bamboo forests, and during the visit, there was no flood plain in front of the embankment.

Site 4: Kalakhowa gaon near Kamalabari ferry ghat [N26°55′20.6″ E94°10′00.4″] Distance from Dibrugarh: 96 km

In 2007, some porcupine structures (Figure A1.17) were raised in the manner of raising a spur about different *satras* (mostly the supreme Vaishnavite estates). Near Kamalabari ghat, eight such spurs were raised (Figure A1.19). It helped resist erosion to some extent. No erosion occurred in this area in 2008. On the way to Kamalabari Chariali (the meeting place of four roads), we could see several erosions displaced people living in the make-shift arrangements. They represented a mixed-type population belonging to different castes and tribal identities.

Site 5: In between Bhogpur Satra and Bengenaati Satra [N26°55′20.6″ E94°10′00.4″] Distance from Dibrugarh: 91.5 km

After 1975, the river shifted inwards by more than 5–6 km. At the site of observation, the river made a convex bend. Several prominent villages, like Upper Sonowal gaon, Kaniajan, and Sukunamukh (etymologically the name came from a dried-up palaeochannel which was the Tuni River that used to divert excess water during the high-flood situation earlier), were completely submerged. Some bamboo porcupines (Figure A.20) were raised. This helped divert the main current, resulting in a braid bar formation ahead in the downstream direction. The sandbar observed near this place had emerged in 2008 itself (Figure A.21).

Site 6: Sumoi Mari gaon [N26°55′59.4″- E94°13′27.0″]

The earlier embankment was eroded by the river, and the riverbank drastically moved in the westward direction. Some remains of the bamboo formed porcupine structures, and a purely temporary means to check erosion were still visible. During dry seasons, the river water receded and the soil became visible. But in the year 2008, the toe-cutting seemed to be much deeper.

Site 7: Dakhinpat [N26°54′38.3″- E94°17′07.2″]

In 1998–1999, the stone spur was made by the Embankment and Development Department (the department that was principally responsible for raising embankments, practically speaking, as a 'one-point programme' to address the problems of flood and erosion. Interestingly, Assam is the state which had the highest density of total line-kilometres of embankments raised in the whole of the country). In 2008, the Brahmaputra Board had raised some concrete porcupine structures. This worked to push the river

FIGURE A1.21
A newly formed braid bar.

towards its midcourse. However, in the downstream areas, erosion increased a lot. Near Aufolamukh ghat, high river dynamism was very much evident from the traces of palaeochannel marks left at many places (Figure A1.22).

In the Dakhinpat area, we could witness a very good example of anthropogenic intervention to tackle flood and erosion problem and their consequences. In 1952, just after the great Assam earthquake of 1950, a huge effort was made prevent the whole of Majuli Island by raising earthen embankments. The earth material was dug out from the adjacent hinterland. This embankment eroded in 1974. Another embankment was raised in 1975 in the same manner.

The dugout portion of the topsoil had become a canal acting as an anabranch of the Brahmaputra River (Figure A.23). The case of the Tuni River was another example of the lopsided anthropogenic intervention. The Tuni River, passing through Majuli Island, was a natural anabranch of the Brahmaputra River that was used to divert excess water during the high-water level condition. By raising an embankment, the Tuni River was blocked. As a result, the previous course of the channel had become a stagnant pool of water, a breeding place for mosquitoes. This was an example of channel metamorphosis due to anthropogenic intervention.

FIGURE A1.22
Palaeochannel marks of very recent origin at Dakhinpat in the middle part of Majuli.

FIGURE A1.23
An artificial anabranch developed near the Dakhinpat area as a consequence of raising an embankment by cutting earth materials of the adjacent area.

During high-flood conditions, river water from the Brahmaputra and the surface runoff from the Mikir Hills through different tributaries, are logged inside the famous Kaziranga national forest and the sanctuary, where besides the one-horn Rhino, hundreds of elephants live. The elephants facing an immense food crisis due to massive deforestation and other habitational problems due to water logging migrating en masse by swimming across the river. During our visit to Dakhinpat, a band of wild elephants took shelter inside the high grasslands, and the villagers looked terrified. This was another aspect of the effect of the anthropogenic intervention on the ecosystem.

Site 8: Sonowal-Kochari Check Dam [N27°04′ 23.7″ E94°25′51.7″]

This place is located in front of the Dikhowmukh on the south bank of the Brahmaputra River. The topsoil located at this place was found sandy up to a considerable depth. Thus, the river action and aggradation were conducted for a considerable length of time. The earthen dam that was raised at this place was covered by boulders stacked inside strong iron nets. As a result, erosion could be arrested considerably. In 2006, long porcupine structures were also raised, and the new siltation helped recover a sufficient amount of land lost during erosion. However, due to lack of knowledge dissemination among the people and poverty, iron components of the said embankment were stolen. During 2008, the place was affected by severe erosion.

Site 9: Halodhibari [N27°05′25.7″-E94°25′29.7″]

At certain points of time in the 1970s, the old course of the Brahmaputra, now called the Kherkutia Suti (it could also be an anabranch), was blocked by a dyke. In 1971, the old dyke gave way during a high flood. In around 1972–1973, another dyke cum embankment was raised in the inner part of the land. The place through which this anabranch joined the main Brahmaputra River was closed at Tekeliphuta. During high-flood situations, the blocked channel received water from the upstream side of the Brahmaputra River. The interesting aspect was that the anthropogenic intervention at different points of time, based mostly on highly localized considerations, changed the river morphology in a much more complex way.

Site 10: Jengraimukh (Kumarbari) [N27°05′01.2″-E94°17′38.0″]

The site was located near the bank of the Lohit River having an actively eroding convex bank. A natural link channel was there linking the Lohit (the local name of the old Brahmaputra) with the Subansiri River. To check erosion, concrete, as well as bamboo porcupines, was raised. These measures act as a partial deterrent. Just 1 day before the day of our visit, slumping took place at the site referred to (Figures A1.24 and A1.25), despite the several porcupine structures and the semi-concretized (frontal part of the sandbag stacks used as lining was cemented) bank lining.

FIGURE A1.24
A case of very recent slumping in the Jengraimukh area located in the western part of Majuli.

FIGURE A1.25
A routine half-hearted measure to check erosion in the western part of Majuli with little success.

Appendix 2: Tables (Additional for Chapter 4)

TABLE A2.1

Elevation, Slope, BAF, and Temporal Variation in Bank-Line Shift (1915–2005) Were Measured in the Upper Reach of the Brahmaputra Valley, Assam (India)

Reach No.	Longitude	Latitude	Elevation (m)	Reach Length (km)	Distance (km)	Plot Points (km)	Slope (m/km)	BAF = (BAL/BAT) × 100	Bank-Line Shift 1975–1915		Bank-Line Shift 2005–1975		Bank-Line Shift 2005–1915	
									Right (N)	Left (S)	Right (N)	Left (S)	Right (N)	Left (S)
	95.433	27.809	124		0	0	0.34	62						
1	95.416	27.795	122	5.84	5.84	2.38	0.34	62	6.15	4.15	0.47	7.48	6.62	11.63
2	95.377	27.764	119	4.66	10.5	7.63	0.52	64	5.68	3.23	0.35	8.22	6.03	11.45
3	95.334	27.739	117	5.03	15.53	12.65	0.52	76	4.67	3.38	0.31	10.6	4.98	13.98
4	95.288	27.714	112	5.39	20.92	17.88	0.51	69	3.4	0.95	0.59	13.2	3.99	14.18
5	95.238	27.694	111	6.37	27.29	23.35	0.51	80	−0.08	0.45	1.54	10.3	1.46	10.75
6	95.178	27.672	110	6.51	33.8	29.79	0.14	81	1.69	1.75	2.33	6.66	4.02	8.41
7	95.117	27.649	107	6.63	40.43	36.32	0.14	80	3.12	1.99	0.71	5.23	3.83	7.22
8	95.05	27.625	107	7.29	47.72	43.4	0.14	76	0.45	3.07	0.08	3.75	0.53	6.82
9	94.986	27.584	107	8.94	56.66	51.19	0.14	72	0.45	3.62	−0.12	1.09	0.33	4.71
10	94.916	27.537	103	7.41	64.07	59.9	0.31	72	0.56	1.06	0.09	0.33	0.65	1.39
11	94.86	27.497	100	7.58	71.65	66.93	0.31	67	1	0.73	0.34	−0.02	1.34	0.71
12	94.809	27.446	100	7.61	79.26	74.56	0.31	59	2.43	0.68	0.65	−0.36	3.08	0.32
13	94.762	27.394	99	7.14	86.4	81.87	0.14	67	3.49	1.34	1.79	−0.52	5.28	0.82
14	94.715	27.34	98	7.34	93.74	89.5	0.14	59	1.98	0.42	0.47	0.44	2.45	0.86
15	94.67	27.286	95	7.37	101.11	96.99	0.38	61	1.03	0.8	−0.02	0.78	1.01	1.58
16	94.625	27.23	92	8.37	109.48	104.53	0.38	53	1.72	1.72	−0.08	0.24	1.64	1.96
17	94.576	27.169	91	8.05	117.53	112.88	0.15	53	1.09	0.69	1.13	0.51	2.22	1.2
18	94.54	27.126	90	4.87	122.4	118.77	0.15	52	−1.42	1.72	1.62	−0.198	0.2	1.522

(*Continued*)

TABLE A2.1 (*Continued*)

Elevation, Slope, BAF, and Temporal Variation in Bank-Line Shift (1915–2005) Were Measured in the Upper Reach of the Brahmaputra Valley, Assam (India)

Reach No.	Longitude	Latitude	Elevation (m)	Reach Length (km)	Distance (km)	Plot Points (km)	Slope (m/km)	BAF= (BAL/BAT) × 100	Bank-Line Shift 1975–1915 Right (N)	Bank-Line Shift 1975–1915 Left (S)	Bank-Line Shift 2005–1975 Right (N)	Bank-Line Shift 2005–1975 Left (S)	Bank-Line Shift 2005–1915 Right (N)	Bank-Line Shift 2005–1915 Left (S)
19	94.508	27.096	88	4.7	127.1	123.6	0.31	47	−0.24	1.19	−0.92	0.16	−1.16	1.35
20	94.448	27.065	86	8.36	135.46	130.39	0.31	48	−0.23	0.47	−0.03	0.32	−0.26	0.79
21	94.375	27.027	85	8.48	143.94	138.79	0.05	39	0.45	0.008	−0.22	0.65	0.23	0.658
22	94.307	26.992	85	6.37	150.31	146.56	0.05	34	0.22	−1.4	−2.3	1.4	−2.08	0
23	94.256	26.966	85	5.41	155.72	152.37	0.05	31	0.17	0.11	−0.54	0.08	−0.37	0.19
24	94.21	26.942	84	5	160.72	157.71	0.27	35	3.57	0.94	−5.5	0.53	−1.93	1.47
25	94.163	26.917	82	5.99	166.71	163.12	0.27	40	4.81	0.84	−5.62	0	−0.81	0.84
26	94.108	26.888	80	6.65	173.36	169.51	0.16	48	1.98	1.05	−2.56	0.02	−0.58	1.066
27	94.054	26.865	80	5.6	178.96	175.37	0.16	51	−0.64	−1.5	2.14	0.66	1.5	−0.84
28	93.998	26.845	73	6.13	185.09	181.36	0.19	62	0.51	−2.04	1.13	−0.13	1.64	−2.17
29	93.94	26.824	73	6.44	191.53	187.52	0.19	72	1.54	−3.5	−0.37	−0.3	1.17	−3.8
30	93.878	26.801	73	5.94	197.47	194.22	0.19	73	0.59	−2.32	0.24	−0.8	0.83	−3.12
31	93.814	26.778	73	6.09	203.56	201.15	0.19	75	2.08	−2.7	0.42	−0.24	2.5	−2.94
32	93.763	26.759	73	6.44	210	206.57	0.19	83	−0.63	−3.8	1.05	2.69	0.42	−1.11
33	93.704	26.754	73	6.17	216.17	212.64	0.19	71	−1.14	−1.44	1.21	2.45	0.07	1.01
34	93.642	26.757	70	6.18	222.35	218.74	0.23	33	0.51	2.08	0.34	0.29	0.85	2.37
35	93.58	26.759	70	6.65	229	225	0.23	33	1.24	0.59	0.13	2.28	1.37	2.87
36	93.513	26.762	70	6.5	235.5	231.6	0.41	43	1.14	2.48	0.24	−0.49	1.38	1.99
37	93.461	26.764	66	3.16	238.66	236.86	0.41	59	0.4	2.2	1.03	−1.58	1.43	0.62

BAF, basin asymmetry factor.

TABLE A2.2

Sinuosity Variation of the Median Path of the Brahmaputra Channel Belt during 1915–2005

Units	Valley Length (km)	Channel Length (km) in 1915	Sinuosity= (CL/VL) in 1915	Channel Length (km) in 1975	Sinuosity= (CL/VL) in 1975	Channel Length (km) in 2005	Sinuosity= (CL/VL) in 2005
1	56.42	58.97	**1.04**	56.63	**1.0**	58.11	1.03
2	65.58	66.82	**1.02**	67.27	**1.03**	66.85	1.02
3	113.15	124.37	**1.1**	123.56	**1.09**	125.58	1.11
Overall 'reach'	228.59	250.16	1.09	247.47	1.08	250.54	1.1

TABLE A2.3

Planar and Temporal Variability of Channel Belt Width (with and without Majuli) in 1915, 1975, and 2005

Distance (km)	Longitude	Latitude	Channel Belt Width (with Majuli) (km)			Channel Belt Width (without Majuli) (km)		
			1915	1975	2005	1915	1975	2005
0	95.433	27.809	4.15	13.05	19.2	4.15	13.05	19.2
1.35	95.424	27.801	2.2	12.44	20.63	2.2	12.44	20.63
2.38	95.416	27.795	2.06	12.78	20.73	2.06	12.78	20.73
3.43	95.408	27.788	2.22	13.3	21.87	2.22	13.3	21.87
4.68	95.399	27.781	2.48	13.18	22.18	2.48	13.18	22.18
5.84	95.39	27.773	3.07	12.84	21.52	3.07	12.84	21.52
6.87	95.382	27.768	3.6	12.7	21	3.6	12.7	21
7.63	95.377	27.764	3.8	8.4	20.45	3.8	8.4	20.45
8.48	95.369	27.758	3.51	12.52	20.58	3.51	12.52	20.58
9.43	95.361	27.752	2.85	10.86	20.92	2.85	10.86	20.92
10.5	95.353	27.748	2.27	10.52	21.29	2.27	10.52	21.29
11.7	95.344	27.742	2.35	10.67	21.21	2.35	10.67	21.21
12.65	95.334	27.739	2.32	10.38	21.24	2.32	10.38	21.24
13.47	95.327	27.735	2.35	10.48	21.35	2.35	10.48	21.35
14.48	95.317	27.73	2.4	10.14	21.53	2.4	10.14	21.53
15.33	95.308	27.726	2.57	9.43	21.13	2.57	9.43	21.13
16.75	95.298	27.721	3.23	8.03	20.87	3.23	8.03	20.87
17.88	95.288	27.714	3.1	6.68	20.63	3.1	6.68	20.63
18.83	95.28	27.712	2.06	5.82	21.26	2.06	5.82	21.26
19.83	95.271	27.707	3.22	5.97	21.02	3.22	5.97	21.02
20.92	95.26	27.703	5.41	6.5	19.58	5.41	6.5	19.58
22.21	95.248	27.698	6.82	7.37	19.2	6.82	7.37	19.2
23.35	95.238	27.694	7.18	8.16	18.83	7.18	8.16	18.83

(*Continued*)

TABLE A2.3 (*Continued*)

Planar and Temporal Variability of Channel Belt Width (with and without Majuli) in 1915, 1975, and 2005

Distance (km)	Longitude	Latitude	Channel Belt Width (with Majuli) (km)			Channel Belt Width (without Majuli) (km)		
			1915	1975	2005	1915	1975	2005
24.59	95.226	27.69	7.22	9.16	18.67	7.22	9.16	18.67
25.97	95.213	27.685	6.94	9.45	18.1	6.94	9.45	18.1
27.29	95.2	27.681	6.4	9.58	18.39	6.4	9.58	18.39
28.68	95.189	27.676	6.31	10.04	19.05	6.31	10.04	19.05
29.79	95.178	27.672	5.63	9.93	18.97	5.63	9.93	18.97
31.33	95.163	27.666	5.15	10.2	17.35	5.15	10.2	17.35
32.41	95.153	27.663	5.47	11.09	17.43	5.47	11.09	17.43
33.8	95.14	27.658	6.18	11.52	18.46	6.18	11.52	18.46
35	95.128	27.653	6.63	11.54	18.04	6.63	11.54	18.04
36.32	95.117	27.649	6.74	11.67	17.83	6.74	11.67	17.83
37.53	95.105	27.645	6.1	11.59	16.59	6.1	11.59	16.59
38.83	95.094	27.64	6.71	11.54	16.83	6.71	11.54	16.83
40.43	95.077	27.634	7.22	11.94	16.59	7.22	11.94	16.59
41.95	95.064	27.63	8.88	11.55	15.85	8.88	11.55	15.85
43.4	95.05	27.625	8.48	11.6	15.82	8.48	11.6	15.82
44.87	95.037	27.617	8.71	11.97	15.45	8.71	11.97	15.45
46.29	95.026	27.61	9.11	12.6	15.11	9.11	12.6	15.11
47.72	95.015	27.602	8.96	12.5	14.68	8.96	12.5	14.68
49.31	95	27.593	8.59	12.87	14.26	8.59	12.87	14.26
51.19	94.986	27.584	8.3	12.86	13.37	8.3	12.86	13.37
52.91	94.972	27.575	7.03	11.25	12.1	7.03	11.25	12.1
54.81	94.956	27.565	8.24	11.36	11.96	8.24	11.36	11.96
56.66	94.941	27.555	8.83	9.98	11.04	8.83	9.98	11.04
58.45	94.926	27.545	8.37	10.03	10.25	8.37	10.03	10.25
59.9	94.916	27.537	7.87	9.98	10.01	7.87	9.98	10.01
61.28	94.904	27.529	7.84	9.41	10.27	7.84	9.41	10.27
62.81	94.892	27.52	7.93	9.64	9.83	7.93	9.64	9.83
64.07	94.883	27.513	7.8	9.41	9.62	7.8	9.41	9.62
65.48	94.872	27.505	7.03	8.56	9.27	7.03	8.56	9.27
66.93	94.86	27.497	6.52	8.17	9.08	6.52	8.17	9.08
68.33	94.848	27.489	6.39	8.29	8.48	6.39	8.29	8.48
69.92	94.838	27.478	6.42	7.87	7.98	6.42	7.87	7.98
71.65	94.827	27.466	5.82	8.16	7.79	5.82	8.16	7.79
73.09	94.818	27.456	4.96	7.98	7.97	4.96	7.98	7.97
74.56	94.809	27.446	4.99	7.58	8.13	4.99	7.58	8.13
75.94	94.8	27.436	4.25	7.77	8.24	4.25	7.77	8.24
77.76	94.789	27.423	4.12	7.84	8.61	4.12	7.84	8.61
79.26	94.779	27.413	4.25	7.1	8.9	4.25	7.1	8.9
80.79	94.769	27.402	3.49	7.68	8.45	3.49	7.68	8.45

(*Continued*)

TABLE A2.3 (*Continued*)

Planar and Temporal Variability of Channel Belt Width (with and without Majuli) in 1915, 1975, and 2005

Distance (km)	Longitude	Latitude	Channel Belt Width (with Majuli) (km)			Channel Belt Width (without Majuli) (km)		
			1915	1975	2005	1915	1975	2005
81.87	94.762	27.394	2.65	9.4	8.91	2.65	9.4	8.91
83.19	94.754	27.385	2.27	10.97	9.98	2.27	10.97	9.98
84.88	94.743	27.373	3.3	10.8	10.48	3.3	10.8	10.48
86.4	94.733	27.362	4.89	8.03	8.09	4.89	8.03	8.09
88.11	94.724	27.351	5.55	7.27	8.11	5.55	7.27	8.11
89.5	94.715	27.34	5.62	8.93	8.59	5.62	8.93	8.59
91.1	94.706	27.328	5.07	9	8.69	5.07	9	8.69
92.34	94.7	27.319	5.05	8.75	9.14	5.05	8.75	9.14
93.74	94.69	27.309	6.05	8.98	9.98	6.05	8.98	9.98
95.16	94.68	27.3	6.79	8.85	10.59	6.79	8.85	10.59
96.99	94.67	27.286	8.11	8.55	11.15	8.11	8.55	11.15
98.41	94.662	27.275	8.53	9.37	9.2	8.53	9.37	9.2
99.62	94.654	27.266	9.16	10.48	10.69	9.16	10.48	10.69
101.11	94.645	27.256	8.91	12.02	12.05	8.91	12.02	12.05
102.9	94.635	27.241	8.67	11.84	11.94	8.67	11.84	11.94
104.53	94.625	27.23	7.8	11.72	11.86	7.8	11.72	11.86
106.36	94.614	27.217	7.26	10.96	11.35	7.26	10.96	11.35
107.91	94.605	27.206	6.29	9.06	9.94	6.29	9.06	9.94
109.48	94.595	27.194	5.89	8.82	9.93	5.89	8.82	9.93
111.4	94.584	27.18	5.58	7.88	10.04	5.58	7.88	10.04
112.88	94.576	27.169	5.6	7.68	9.35	5.6	7.68	9.35
114.02	94.568	27.161	5.68	7.03	8.79	5.68	7.03	8.79
115.66	94.558	27.149	6.5	6.4	8.35	6.5	6.4	8.35
117.53	94.547	27.136	7.11	6.79	8.79	7.11	6.79	8.79
118.77	94.54	27.126	6.42	7.29	9.04	6.42	7.29	9.04
120.28	94.532	27.115	6.95	7.45	7.98	6.95	7.45	7.98
122.4	94.518	27.1	9.2	8.42	9.06	9.2	8.42	9.06
123.6	94.508	27.096	9.88	11.31	9.46	4.42	8.34	9.46
125.45	94.492	27.088	9.08	9.8	10.36	4.17	7.9	10.36
127.1	94.477	27.08	10.31	10.36	11.25	3.46	7.03	11.25
128.65	94.464	27.073	10.97	12.02	12.42	1.95	6.26	9.51
130.39	94.448	27.065	11.22	11.28	11.26	2.86	5.81	7.88
131.9	94.435	27.057	11.75	11.83	11.83	4.57	6.84	8.98
133.8	94.419	27.05	13.58	13.81	14.29	6.57	8.46	8.96
135.46	94.404	27.042	14.89	14.79	15.03	7.1	9.86	10.83
137.31	94.39	27.034	14.8	15.3	15.48	5.89	9.72	10.22
138.79	94.375	27.027	15.19	14.84	15.5	5.33	8.93	9.27
140.46	94.361	27.02	17.49	17.41	18.2	4.79	8.91	8.69
142	94.347	27.012	17.04	18.84	19.86	5.57	9.25	9.35

(Continued)

TABLE A2.3 (*Continued*)

Planar and Temporal Variability of Channel Belt Width (with and without Majuli) in 1915, 1975, and 2005

Distance (km)	Longitude	Latitude	Channel Belt Width (with Majuli) (km)			Channel Belt Width (without Majuli) (km)		
			1915	1975	2005	1915	1975	2005
143.94	94.33	27.005	21.98	21.34	22.63	6.87	8.21	9.64
145.29	94.318	26.997	22.9	22.18	23.43	7.13	7.16	9.96
146.56	94.307	26.992	23.75	21.68	21.66	6.9	5.74	9.7
147.66	94.298	26.986	23.75	21.73	21.87	4	5.54	8.79
149.04	94.285	26.981	23.16	22.74	23.13	3.89	3.09	5.81
150.31	94.274	26.975	21.82	21.1	22.26	3.75	2.62	3.89
151.36	94.265	26.97	20.23	20.65	19.13	3.73	2.9	3.35
152.37	94.256	26.966	20.37	19.95	18.97	3.76	4.38	3.33
153.35	94.25	26.962	20.95	20.73	19.86	3.7	5.07	3.91
154.43	94.239	26.957	21.4	23.27	21.16	4.34	5.66	5.55
155.72	94.227	26.951	21.24	23.3	23.67	5.33	6.58	7.16
156.8	94.219	26.946	21.74	21.03	20.37	7.27	8.43	9.88
157.71	94.21	26.942	17.16	18.67	18.72	7.8	8.32	10.4
158.83	94.2	26.937	17.52	17.57	17.46	6.97	7.64	9.88
159.71	94.19	26.93	18.56	17.83	17.32	6.74	7.79	9.38
160.72	94.183	26.928	18.94	19.67	19.12	6.82	7.95	9.32
161.87	94.174	26.923	19.42	19.65	19.13	6.98	7.39	8.98
163.12	94.163	26.917	18.44	19.28	15.61	6.77	8.03	9.11
164.47	94.151	26.912	15.82	17.7	17.22	6.85	9.2	10.2
165.5	94.142	26.906	16.04	17.65	17.67	6.87	10.91	10.44
166.71	94.132	26.901	16.01	17.18	17.33	6.05	12.17	12.52
168.06	94.119	26.895	16.24	17.6	16.48	4.94	12.76	12.37
169.51	94.108	26.888	17.56	18.25	18.18	5.49	12.04	11.63
170.65	94.098	26.883	16.82	16.91	16.14	4.92	10.52	10.5
171.86	94.086	26.877	16.72	17.12	16.88	5.45	10.28	9.57
173.36	94.073	26.872	16.1	17.36	17.86	6.03	9.82	9.83
174.31	94.064	26.87	16.25	16.35	18.41	6.29	9.24	10.88
175.37	94.054	26.865	18.09	12.5	18.97	6.61	9.14	11.68
176.58	94.043	26.861	18.15	14.58	17.81	6.73	9.08	12
177.62	94.033	26.858	19.12	14.53	15.77	6.89	9.09	11.2
178.96	94.021	26.853	17.14	14.63	15.53	7.1	8.87	10.72
180.42	94.007	26.849	17.09	15.16	16.25	6.55	6.24	9.8
181.36	93.998	26.845	16.66	15.3	17.07	5.73	5.89	10.36
182.48	93.988	26.841	15.72	15.18	15.66	4.99	6.18	12.23
183.9	93.975	26.837	15.45	13.95	14.98	5.29	6.45	11.41
185.09	93.963	26.832	16.67	14.47	15.37	5.49	7.61	10.73
186.22	93.953	26.828	17.86	15.93	15.7	5.87	8.48	10.96
187.52	93.94	26.824	18.44	16.8	14.7	7.84	9.25	11.81
189.26	93.925	26.818	18.97	16.2	14.69	8.34	7.87	11.8

(*Continued*)

TABLE A2.3 (*Continued*)

Planar and Temporal Variability of Channel Belt Width (with and without Majuli) in 1915, 1975, and 2005

Distance (km)	Longitude	Latitude	Channel Belt Width (with Majuli) (km)			Channel Belt Width (without Majuli) (km)		
			1915	1975	2005	1915	1975	2005
190.45	93.913	26.814	17.27	15.61	15.3	7.68	6.37	11.92
191.53	93.904	26.811	17.03	15.27	15.96	8.34	6.74	11.72
192.94	93.891	26.806	15.24	13.86	15.27	7.8	8.03	10.48
194.22	93.878	26.801	15.72	13.9	12.7	7.35	8.5	11.04
195.28	93.869	26.798	15.87	13.65	12.25	7.11	7.9	11.7
196.32	93.858	26.794	16.19	13.87	12.39	7.84	8.71	12.29
197.47	93.848	26.79	16.19	13.86	12.41	9.01	9.46	12.41
198.62	93.838	26.785	13.01	13.05	12.25	8.75	11.7	12.25
199.88	93.826	26.782	11.34	11.67	12.58	9.12	12.65	12.58
201.15	93.814	26.778	11.86	12.6	13.7	7.95	12.81	13.69
202.44	93.802	26.774	13.26	12.76	14	6.85	12.76	14
203.56	93.792	26.77	12.58	10.77	13.4	8.25	10.77	13.4
205.08	93.778	26.764	13.79	8.82	12.34	5.91	8.82	12.34
206.57	93.763	26.759	9.43	6.42	8.8	6.97	6.42	8.8
207.72	93.753	26.755	9.37	4.6	8.87	6.84	4.6	8.87
209.1	93.74	26.753	11.62	4.17	10.83	6.71	4.17	10.83
210	93.729	26.753	11.59	3.7	11.46	7.08	4.71	11.46
211.31	93.716	26.754	11.47	5.37	10.86	6.9	5.37	10.86
212.64	93.704	26.754	8.75	7.72	11.92	7.43	7.72	11.92
213.75	93.693	26.755	9.61	10.25	11.76	7.47	10.25	11.76
214.84	93.681	26.755	10.51	10.96	11.59	6.9	10.96	11.59
216.17	93.668	26.756	9.12	10.03	12.02	8.22	10.03	12.02
217.61	93.654	26.757	8.58	10.88	10.73	8.58	10.88	10.73
218.74	93.642	26.757	8.58	12.47	11.92	8.58	12.47	11.92
219.82	93.631	26.758	8.51	11.97	12.94	8.51	11.97	12.94
221.38	93.617	26.758	8.32	10.11	11.99	8.32	10.11	11.99
222.35	93.606	26.759	7.8	9.41	10.8	7.8	9.41	10.8
223.59	93.593	26.759	6.98	8.77	9.61	6.98	8.77	9.61
225	93.58	26.759	6.53	8.06	11.1	6.53	8.06	11.1
226.29	93.567	26.759	5.92	6.37	10.25	5.92	6.37	10.25
227.87	93.552	26.761	3.86	6.32	9.32	3.86	6.32	9.32
229	93.538	26.761	2.77	6.11	6.97	2.77	6.11	6.97
230.42	93.526	26.761	2.46	6.66	6.08	2.46	6.66	6.08
231.6	93.513	26.762	2.64	6.69	5.58	2.64	6.69	5.58
232.9	93.5	26.762	2.43	5.76	5.37	2.43	5.76	5.37
234.19	93.487	26.763	2.56	5.5	5.66	2.56	5.5	5.66
235.5	93.474	26.763	2.11	5.76	5.36	2.11	5.76	5.36
236.86	93.461	26.764	2.7	5.34	4.79	2.7	5.34	4.79
238.66	93.443	26.765	3.36	5.04	3.97	3.36	5.04	3.97

TABLE A2.4

Relative Variability in Width (%) of the Brahmaputra River in the Upper Reach of the Brahmaputra Valley in 1915, 1975, and 2005

Distance (km)	Longitude	Latitude	Relative Variability in Width (%) (with Majuli)			Relative Variability in Width (%) (without Majuli)		
			1915	1975	2005	1915	1975	2005
0	95.433	27.809	−57.9	11.2	34.8	−29.5	47.0	60.5
1.35	95.424	27.801	−77.7	6.0	44.9	−62.6	40.1	72.5
2.38	95.416	27.795	−79.1	8.9	45.6	−65.0	43.9	73.3
3.43	95.408	27.788	−77.5	13.3	53.6	−62.3	49.8	82.9
4.68	95.399	27.781	−74.8	12.3	55.8	−57.9	48.4	85.5
5.84	95.39	27.773	−68.9	9.4	51.1	−47.9	44.6	79.9
6.87	95.382	27.768	−63.5	8.2	47.5	−38.9	43.0	75.6
7.63	95.377	27.764	−61.5	−28.4	43.6	−35.5	−5.4	71.0
8.48	95.369	27.758	−64.4	6.6	44.5	−40.4	41.0	72.1
9.43	95.361	27.752	−71.1	−7.5	46.9	−51.6	22.3	74.9
10.5	95.353	27.748	−77.0	−10.4	49.5	−61.5	18.5	78.0
11.7	95.344	27.742	−76.2	−9.1	48.9	−60.1	20.2	77.3
12.65	95.334	27.739	−76.5	−11.6	49.2	−60.6	16.9	77.6
13.47	95.327	27.735	−76.2	−10.7	49.9	−60.1	18.0	78.5
14.48	95.317	27.73	−75.7	−13.6	51.2	−59.3	14.2	80.0
15.33	95.308	27.726	−73.9	−19.7	48.4	−56.4	6.2	76.7
16.75	95.298	27.721	−67.2	−31.6	46.6	−45.2	−9.6	74.5
17.88	95.288	27.714	−68.6	−43.1	44.9	−47.4	−24.8	72.5
18.83	95.28	27.712	−79.1	−50.4	49.3	−65.0	−34.5	77.8
19.83	95.271	27.707	−67.3	−49.1	47.6	−45.3	−32.8	75.8
20.92	95.26	27.703	−45.1	−44.6	37.5	−8.1	−26.8	63.7
22.21	95.248	27.698	−30.8	−37.2	34.8	15.8	−17.0	60.5
23.35	95.238	27.694	−27.2	−30.5	32.2	21.9	−8.1	57.4
24.59	95.226	27.69	−26.8	−22.0	31.1	22.6	3.2	56.1
25.97	95.213	27.685	−29.6	−19.5	27.1	17.8	6.4	51.3
27.29	95.2	27.681	−35.1	−18.4	29.1	8.7	7.9	53.8
28.68	95.189	27.676	−36.0	−14.5	33.8	7.1	13.1	59.3
29.79	95.178	27.672	−42.9	−15.4	33.2	−4.4	11.8	58.6
31.33	95.163	27.666	−47.8	−13.1	21.8	−12.6	14.9	45.1
32.41	95.153	27.663	−44.5	−5.5	22.4	−7.1	24.9	45.7
33.8	95.14	27.658	−37.3	−1.9	29.6	4.9	29.7	54.3
35	95.128	27.653	−32.8	−1.7	26.7	12.6	30.0	50.8
36.32	95.117	27.649	−31.6	−0.6	25.2	14.4	31.4	49.1
37.53	95.105	27.645	−38.1	−1.3	16.5	3.6	30.5	38.7
38.83	95.094	27.64	−31.9	−1.7	18.2	13.9	30.0	40.7
40.43	95.077	27.634	−26.8	1.7	16.5	22.6	34.5	38.7

(Continued)

TABLE A2.4 (*Continued*)

Relative Variability in Width (%) of the Brahmaputra River in the Upper Reach of the Brahmaputra Valley in 1915, 1975, and 2005

Distance (km)	Longitude	Latitude	Relative Variability in Width (%) (with Majuli)			Relative Variability in Width (%) (without Majuli)		
			1915	1975	2005	1915	1975	2005
41.95	95.064	27.63	−9.9	−1.6	11.3	50.8	30.1	32.5
43.4	95.05	27.625	−14.0	−1.2	11.1	44.0	30.6	32.3
44.87	95.037	27.617	−11.7	2.0	8.5	47.9	34.8	29.2
46.29	95.026	27.61	−7.6	7.3	6.1	54.7	41.9	26.3
47.72	95.015	27.602	−9.1	6.5	3.1	52.1	40.8	22.7
49.31	95	27.593	−12.9	9.6	0.1	45.8	44.9	19.2
51.19	94.986	27.584	−15.8	9.5	−6.1	40.9	44.8	11.8
52.91	94.972	27.575	−28.7	−4.2	−15.0	19.4	26.7	1.2
54.81	94.956	27.565	−16.4	−3.2	−16.0	39.9	27.9	0.0
56.66	94.941	27.555	−10.4	−15.0	−22.5	49.9	12.4	−7.7
58.45	94.926	27.545	−15.1	−14.6	−28.0	42.1	13.0	−14.3
59.9	94.916	27.537	−20.2	−15.0	−29.7	33.6	12.4	−16.3
61.28	94.904	27.529	−20.5	−19.8	−27.9	33.1	6.0	−14.1
62.81	94.892	27.52	−19.6	−17.9	−31.0	34.6	8.6	−17.8
64.07	94.883	27.513	−20.9	−19.8	−32.4	32.4	6.0	−19.6
65.48	94.872	27.505	−28.7	−27.1	−34.9	19.4	−3.6	−22.5
66.93	94.86	27.497	−33.9	−30.4	−36.2	10.7	−8.0	−24.1
68.33	94.848	27.489	−35.2	−29.4	−40.4	8.5	−6.6	−29.1
69.92	94.838	27.478	−34.9	−33.0	−44.0	9.0	−11.4	−33.3
71.65	94.827	27.466	−41.0	−30.5	−45.3	−1.2	−8.1	−34.9
73.09	94.818	27.456	−49.7	−32.0	−44.0	−15.8	−10.1	−33.4
74.56	94.809	27.446	−49.4	−35.4	−42.9	−15.3	−14.6	−32.0
75.94	94.8	27.436	−56.9	−33.8	−42.1	−27.8	−12.5	−31.1
77.76	94.789	27.423	−58.2	−33.2	−39.5	−30.1	−11.7	−28.0
79.26	94.779	27.413	−56.9	−39.5	−37.5	−27.8	−20.0	−25.6
80.79	94.769	27.402	−64.6	−34.6	−40.7	−40.7	−13.5	−29.3
81.87	94.762	27.394	−73.1	−19.9	−37.4	−55.0	5.9	−25.5
83.19	94.754	27.385	−77.0	−6.6	−29.9	−61.5	23.5	−16.6
84.88	94.743	27.373	−66.5	−8.0	−26.4	−44.0	21.6	−12.4
86.4	94.733	27.362	−50.4	−31.6	−43.2	−17.0	−9.6	−32.4
88.11	94.724	27.351	−43.7	−38.1	−43.0	−5.8	−18.1	−32.2
89.5	94.715	27.34	−43.0	−23.9	−39.7	−4.6	0.6	−28.2
91.1	94.706	27.328	−48.6	−23.3	−39.0	−13.9	1.4	−27.3
92.34	94.7	27.319	−48.8	−25.5	−35.8	−14.3	−1.5	−23.6
93.74	94.69	27.309	−38.6	−23.5	−29.9	2.7	1.1	−16.6
95.16	94.68	27.3	−31.1	−24.6	−25.6	15.3	−0.3	−11.5

(*Continued*)

TABLE A2.4 (*Continued*)

Relative Variability in Width (%) of the Brahmaputra River in the Upper Reach of the Brahmaputra Valley in 1915, 1975, and 2005

Distance (km)	Longitude	Latitude	Relative Variability in Width (%) (with Majuli)			Relative Variability in Width (%) (without Majuli)		
			1915	1975	2005	1915	1975	2005
96.99	94.67	27.286	−17.7	−27.2	−21.7	37.7	−3.7	−6.8
98.41	94.662	27.275	−13.5	−20.2	−35.4	44.8	5.5	−23.1
99.62	94.654	27.266	−7.1	−10.7	−24.9	55.5	18.0	−10.6
101.11	94.645	27.256	−9.6	2.4	−15.4	51.3	35.4	0.8
102.9	94.635	27.241	−12.1	0.9	−16.2	47.2	33.3	−0.2
104.53	94.625	27.23	−20.9	−0.2	−16.7	32.4	32.0	−0.8
106.36	94.614	27.217	−26.4	−6.6	−20.3	23.3	23.4	−5.1
107.91	94.605	27.206	−36.2	−22.8	−30.2	6.8	2.0	−16.9
109.48	94.595	27.194	−40.3	−24.9	−30.3	0.0	−0.7	−17.0
111.4	94.584	27.18	−43.4	−32.9	−29.5	−5.3	−11.3	−16.1
112.88	94.576	27.169	−43.2	−34.6	−34.3	−4.9	−13.5	−21.8
114.02	94.568	27.161	−42.4	−40.1	−38.3	−3.6	−20.8	−26.5
115.66	94.558	27.149	−34.1	−45.5	−41.4	10.4	−27.9	−30.2
117.53	94.547	27.136	−27.9	−42.2	−38.3	20.7	−23.5	−26.5
118.77	94.54	27.126	−34.9	−37.9	−36.5	9.0	−17.9	−24.4
120.28	94.532	27.115	−29.5	−36.5	−44.0	18.0	−16.1	−33.3
122.4	94.518	27.1	−6.7	−28.3	−36.4	56.2	−5.2	−24.2
123.6	94.508	27.096	0.2	−3.7	−33.6	−25.0	−6.1	−20.9
125.45	94.492	27.088	−7.9	−16.5	−27.2	−29.2	−11.0	−13.4
127.1	94.477	27.08	4.6	−11.8	−21.0	−41.3	−20.8	−5.9
128.65	94.464	27.073	11.3	2.4	−12.8	−66.9	−29.5	−20.5
130.39	94.448	27.065	13.8	−3.9	−20.9	−51.4	−34.6	−34.1
131.9	94.435	27.057	19.2	0.8	−16.9	−22.4	−23.0	−24.9
133.8	94.419	27.05	37.7	17.6	0.4	11.5	−4.7	−25.1
135.46	94.404	27.042	51.0	26.0	5.5	20.5	11.0	−9.4
137.31	94.39	27.034	50.1	30.3	8.7	0.0	9.5	−14.5
138.79	94.375	27.027	54.1	26.4	8.8	−9.5	0.6	−22.5
140.46	94.361	27.02	77.4	48.3	27.8	−18.7	0.3	−27.3
142	94.347	27.012	72.8	60.5	39.5	−5.4	4.2	−21.8
143.94	94.33	27.005	122.9	81.8	58.9	16.6	−7.5	−19.4
145.29	94.318	26.997	132.3	88.9	64.5	21.1	−19.4	−16.7
146.56	94.307	26.992	140.9	84.7	52.1	17.1	−35.4	−18.9
147.66	94.298	26.986	140.9	85.1	53.6	−32.1	−37.6	−26.5
149.04	94.285	26.981	134.9	93.7	62.4	−34.0	−65.2	−51.4
150.31	94.274	26.975	121.3	79.7	56.3	−36.3	−70.5	−67.5
151.36	94.265	26.97	105.2	75.9	34.3	−36.7	−67.3	−72.0

(*Continued*)

TABLE A2.4 (*Continued*)

Relative Variability in Width (%) of the Brahmaputra River in the Upper Reach of the Brahmaputra Valley in 1915, 1975, and 2005

Distance (km)	Longitude	Latitude	Relative Variability in Width (%) (with Majuli)			Relative Variability in Width (%) (without Majuli)		
			1915	1975	2005	1915	1975	2005
152.37	94.256	26.966	106.6	69.9	33.2	−36.2	−50.7	−72.2
153.35	94.25	26.962	112.5	76.6	39.5	−37.2	−42.9	−67.3
154.43	94.239	26.957	117.0	98.2	48.6	−26.3	−36.3	−53.6
155.72	94.227	26.951	115.4	98.5	66.2	−9.5	−25.9	−40.1
156.8	94.219	26.946	120.5	79.1	43.0	23.4	−5.1	−17.4
157.71	94.21	26.942	74.0	59.0	31.5	32.4	−6.3	−13.0
158.83	94.2	26.937	77.7	49.7	22.6	18.3	−14.0	−17.4
159.71	94.19	26.93	88.2	51.9	21.6	14.4	−12.3	−21.6
160.72	94.183	26.928	92.1	67.5	34.3	15.8	−10.5	−22.1
161.87	94.174	26.923	97.0	67.4	34.3	18.5	−16.8	−24.9
163.12	94.163	26.917	87.0	64.2	9.6	14.9	−9.6	−23.8
164.47	94.151	26.912	60.4	50.8	20.9	16.3	3.6	−14.7
165.5	94.142	26.906	62.7	50.3	24.1	16.6	22.9	−12.7
166.71	94.132	26.901	62.4	46.3	21.7	2.7	37.0	4.7
168.06	94.119	26.895	64.7	49.9	15.7	−16.1	43.7	3.4
169.51	94.108	26.888	78.1	55.5	27.7	−6.8	35.6	−2.8
170.65	94.098	26.883	70.6	44.0	13.3	−16.5	18.5	−12.2
171.86	94.086	26.877	69.6	45.8	18.5	−7.5	15.8	−20.0
173.36	94.073	26.872	63.3	47.9	25.4	2.4	10.6	−17.8
174.31	94.064	26.87	64.8	39.3	29.3	6.8	4.1	−9.0
175.37	94.054	26.865	83.5	6.5	33.2	12.2	2.9	−2.3
176.58	94.043	26.861	84.1	24.2	25.1	14.3	2.3	0.3
177.62	94.033	26.858	93.9	23.8	10.7	17.0	2.4	−6.4
178.96	94.021	26.853	73.8	24.6	9.1	20.5	−0.1	−10.4
180.42	94.007	26.849	73.3	29.1	14.1	11.2	−29.7	−18.1
181.36	93.998	26.845	69.0	30.3	19.9	−2.7	−33.7	−13.4
182.48	93.988	26.841	59.4	29.3	10.0	−15.3	−30.4	2.3
183.9	93.975	26.837	56.7	18.8	5.2	−10.2	−27.4	−4.6
185.09	93.963	26.832	69.1	23.3	7.9	−6.8	−14.3	−10.3
186.22	93.953	26.828	81.1	35.7	10.3	−0.3	−4.5	−8.4
187.52	93.94	26.824	87.0	43.1	3.2	33.1	4.2	−1.3
189.26	93.925	26.818	92.4	38.0	3.2	41.6	−11.4	−1.3
190.45	93.913	26.814	75.2	33.0	7.4	30.4	−28.3	−0.3
191.53	93.904	26.811	72.7	30.1	12.1	41.6	−24.1	−2.0
192.94	93.891	26.806	54.6	18.1	7.2	32.4	−9.6	−12.4
194.22	93.878	26.801	59.4	18.4	−10.8	24.8	−4.3	−7.7

(*Continued*)

TABLE A2.4 *(Continued)*

Relative Variability in Width (%) of the Brahmaputra River in the Upper Reach of the Brahmaputra Valley in 1915, 1975, and 2005

Distance (km)	Longitude	Latitude	Relative Variability in Width (%) (with Majuli)			Relative Variability in Width (%) (without Majuli)		
			1915	1975	2005	1915	1975	2005
195.28	93.869	26.798	61.0	16.3	−14.0	20.7	−11.0	−2.2
196.32	93.858	26.794	64.2	18.1	−13.0	33.1	−1.9	2.8
197.47	93.848	26.79	64.2	18.1	−12.9	53.0	6.5	3.8
198.62	93.838	26.785	31.9	11.2	−14.0	48.6	31.8	2.4
199.88	93.826	26.782	15.0	−0.6	−11.7	54.8	42.5	5.2
201.15	93.814	26.778	20.3	7.3	−3.8	35.0	44.3	14.5
202.44	93.802	26.774	34.5	8.7	−1.7	16.3	43.7	17.1
203.56	93.792	26.77	27.6	−8.3	−5.9	40.1	21.3	12.0
205.08	93.778	26.764	39.9	−24.9	−13.3	0.3	−0.7	3.2
206.57	93.763	26.759	−4.4	−45.3	−38.2	18.3	−27.7	−26.4
207.72	93.753	26.755	−5.0	−60.8	−37.7	16.1	−48.2	−25.8
209.1	93.74	26.753	17.8	−64.5	−23.9	13.9	−53.0	−9.4
210	93.729	26.753	17.5	−68.5	−19.5	20.2	−47.0	−4.2
211.31	93.716	26.754	16.3	−54.3	−23.7	17.1	−39.5	−9.2
212.64	93.704	26.754	−11.3	−34.2	−16.3	26.1	−13.1	−0.3
213.75	93.693	26.755	−2.5	−12.7	−17.4	26.8	15.4	−1.7
214.84	93.681	26.755	6.6	−6.6	−18.6	17.1	23.4	−3.1
216.17	93.668	26.756	−7.5	−14.6	−15.6	39.6	13.0	0.5
217.61	93.654	26.757	−13.0	−7.3	−24.6	45.7	22.5	−10.3
218.74	93.642	26.757	−13.0	6.2	−16.3	45.7	40.4	−0.3
219.82	93.631	26.758	−13.7	2.0	−9.1	44.5	34.8	8.2
221.38	93.617	26.758	−15.6	−13.9	−15.8	41.3	13.9	0.3
222.35	93.606	26.759	−20.9	−19.8	−24.2	32.4	6.0	−9.7
223.59	93.593	26.759	−29.2	−25.3	−32.5	18.5	−1.2	−19.6
225	93.58	26.759	−33.8	−31.3	−22.1	10.9	−9.2	−7.2
226.29	93.567	26.759	−40.0	−45.7	−28.0	0.5	−28.3	−14.3
227.87	93.552	26.761	−60.9	−46.2	−34.6	−34.5	−28.8	−22.1
229	93.538	26.761	−71.9	−48.0	−51.1	−53.0	−31.2	−41.7
230.42	93.526	26.761	−75.1	−43.3	−57.3	−58.2	−25.0	−49.2
231.6	93.513	26.762	−73.2	−43.0	−60.8	−55.2	−24.7	−53.3
232.9	93.5	26.762	−75.4	−50.9	−62.3	−58.7	−35.1	−55.1
234.19	93.487	26.763	−74.0	−53.2	−60.3	−56.5	−38.1	−52.7
235.5	93.474	26.763	−78.6	−50.9	−62.4	−64.2	−35.1	−55.2
236.86	93.461	26.764	−72.6	−54.5	−66.4	−54.2	−39.9	−59.9
238.66	93.443	26.765	−65.9	−57.1	−72.1	−43.0	−43.2	−66.8

TABLE A2.5

Measured Values of Different Morphologic Parameters in 37 Smaller Reaches in the Upper Reach of the Brahmaputra Valley in Assam at Different Times in 1915, 1975, and 2005

Reach No.	Distance (km)	Longitude	Latitude	Width Average (km)			CHB (km²)			Interfluves/Older Flood Plains (km²)			BB Area (km²)			CH (CHB-(IF+BB)) (km²)			BB/CH		
				1915	1975	2005	1915	1975	2005	1915	1975	2005	1915	1975	2005	1915	1975	2005	1915	1975	2005
1	2.38	95.416	27.795	2.62	12.95	20.92	16.45	85.29	140.30	0.00	28.70	23.72	7.95	41.34	67.81	8.50	15.25	32.01	0.936	2.710	2.118
2	7.63	95.377	27.764	3.37	11.46	20.89	17.07	62.91	111.65	0.00	32.32	35.69	10.00	10.96	39.99	7.07	19.63	23.00	1.414	0.558	1.739
3	12.65	95.334	27.739	2.34	10.44	21.32	13.18	55.89	131.23	0.00	21.50	55.45	5.52	19.53	44.94	7.67	14.87	24.66	0.720	1.314	1.822
4	17.88	95.288	27.714	2.84	7.19	20.98	19.14	40.33	141.75	0.00	0.75	70.24	10.44	23.02	44.50	8.70	16.55	24.55	1.199	1.391	1.812
5	23.35	95.238	27.694	6.71	8.13	18.88	50.04	62.60	160.37	0.00	0.00	63.97	38.41	45.17	52.21	11.63	17.43	44.19	3.303	2.591	1.182
6	29.79	95.178	27.672	5.79	10.17	18.24	41.36	74.75	134.24	0.00	0.00	12.46	28.05	60.22	91.53	13.31	14.53	30.25	2.107	4.144	3.026
7	36.32	95.117	27.649	6.47	11.57	17.55	48.28	86.71	120.18	0.00	0.00	0.00	31.18	68.95	94.82	17.09	17.77	25.36	1.824	3.880	3.740
8	43.4	95.05	27.625	8.48	11.93	15.76	71.22	95.70	117.43	0.00	0.00	0.00	55.27	76.35	87.41	15.95	19.35	30.02	3.464	3.946	2.912
9	51.19	94.986	27.584	8.22	12.17	13.27	82.18	120.51	127.40	0.00	0.00	0.00	57.50	94.30	86.17	24.68	26.21	41.23	2.329	3.598	2.090
10	59.9	94.916	27.537	8.17	9.81	10.28	66.69	79.59	83.22	0.00	0.00	0.00	48.15	63.22	60.09	18.54	16.37	23.13	2.596	3.862	2.598
11	66.93	94.86	27.497	6.83	8.46	8.89	57.11	73.24	75.34	0.00	0.00	0.00	40.53	56.46	51.15	16.58	16.78	24.19	2.445	3.364	2.115
12	74.56	94.809	27.446	4.83	7.87	8.15	39.21	66.74	74.02	0.00	0.00	0.00	27.51	44.86	50.79	11.71	21.89	23.23	2.350	2.050	2.186
13	81.87	94.762	27.394	3.19	9.19	9.34	25.64	74.49	74.57	0.00	0.00	0.00	14.81	55.97	52.40	10.83	18.52	22.17	1.368	3.022	2.363
14	89.5	94.715	27.34	5.24	8.4	8.52	45.17	69.67	71.56	0.00	0.00	0.00	32.40	48.82	46.67	12.77	20.85	24.89	2.538	2.342	1.875
15	96.99	94.67	27.286	7.73	9.25	10.32	66.72	81.38	87.57	0.00	0.00	0.00	51.90	62.94	62.39	14.81	18.44	25.17	3.503	3.413	2.478
16	104.53	94.625	27.23	7.79	11.12	11.43	70.16	103.31	106.53	0.00	0.00	0.00	53.87	78.94	78.86	16.29	24.37	27.66	3.307	3.239	2.851
17	112.88	94.576	27.169	5.85	7.56	9.29	54.13	66.69	83.16	0.00	0.00	0.00	36.96	49.29	56.33	17.17	17.40	26.83	2.152	2.832	2.099
18	118.77	94.54	27.126	6.83	7.18	8.6	35.92	41.26	47.50	2.38	0.00	0.00	22.95	23.36	35.38	10.59	17.90	12.12	2.166	1.305	2.919
19	123.6	94.508	27.096	9.34	9.84	9.63	31.60	37.06	33.28	14.24	6.84	0.00	10.08	21.60	23.59	7.28	8.62	9.69	1.384	2.505	2.436

(Continued)

TABLE A2.5 (Continued)

Measured Values of Different Morphologic Parameters in 37 Smaller Reaches in the Upper Reach of the Brahmaputra Valley in Assam at Different Times in 1915, 1975, and 2005

Reach No.	Distance (km)	Longitude	Latitude	Width Average (km)			CHB (km²)			Interfluves/Older Flood Plains (km²)			BB Area (km²)			CH (CHB-(IF+BB)) (km²)			BB/CH		
				1915	1975	2005	1915	1975	2005	1915	1975	2005	1915	1975	2005	1915	1975	2005	1915	1975	2005
20	130.39	94.448	27.065	11.57	11.86	12.21	105.15	108.88	114.14	63.38	47.45	21.94	29.32	38.72	67.73	12.46	22.71	24.48	2.353	1.705	2.767
21	138.79	94.375	27.027	15.88	16.24	16.81	141.93	142.45	148.82	86.01	80.96	59.44	39.83	34.34	57.86	16.08	27.14	31.52	2.477	1.265	1.836
22	146.56	94.307	26.992	23.11	21.93	22.54	175.81	170.99	172.96	126.31	112.53	108.73	31.31	22.22	42.84	18.18	36.23	21.39	1.722	0.613	2.002
23	152.37	94.256	26.966	20.95	21.14	20.28	125.74	126.62	119.35	97.56	95.67	97.25	11.91	13.36	11.94	16.27	17.59	10.15	0.732	0.760	1.176
24	157.71	94.21	26.942	19.24	19.68	19.51	120.02	122.71	121.65	63.20	72.78	69.00	41.75	28.62	37.94	15.07	21.32	14.71	2.770	1.343	2.579
25	163.12	94.163	26.917	17.73	18.79	17.75	113.21	121.03	114.81	54.67	50.94	46.05	45.74	46.46	47.06	12.79	23.62	21.70	3.575	1.967	2.168
26	169.51	94.108	26.888	16.67	17.41	17	115.38	124.94	121.78	53.82	33.15	50.97	38.31	58.17	49.99	23.26	33.62	20.82	1.647	1.730	2.400
27	175.37	94.054	26.865	17.54	15.06	17.76	120.36	103.37	125.15	46.59	30.07	39.76	52.99	30.64	59.34	20.77	42.66	26.06	2.551	0.718	2.277
28	181.36	93.998	26.845	16.41	14.84	15.9	121.63	113.26	119.45	63.79	47.71	15.00	41.31	22.61	79.51	16.52	42.94	24.94	2.500	0.527	3.188
29	187.52	93.94	26.824	17.84	15.8	15.15	124.32	98.86	98.83	58.46	24.55	0.08	48.98	48.85	68.30	16.89	25.46	30.46	2.900	1.919	2.242
30	194.22	93.878	26.801	16.01	14.11	13.71	104.64	92.57	92.51	43.30	31.34	0.00	44.75	28.85	71.90	16.58	32.37	20.62	2.700	0.891	3.487
31	201.15	93.814	26.778	13.13	12.79	12.99	93.21	88.91	87.64	16.52	6.50	0.00	58.02	53.87	68.17	18.67	28.54	19.48	3.107	1.887	3.500
32	206.57	93.763	26.759	11.36	6.96	10.85	86.56	54.70	80.34	0.00	0.00	0.00	63.89	29.97	55.94	22.66	24.73	24.40	2.819	1.212	2.293
33	212.64	93.704	26.754	10.34	7.6	11.52	66.46	44.94	72.29	0.00	0.00	0.00	47.97	24.32	55.94	18.49	20.62	16.34	2.594	1.180	3.423
34	218.74	93.642	26.757	8.62	11.09	11.92	62.47	81.33	84.93	0.00	0.00	0.00	43.82	59.26	67.05	18.65	22.07	17.87	2.350	2.685	3.752
35	225	93.58	26.759	6.22	7.79	10.22	47.19	60.32	80.86	0.00	0.00	0.00	24.01	40.48	64.08	23.18	19.84	16.78	1.036	2.040	3.818
36	231.6	93.513	26.762	2.57	6.14	5.93	19.11	43.64	45.04	0.00	0.00	0.00	9.04	27.07	32.71	10.08	16.58	12.33	0.897	1.633	2.653
37	236.86	93.461	26.764	2.72	5.38	4.71	12.90	26.70	25.38	0.00	0.00	0.00	6.50	13.23	15.95	6.40	13.47	9.43	0.983	1.016	1.692

CHB, channel belt area; BB, braid bar; CH, channel area.

TABLE A2.6

Changes in Different Morphologic Parameters of the Upper Brahmaputra within Two Different Time Frames 1915–1975 and 1975–2005

Reach No.	Longitude	Latitude	Distance (km)	Net Change in Average Width (km)		Change in CHB (km²)		Temporal Change in BB Area (km²)		Temporal Change in CH (km²)		Temporal Change in BB/CH Ratio	
				1915–1975	1975–2005	1915–1975	1975–2005	1915–1975	1975–2005	1915–1975	1975–2005	1915–1975	1975–2005
1	95.416	27.795	2.38	10.3	7.95	68.84	55.01	33.38	26.47	6.76	16.76	1.77	-0.59
2	95.377	27.764	7.63	8.91	8.57	45.84	48.74	0.96	29.03	12.56	3.37	-0.86	1.18
3	95.334	27.739	12.65	8.05	10.91	42.71	75.34	14.01	25.41	7.20	9.79	0.59	0.51
4	95.288	27.714	17.88	4.35	13.82	21.19	101.42	12.59	21.47	7.85	8.00	0.19	0.42
5	95.238	27.694	23.35	0.37	11.84	12.56	97.77	6.76	7.04	5.80	26.75	-0.71	-1.41
6	95.178	27.672	29.79	3.44	8.99	33.38	59.49	32.17	31.31	1.22	15.72	2.04	-1.12
7	95.117	27.649	36.32	5.11	5.94	38.44	33.46	37.76	25.87	0.67	7.59	2.06	-0.14
8	95.05	27.625	43.4	3.52	3.83	24.48	21.73	21.08	11.06	3.39	10.67	0.48	-1.03
9	94.986	27.584	51.19	4.07	0.97	38.33	6.89	36.80	-8.13	1.53	15.02	1.27	-1.51
10	94.916	27.537	59.9	1.62	0.42	12.90	3.63	15.07	-3.13	-2.18	6.76	1.27	-1.26
11	94.86	27.497	66.93	1.73	0.32	16.14	2.10	15.93	-5.31	0.21	7.41	0.92	-1.25
12	94.809	27.446	74.56	3.11	0.29	27.53	7.28	17.35	5.93	10.18	1.35	-0.30	0.14
13	94.762	27.394	81.87	4.83	1.27	48.85	0.08	41.15	-3.57	7.69	3.65	1.65	-0.66
14	94.715	27.34	89.5	2.4	0.91	24.50	1.89	16.42	-2.15	8.08	4.04	-0.20	-0.47
15	94.67	27.286	96.99	1.83	0.76	14.66	6.19	11.03	-0.54	3.63	6.73	-0.09	-0.93
16	94.625	27.23	104.53	3.44	0.16	33.15	3.21	25.07	-0.08	8.08	3.29	-0.07	-0.39
17	94.576	27.169	112.88	1.78	1.64	12.56	16.47	12.33	7.04	0.23	9.43	0.68	-0.73
18	94.54	27.126	118.77	0.3	1.422	5.34	6.24	0.41	12.02	7.30	-5.78	-0.86	1.61
19	94.508	27.096	123.6	0.95	-0.76	5.46	-3.78	11.53	1.99	1.35	1.06	1.12	-0.07

(Continued)

TABLE A2.6 (*Cotinued*)

Changes in Different Morphologic Parameters of the Upper Brahmaputra within Two Different Time Frames 1915–1975 and 1975–2005

Reach No.	Longitude	Latitude	Distance (km)	Net Change in Average Width (km)		Change in CHB (km²)		Temporal Change in BB Area (km²)		Temporal Change in CH (km²)		Temporal Change in BB/CH Ratio	
				1915–1975	1975–2005	1915–1975	1975–2005	1915–1975	1975–2005	1915–1975	1975–2005	1915–1975	1975–2005
20	94.448	27.065	130.39	0.24	0.29	3.73	5.26	9.40	29.01	10.26	1.76	-0.65	1.06
21	94.375	27.027	138.79	0.458	0.43	0.52	6.37	-5.49	23.52	11.06	4.38	-1.21	0.57
22	94.307	26.992	146.56	-1.18	-0.9	-4.82	1.97	-9.09	20.62	18.05	-14.84	-1.11	1.39
23	94.256	26.966	152.37	0.28	-0.46	0.88	-7.28	1.45	-1.42	1.32	-7.43	0.03	0.42
24	94.21	26.942	157.71	4.51	-4.97	2.69	-1.06	-13.13	9.32	6.24	-6.60	-1.43	1.24
25	94.163	26.917	163.12	5.65	-5.62	7.82	-6.22	0.73	0.60	10.83	-1.92	-1.61	0.20
26	94.108	26.888	169.51	3.03	-2.544	9.56	-3.16	19.87	-8.18	10.36	-12.79	0.08	0.67
27	94.054	26.865	175.37	-2.14	2.8	-16.99	21.78	-22.35	28.70	21.89	-16.60	-1.83	1.56
28	93.998	26.845	181.36	-1.53	1	-8.37	6.19	-18.70	56.90	26.42	-18.00	-1.97	2.66
29	93.94	26.824	187.52	-1.96	-0.67	-25.46	-0.03	-0.13	19.45	8.57	5.00	-0.98	0.32
30	93.878	26.801	194.22	-1.73	-0.56	-12.07	-0.05	-15.90	43.05	15.80	-11.76	-1.81	2.60
31	93.814	26.778	201.15	-0.62	0.18	-4.30	-1.27	-4.14	14.30	9.87	-9.06	-1.22	1.61
32	93.763	26.759	206.57	-4.43	3.74	-31.86	25.64	-33.93	25.98	2.07	-0.34	-1.61	1.08
33	93.704	26.754	212.64	-2.58	3.66	-21.52	27.35	-23.65	31.62	2.12	-4.27	-1.41	2.24
34	93.642	26.757	218.74	2.59	0.63	18.86	3.60	15.44	7.80	3.42	-4.20	0.34	1.07
35	93.58	26.759	225	1.83	2.41	13.13	20.54	16.47	23.59	-3.34	-3.06	1.00	1.78
36	93.513	26.762	231.6	3.62	-0.25	24.53	1.40	18.03	5.65	6.50	-4.25	0.74	1.02
37	93.461	26.764	236.86	2.6	-0.55	13.80	-1.32	6.73	2.72	7.07	-4.04	-0.03	0.71

CHB, channel belt area; BB, braid bar; CH, channel area.

TABLE A2.7

Gross Temporal Variation in the Geomorphic Parameters (Relic Islands like Majuli and Dibru-Saikhowa Excluded from the Calculation) in Three Different Units of the Brahmaputra River in the Upper Reach of the Brahmaputra Valley in 1915, 1975, and 2005

Segments	Geomorphological Parameters	1915	1975	2005
Unit 1	Average channel belt width (in km)	5.28	10.65 (+102%)	18.48 (+250%)
	CHB (in km²)	358.92	684.69 (+90.8%)	1,184.55 (+230%)
	CH (in km²)	114.61	161.59 (+41%)	275.26 (+140%)
	BB area (in km²)	244.31	439.83 (+80%)	609.37 (+149%)
	CH/CHB	0.32	0.24 (−25%)	0. 23 (−28%)
	BB area/CH	2.13	2.72 (+27.7%)	2.21 (+3.75%)
Unit 2	Average channel belt width (in km)	6.25	8.80 (+40%)	9.42 (+51%)
	CHB (in km²)	460.76	656.38 (+42.4%)	703.46 (+52.7%)
	CH (in km²)	129.29	172.52 (+33%)	209.4 (+62%)
	BB area (in km²)	329.08	483.86 (+47%)	494.06 (+50%)
	CH/CHB	0.28	0.26 (−7%)	0.30 (+7%)
	BB area/CH	2.55	2.80 (+9.8%)	2.36 (−7.4%)
Unit 3	Average channel belt width (in km)	13.93	13.70 (−1.6%)	14.38 (+3.26%)
	CHB (in km²)	1,787.68	1,763.28 (−1.36%)	1,859.22 (+4%)
	CH (in km²)	310.28	480.13 (+54.7%)	373.16 (+20.3%)
	BB area (in km²)	689.53	642.65 (−6.8%)	977.85 (+41.8%)
	CH/CHB	0.17	0.27 (+58. 8%)	0.20 (+17.6%)
	BB area/CH	2.22	1.34 (−39.6%)	2.62 (+18%)

Positive numbers inside the brackets indicate growth over the 1915 status.
CHB, channel belt area; BB, braid bar; CH, channel area.

TABLE A2.8

Temporal Variation in the Parameters like Average Channel Belt Widths, Areas of the Channel Belts, Channels, Sandbars, and the Ratios of the Channel and CHBs, as well as BB Areas and CHs, Are Shown for Three Units (Majuli included) of the Brahmaputra River in the Upper Reach of the Brahmaputra Valley

Units	Geomorphological Parameters	1915	1975	2005	Net Change (1915–1975)	Net Change (1915–2005)
Unit 1	Average channel belt width (in km)	5.28	10.65	18.48	5.37 (+102%)	13.20 (+250%)
	CHB (in km²)	358.92	678.39	1,186.27	319.68 (+89.12%)	827.56 (+230.7%)
	CH (in km²)	122.04	144.13	270.24	22.09 (+18.1%)	148.2 (+121.4%)
	Sandbar area (in km²)	234.08	534.26	916.03	300.18 (+128.24%)	681.95 (+291.32%)
	CH/CHB	0.35	0.21	0.23	−0.14 (−40%)	−0.12 (−34.3%)
	Sandbar area/CH	1.92	3.71	3.39	1.79 (+93.2%)	1.47 (+76.6%)
Unit 2	Average channel belt width (in km)	6.25	8.80	9.42	2.55 (+40 %)	3.17 (+51 %)
	CHB (in km²)	460.32	655.47	698.42	195.15 (+42.4%)	238.1 (+51.72%)
	CH (in km²)	128.33	164.49	205.64	36.16 (+28.3%)	77.31 (+60.2%)
	Sandbar area (in km²)	331.98	490.98	492.77	159 (+47.9%)	160.79 (+48.4%)
	CH/CHB	0.28	0.25	0.29	−0.03 (−10.7%)	0.01 (+3.57%)
	Sandbar area/CH	2.59	2.98	2.40	0.39 (+15.06%)	−0.19 (−7.33%)
Unit 3	Average channel belt width (in km)	13.93	13.70	14.38	−0.23 (−1.6%)	0.45 (+3.26%)
	CHB (in km²)	1,789.16	1,756.32	1,855.49	−32.84 (−1.83%)	66.33 (+3.71%)
	CH (in km²)	300.18	473.22	365.55	173.04 (+57.64%)	65.37 (+21.78%)
	Sandbar area (in km²)	1,489.0	1,283.0	1,490.0	−206 (−13.8%)	1 (+0.06%)
	CH/CHB	0.17	0.27	0.20	0.10 (+58.8%)	0.03 (+15%)
	Sandbar area/CH	4.96	2.71	4.07	−2.25 (−45.36%)	−0.89 (−17.9%)

CHB, channel belt area; BB, braid bar; CH, channel area.

TABLE A2.9

Slope Variation at Different Stretches of the Burhi Dihing River and Temporal Sinuosity Variation for the Corresponding Stretches from 1915 to 2005

Stretch No.	Longitude	Latitude	Elevation AMSL (m)	Stretch Length (km)	Cumulative Stretch Length (km)	Slope (m/km)	Sinuosity (1915)	Sinuosity (2005)
1	95.66	27.28	150.9	0	0			
1	95.56	27.29	141.4	9.96	9.96	0.954	1.41	1.46
2	95.49	27.24	140.2	9.46	19.42	0.127	1.14	1.18
3	95.42	27.25	146.9	7.35	26.77	−0.912	1.91	2
4	95.39	27.29	128.93	4.86	31.63	3.698	1.04	1.03
5	95.37	27.28	123.4	1.87	33.5	2.957	1.05	1.4
6	95.34	27.29	120.7	3.79	37.29	0.712	2.5	1.2
7	95.33	27.32	120.4	3.3	40.59	0.091	1.1	1
8	95.29	27.3	119.8	4.02	44.61	0.149	1.03	1.03
9	95.19	27.35	116.13	11.63	56.24	0.316	1.41	1.09
10	95.14	27.33	113.7	6.55	62.79	0.371	1.55	1.43
11	95.07	27.31	109.1	6.97	69.76	0.660	2.14	1.1
12	95.01	27.38	106.9	9.98	79.74	0.220	1.37	1.51
13	94.92	27.31	103	10.91	90.65	0.357	1.84	1.07
14	94.85	27.28	104.2	8.38	99.03	−0.143	1.85	1.33
15	94.79	27.25	100.3	6.36	105.39	0.6131	1.99	1.9
16	94.71	27.26	96.32	8.64	114.03	0.461	2.53	2.2

Correlation coefficient between the old and the new sinuosity signatures of the Burhi Dihing River: **0.507**. AMSL, above the mean sea level.

TABLE A2.10

Slope Variation at Different Stretches of the Disang River and Temporal Sinuosity Variation for the Corresponding Stretches from 1915 to 2005

Stretch No.	Longitude	Latitude	Elevation AMSL (m)	Stretch Length (km)	Cumulative Stretch Length (km)	Slope (m/km)	Sinuosity (1915)	Sinuosity (2005)
	95.38	27	212.4	0	0			
1	95.37	27.04	202	4.23	4.23	2.459	1.94	2.38
2	95.39	27.05	195.7	2.22	6.45	2.839	1.06	1.15
3	95.38	27.09	164.6	4.07	10.52	7.641	1.13	1.48
4	95.37	27.14	150.6	5.92	16.44	2.365	1.21	1.43
5	95.35	27.17	119.8	3.99	20.43	7.719	1.09	1.14
6	95.32	27.19	117.3	3.22	23.65	0.776	1.11	1.65
7	95.3	27.19	117	2.8	26.45	0.107	1.34	1.46
8	95.27	27.19	116.4	2.96	29.41	0.203	1.79	2.08
9	95.2	27.16	114.9	6.9	36.31	0.217	2	2.43
10	95.16	27.13	111.9	5.31	41.62	0.565	1.94	2.74
11	95.12	27.13	107.3	3.76	45.38	1.223	1.87	1.87
12	95.06	27.09	106.4	7.14	52.52	0.126	2.17	2.65
13	95.03	27.07	107.9	4.12	56.64	-0.365	1.49	1.84
14	94.97	27.06	103.6	6.45	63.09	0.667	1.6	1.95
15	94.95	27.05	99.7	2.57	65.66	1.518	3.33	3.35
16	94.91	27.04	99.1	4.18	69.84	0.144	2.12	2.9
17	94.87	27.04	100.6	3.78	73.62	-0.397	1.74	2.24
18	94.8	26.99	97.5	8.63	82.25	0.359	1.62	1.94
19	94.77	26.98	95.4	3.12	85.37	0.673	1.76	2.09
20	94.74	27	93.9	3.67	89.04	0.409	1.37	1.49
21	94.73	27.04	93.6	4.57	93.61	0.066	1.59	2.02
22	94.67	27.04	96.6	5.39	99	-0.557	2.05	2.29
23	94.59	27.08	92	9.2	108.2	0.500	1.27	1.31
24	94.56	27.08	91.7	2.8	111	0.107	1.88	2.14

The correlation coefficient between the old and the new sinuosity signatures of the Disang River: **0.925**. AMSL, above the mean sea level.

TABLE A2.11

Slope Variation at Different Stretches of the Dikhau River and Temporal Sinuosity Variation for the Corresponding Stretches from 1915 to 2005

Stretch No.	Longitude	Latitude	Elevation AMSL (m)	Stretch Length (km)	Cumulative Stretch Length (km)	Slope (m/km)	Sinuosity (1915)	Sinuosity (2005)
	94.72	26.62	292.6	0	0			
1	94.78	26.66	261.8	7.19	7.19	4.284	1.05	
2	94.78	26.7	299	4.47	11.66	-8.322	1.18	
3	94.84	26.72	179.2	5.81	17.47	20.620	1.31	
4	94.81	26.77	231.6	6.13	23.6	-8.548	1.22	
5	94.81	26.79	178	2.75	26.35	19.491	2.35	
6	94.8	26.86	93.3	7.63	33.98	11.101	1.95	1.97
7	94.74	26.92	97.2	8.88	42.86	-0.440	2.2	2.4
8	94.67	26.98	97.2	8.82	51.68	0	1.7	1.68
9	94.62	26.98	105.2	5.52	57.2	-1.450	1.4	1.26
10	94.57	26.98	96.9	4.71	61.91	1.762	2.05	1.57
11	94.57	26.96	93	2.12	64.03	1.840	2.68	2.6
12	94.54	26.95	89.6	3.48	67.51	0.977	1.43	1.58
13	94.47	27	86.3	9.03	76.54	0.365	2.83	2.39

The correlation coefficient between the old and the new sinuosity signatures of the Dikhau River: **0.883**.
AMSL, above the mean sea level.

TABLE A2.12

Slope Variation at Different Stretches of the Jiya Dhol River and Temporal Sinuosity Variation for the Corresponding Stretches from 1915 to 2005

Stretch No.	Longitude	Latitude	Elevation AMSL (m)	Stretch Length (km)	Cumulative Stretch Length (km)	Slope (m/km)	Length (1915) (km)	Length (2005) (km)	Sinuosity (1915)	Sinuosity (2005)
	94.29	27.61	409.96	0	0					
1	94.34	27.62	313.03	5.08	5.08	19.081	5.87	6	1.156	1.181
2	94.42	27.64	213.97	8.53	13.61	11.613	8.98	12.62	1.053	1.480
3	94.44	27.61	261.52	3.41	17.02	−13.944	3.62	5.78	1.062	1.695
4	94.45	27.57	139.3	4.62	21.64	26.455	4.71	5.08	1.020	1.100
5	94.49	27.45	102.1	14.21	35.85	2.618	22.18	15.58	1.561	1.100
6	94.44	27.31	94.49	15.92	51.77	0.478	27.79	34.83	1.746	2.188
7	94.41	27.26	93.57	6.26	58.03	0.147	9.04	7.22	1.444	1.153
8	94.43	27.25	89	1.87	59.9	2.444	3.54	4.1	1.893	2.193
9	94.31	27.16	85	15.05	74.95	0.266	35	28.36	2.326	1.884
10	94.3	27.09	86.9	7.95	82.9	−0.239	13.24	11.33	1.665	1.425

Correlation coefficient between the old (1915) and the new (2005) sinuosity signatures of the Jiya Dhol River: 0.580.
AMSL, above the mean sea level.

TABLE A2.13

Slope Variation at Different Stretches of the Sisi River and Temporal Sinuosity Variation for the Corresponding Stretches from 1915 to 2005

Stretch No.	Longitude	Latitude	Elevation AMSL (m)	Stretch Length (km)	Cumulative Stretch Length (km)	Slope (m/km)	Length 1915 (km)	Length 2005 (km)	Sinuosity (1915)	Sinuosity (2005)
0	94.71	27.75	344.4	0	0					
1	94.68	27.71	236.2	5.45	5.45	19.853	5.65	7.79	1.037	1.430
2	94.69	27.66	140.5	5.99	11.44	15.977	7.06	12.02	1.179	2.007
3	94.63	27.43	99.67	25.64	37.08	1.592	42.89	31.61	1.673	1.233
4	94.54	27.27	93.88	20.12	57.2	0.288	34.63	27.45	1.721	1.364
5	94.48	27.24	91.74	7.02	64.22	0.305	12.9	17.88	1.838	2.547
6	94.41	27.18	89	9.09	73.31	0.301	10.46	12.86	1.151	1.415
7	94.36	27.16	86.26	5.87	79.18	0.467	9.59	7.82	1.634	1.332
8	94.31	27.16	85.34	4.51	83.69	0.204	7.88	7.9	1.747	1.752

The correlation coefficient between the old (1915) and the new (2005) sinuosity signatures of the Sisi River: **0.220**. AMSL, above the mean sea level.

TABLE A2.14

Slope Variation at Different Stretches of the Simen River and Temporal Sinuosity Variation for the Corresponding Stretches from 1915 to 2005

Stretch No.	Longitude	Latitude	Elevation AMSL (m)	Stretch Length (km)	Cumulative Stretch Length (km)	Slope (m/km)	Length (1915) (km)	Length (2005) (km)	Sinuosity (1915)	Sinuosity (2005)
0	94.79	27.98	1,119.2	0	0					
1	94.88	28.00	707.1	8.87	8.87	46.460	9.86	12.67	1.112	1.428
2	94.96	27.98	329.2	7.58	16.45	49.855	8.87	9.64	1.170	1.272
3	94.92	27.89	231.9	10.72	27.17	9.077	11.52	11.8	1.075	1.101
4	94.92	27.85	210.3	3.78	30.95	5.714	7.55	8.51	1.997	2.251
5	94.87	27.74	147.8	13.76	44.71	4.542	17.4	20.34	1.265	1.478
6	94.87	27.64	114.3	10.89	55.6	3.076	24.27	14.06	2.229	1.291
7	94.84	27.58	107.3	7.14	62.74	0.980	13.08	9.32	1.832	1.305
8	94.82	27.53	103.3	6	68.74	0.667	7.19	7.55	1.198	1.258

The correlation coefficient between the old (1915) and the new (2005) sinuosity signatures of the Simen River: 0.439. AMSL, above the mean sea level.

TABLE A2.15

Slope Variation at Different Stretches of the Dikari River and Temporal Sinuosity Variation for the Corresponding Stretches from 1915 to 2005

Stretch No.	Longitude	Latitude	Elevation AMSL (m)	Stretch Length (km)	Cumulative Stretch Length (km)	Slope (m/km)	Length (1915) (km)	Length (2005) (km)	Sinuosity (1915)	Sinuosity (2005)
0	95.09	27.86	347.5	0	0					
1	95.11	27.83	149.3	2.65	2.65	74.792	2.85	3.85	1.075	1.453
2	95.11	27.82	132	1.46	4.11	11.849	1.61	1.88	1.103	1.288
3	95.1	27.8	131.4	1.64	5.75	0.366	2.2	1.98	1.341	1.207
4	95.1	27.78	126.2	2.2	7.95	2.364	2.69	3.35	1.223	1.523
5	95.1	27.75	118.3	3.49	11.44	2.264	3.96	6	1.135	1.720
6	95.12	27.72	114	4.38	15.82	0.982	10.85	4.71	2.477	1.075

The correlation coefficient between the old (1915) and the new (2005) sinuosity signatures of the Depi River: **−0.68226**. AMSL, above the mean sea level.

TABLE A2.16

Slope Variation at Different Stretches of the Depi River and Temporal Sinuosity Variation for the Corresponding Stretches from 1915 to 2005

Stretch No.	Longitude	Latitude	Elevation AMSL (m)	Stretch Length (km)	Cumulative Stretch Length (km)	Slope (m/km)	Length (1915) (km)	Length (2005) (km)	Sinuosity (1915)	Sinuosity (2005)
0	95.17	28	370.9	0	0					
1	95.18	27.96	282.8	3.86	3.86	22.824	4.12	4.9	1.067	1.269
2	95.24	27.97	182	6.13	9.99	16.444	6.5	7	1.060	1.142
3	95.29	27.9	140.2	8.59	18.58	4.866	10.15	9.67	1.182	1.126
4	95.33	27.86	130.4	5.68	24.26	1.725	8.54	7.21	1.504	1.270
5	95.34	27.83	128	3.72	27.98	0.645	4.87	5.25	1.309	1.411

Correlation coefficient between the old (1915) and the new (2005) sinuosity signatures of the Remi River: 0.453103.
AMSL, above the mean sea level.

TABLE A2.17

Slope Variation at Different Stretches of the Silli or Leko Jan River and Temporal Sinuosity Variation for the corresponding Stretches from 1915 to 2005

Stretch No.	Longitude	Latitude	Elevation AMSL (m)	Stretch Length (km)	Cumulative Stretch Length (km)	Slope (m/km)	Length (1915) (km)	Length (2005) (km)	Sinuosity (1915)	Sinuosity (2005)
0	95.09	27.95	446.8	0	0					
1	95.14	27.93	292.6	5.44	5.44	28.346	7.26	5.89	1.335	1.083
2	95.18	27.92	235.9	4.22	9.66	13.436	4.46	4.94	1.057	1.171
3	95.22	27.91	184.1	4.71	14.37	10.998	5.7	6.08	1.210	1.291
4	95.27	27.87	135	6.73	21.1	7.296	8.64	7.72	1.284	1.147
5	95.26	27.84	126.8	2.73	23.83	3.004	4.07	5.42	1.491	1.985
6	95.28	27.82	125.3	2.78	26.61	0.540	3	4.02	1.079	1.446
7	95.28	27.79	120.4	3.46	30.07	1.416	3.91	4.6	1.130	1.329
8	95.25	27.76	117.6	3.8	33.87	0.737	5.74	5.86	1.511	1.542
9	95.23	27.76	115.8	1.98	35.85	0.909	2.04	2.19	1.030	1.106
10	95.21	27.74	114.9	3.36	39.21	0.268	5.25	5.26	1.563	1.565

The correlation coefficient between the old (1915) and the new (2005) sinuosity signatures of the Silli or Leko Jan River: **0.626533**. AMSL, above the mean sea level.

TABLE A2.18

Slope Variation at Different Stretches of the Remi River and Temporal Sinuosity Variation for the Corresponding Stretches from 1915 to 2005

Stretch No.	Longitude	Latitude	Elevation AMSL (m)	Stretch Length (km)	Cumulative Stretch Length (km)	Slope (m/km)	Length (1915) (km)	Length (2005) (km)	Sinuosity (1915)	Sinuosity (2005)
0	95.05	27.96	446.8	0	0					
1	95.07	27.95	366.4	2.73	2.73	29.451	3.64	4.22	1.333	1.546
2	95.06	27.93	308.4	2.8	5.53	20.714	3.59	4.15	1.282	1.482
3	95.06	27.9	252.9	3.22	8.75	17.236	3.85	5.74	1.1957	1.783
4	95.02	27.86	208.8	5.31	14.06	8.305	7.02	8.43	1.322	1.588
5	95.04	27.81	133.8	6.66	20.72	11.261	8.98	13.23	1.348	1.986
6	95.04	27.76	120.7	5.25	25.97	2.495	8.32	6.1	1.585	1.162
7	95.01	27.71	108.2	6.45	32.42	1.938	12.17	7.87	1.887	1.220
8	94.99	27.68	107.9	3.83	36.25	0.078	4.76	5.49	1.243	1.433
9	94.97	27.65	107.3	4.02	40.27	0.149	6.77	5.78	1.684	1.438

The correlation coefficient between the old (1915) and the new (2005) sinuosity signatures of the Dikari River: −0.62808. AMSL, above the mean sea level.

TABLE A2.19

Slope Variation at Different Stretches of the Subansiri River and Temporal Sinuosity Variation for the Corresponding Stretches from 1915 to 2005

Stretch No.	Longitude	Latitude	Elevation AMSL (m)	Stretch Length (km)	Cumulative Stretch Length (km)	Slope (m/km)	Channel Length (km)			Sinuosity (Channel Length/Valley Length)		
							1915	1975	2005	1915	1975	2005
0	94.22	27.99	238	0	0							
1	94.26	27.99	230	3.96	3.96	2.02	3.96	4.10	4.75	1	1.04	1.20
2	94.31	27.93	223	8.46	12.42	0.83	8.53	9.44	11.30	1.01	1.12	1.33
3	94.32	27.86	220	7.55	19.97	0.40	8.45	7.93	7.95	1.12	1.05	1.05
4	94.34	27.81	216	6.10	26.07	0.66	7.50	7.74	8.09	1.23	1.27	1.33
5	94.36	27.73	178	8.79	34.85	4.33	13.55	13.72	14.05	1.54	1.56	1.60
6	94.31	27.69	165	6.08	40.93	2.14	6.82	7.03	6.77	1.12	1.16	1.11
7	94.23	27.63	146	11.21	52.15	1.69	11.33	11.46	11.49	1.01	1.02	1.02
8	94.23	27.6	132	3.23	55.38	4.33	4.97	4.81	5.00	1.54	1.49	1.55
9	94.24	27.57	121	3.28	58.66	3.35	4.28	4.25	4.30	1.30	1.29	1.31
10	94.26	27.54	120	4.25	62.91	0.24	4.60	4.63	4.71	1.08	1.09	1.11
11	94.28	27.54	99	9.93	72.84	2.12	11.20	10.36	10.73	1.13	1.04	1.08
12	94.26	27.37	97	9.04	81.88	0.22	16.54	10.14	9.88	1.83	1.12	1.09
13	94.27	27.29	91	8.74	90.62	0.69	10.83	9.20	11.10	1.24	1.05	1.27
14	94.26	27.21	89	8.50	99.11	0.24	16.44	14.43	9.30	1.94	1.70	1.09
15	94.21	27.13	86	9.03	108.14	0.33	11.52	13.79	9.03	1.28	1.53	1.00
16	94.15	27.08	81	7.27	115.41	0.69	11.34	18.17	7.55	1.56	2.50	1.04
17	94.12	27.03	85	6.68	122.09	-0.60	18.04	19.74	7.00	2.70	2.96	1.05
18	94.07	26.98	83	7.29	129.38	0.27	10.22	14.09	7.56	1.40	1.93	1.04
19	94.02	26.95	82	5.92	135.30	0.17	10.99	7.71	7.06	1.86	1.30	1.19
20	93.97	26.93	81	5.71	141.01	0.18	9.93	7.35	5.71	1.74	1.29	1.00
21	93.85	26.9	80	8.48	149.49	0.12	8.93	12.02	12.94	1.05	1.42	1.53
22	93.84	26.86	79	6.45	155.94	0.15	7.71	7.35	10.12	1.19	1.14	1.57

AMSL, above the mean sea level.

TABLE A2.20

Some of the Earthquakes of Magnitude 5+ Which Took Place within the Study Period 1915 to 2005 and the Epicentres of Which Were Located Very Close to the Upper Reach of the Brahmaputra Valley Area

Serial No.	Longitude	Latitude	Elevation (m)	AMSL	Date of Occurrence	Place	Focal Depth (km)	Magnitude (*Mw*)
1	96.700	28.500	4,310	15 August 1950	Arunachal Pradesh	25.0	8.7
2	95.921	28.642	1,352	01 March 1987	Arunachal Pradesh	33.0	5.1
3	96.036	28.323	1,792	21 February 1985	Eastern Xizang–India border region	33.0	5.4
4	95.900	28.231	2,499	10 June 1975	Arunachal Pradesh	26.0	5.1
5	95.100	28.600	1,129	08 October 1963	Arunachal Pradesh	24.0	5.4
6	94.919	27.761	136	26 November 1982	Arunachal Pradesh–Assam region	33.0	5.1
7	94.367	28.434	2,479	11 March 1967	Arunachal Pradesh	12.0	5.3
8	94.318	28.458	3,590	14 March 1967	Arunachal Pradesh	12.0	5.8
9	94.000	28.500	3,060	29 July 1947	Arunachal Pradesh	-	7.9
10	93.800	28.100	1,600	21 October 1964	Eastern Xizang–India border region	37.0	5.9

(Continued)

TABLE A2.20 (*Continued*)

Some of the Earthquakes of Magnitude 5+ Which Took Place within the Study Period 1915 to 2005 and the Epicentres of Which Were Located Very Close to the Upper Reach of the Brahmaputra Valley Area

Serial No.	Longitude	Latitude	Elevation (m) AMSL	Date of Occurrence	Place	Focal Depth (km)	Magnitude (Mw)
11	93.990	27.396	917	19 February 1970	Arunachal Pradesh	18.0	5.5
12	92.893	26.894	89	20 February 1983	Assam	33.0	5.2
13	93.187	26.637	68	23 June 1991	Assam	33.20	5.3
14	93.414	26.689	81	06 September 1987	Assam	43.80	5.1
15	93.114	26.471	142	13 November 1972	Assam	33.0	5.0
16	93.231	26.469	282	17 July 1971	Assam	49.0	5.3
17	93.318	26.437	608	30 April 1990	Assam	33.0	5.3
18	93.499	26.345	310	02 December 1998	Assam	33.0	5.0
19	95.045	26.864	142	23 July 1986	Myanmar–India border region	33.0	5.4

Source: USGS.

These earthquakes are supposed to restructure the mountainous reach as well as regulate the sediment yield for the mountain-fed rivers. AMSL, above the mean sea level.

Glossary

Accommodation space: The space available for the potential deposition of sediments. In the marine condition, it is a combined effect due to the sea surface (rise or fall), seafloor (upliftment or subsidence), and changes in the rates of sediment accumulation. For the continents, this is the space available for sediment deposition controlled by the regional base level and the local base level. Changing slopes, discharge, and sediment load of the rivers play significant roles in determining the potential accommodation space. For landforms having a high degree of tectonic control, the accommodation space can also be related to morphotectonic variability.

Accretionary complex: Usually along the convergent plate margins, sediments keep on piling up, which are scraped off from the subducting plate, and it geometrically appears to resemble a wedge or a prism and is constituted of diverse materials like turbidites of terrestrial origin as well as sediments of marine origin and thus forms a complex.

Amplitude standout: Geophysical response due to a targeted body (which needs to be located or mapped) is mostly associated with unwanted signals from different sources of noise. The difference in the amplitudes of the targeted body and the background noise is usually referred to as the amplitude standout. Sometimes when the amplitude standout is too small, repetitive data acquisition is carried out. If the signal is having phase consistency, an additive method is applied, which amplifies the in-phase signals and cancels the out-of-phase noise. As a result, amplitude standout increases.

Anabranching: When rivers in certain reaches develop branches to join subsequently in the downstream side of the main flow, the process is called anabranching. For braided rivers, this process helps develop islands having different degrees of stability.

Anastomosing pattern: This is a type of anabranching that happens in low-energy conditions, where channels mostly carry suspended load. If the valley relief is constituted of fairly compact tracts, due to lower stream power, a channel, instead of eroding the older flood plain, is observed to flow through multiple weaker paths and form a netlike pattern.

Antecedent: An old river with considerably high stream power while interacting with an uplifting landform; it keeps on maintaining its old course by continuously cutting it, and we get an antecedent river. In the present-day relief, these types of channels are seen to cut through the younger mountains. Siang is an antecedent river.

Anticlinal ridges: Surface expression of deep structural folding in the form of anticlinal hills.

Assam–Arakan basin: Subsurface geophysical data show that the upper reach of the Brahmaputra valley, Naga-Patkai Schuppen belt, and the Arakan–Yoma fold belt shared a continuity in the past (till early Miocene or say around 20 million years back when the Himalayas started rising very fast), which can be called as the shelf-slope-basinal system, which is commonly referred by the geoscientists in the oil industries as the Assam–Arakan basin. There is no such connectivity in the present-day surface relief.

Avulsion: When a river suddenly changes its course, it is called avulsion, which can be co-seismic due to a sudden change in the tilt of the valley floor or other reasons involving changes in the base level of the valley.

Bank-line shift: When a river erodes its bank, what is observed in the plan-form is bank-line shift.

Basement complex: Sedimentary basins are usually supposed to have different layers of soft rocks and the fluids inside the pore spaces stacked over the crystalline metamorphic hard rocks called basement, which for practical purposes is very compact having an insignificant amount of fluids. Since the origin of the basement can be diverse, its composition is also highly variable, hence the suffix *complex*.

Basin-forming tectonics: The sediment architecture of a basin is controlled by the kind of tectonic framework it belongs to, which is broadly either convergent or divergent. Convergent and divergent margins are found on both the continental and oceanic crustal plates. Next the location of a basin on the plate (at the interior or margin of the plate) and the type of structural movements (like sagging, normal faulting, or wrench faulting). These issues give a few type of situations interpreted by geoscientists principally from the seismic sections and the geophysical log data to narrate the history of the evolution of different basins in terms of eustasy, rate of sediment influx, and the tectonic controls. All these factors are clubbed up as basin-forming tectonics.

Basin-modifying tectonics: Above the basement up to the surface, a basin can either keep on forming continuously, or at different levels, it can have unconformities. Across the unconformities, basins seem to follow different histories of evolution. Accordingly, basins can have mono-history or poly-history. Since unconformities represent time gaps due to prolonged non-deposition or erosion of sediments, it can be generalized that a basin was structurally modifying itself so that the next stage of basin formation could start. This is known as basin-modifying tectonics. Kingston et al. (1983) suggested three types of basin-modifying tectonics: episodic wrenches, adjacent fold belts, and complete folding of a basin area.

Bed–bank relationship: The average elevation of the principal river bed of a valley can be treated as the local base level for the given reach. Depending upon the slope of the river bed, load carried by the river, and its stream power, sedimentary layers form *continuously* on the river bed. Banks representing flood plains form *episodically* during floods only. Depending upon the thickness differences of the new layers deposited on the older river bed and the bank, availability of the accommodation space and the corresponding bed–bank relationship can have highly variable possibilities.

Blind fault-related folds: During the development of blind faults, if the top layers are sufficiently ductile, first bending and then folding take place.

Blind faults: Subsurface thrust faults that are yet to reach the surface to become visible are called blind faults. As these faults remain 'buried', during field visits, geologists fail to identify them. Sometimes, by indirect evidence like the flow geometry of the rivers, blind faults can be suspected and then by conducting geophysical investigations validated.

Bouguer gravity anomaly: Gravity anomaly measured by using a gravimeter for a station involves a cumulative anomaly due to different factors like latitude, elevation, and unevenness of terrain in addition to the target, the principal object. The corrections made to the measured gravity anomaly for all the factors, except the target, is called Bouguer gravity anomaly or simply Bouguer gravity.

Cascade flow: Flow of fluids through a uniform slope can be treated equivalent to a succession of steps of very small widths, which is advantageous to apply some of the mathematical tools.

Catchment: The area on the landscape where all natural sources of water are collected. Generally, the water of a catchment never crosses the catchment boundary.

Channel capture: In geomorphology, this is also known as river capture, stream capture, river piracy, or stream piracy. This happens when a channel or a channel system or a watershed is diverted from its bed and starts flowing through the bed of a neighbouring bed. This can happen due to many reasons, like tectonic tilt after a major earthquake, large-scale glacier retreat and formation of large lakes in the high-altitude areas, and, subsequently, glacial lake outburst-related floods, rapid land erosion caused by rivers due to faster rate of aggradation, etc.

Channel cut-offs: Channels, particularly of meandering type, are occasionally observed to follow short-cut paths when discharge volume increases, or due to very recent tectonic activities, valley slope changes, which are called cut-offs. This process leaves behind oxbow lakes.

Channel-belt: Braided rivers seem to flow through several inter-connected channels, which during the bank-full condition only, seems to flow as one. Thus, the width of a highly braided river as seen in the satellite imageries can be treated as the width of the channel belt, which is equal to the width of the river under bank-full conditions. Thalweg of such a highly braided channel belt can be assumed to be represented by the median path of the widest channel (not the median path of the river under bank-full conditions), which can also be treated as the deepest.

Coda waves: Coda waves in the seismograms are usually referred to as the tail parts of the prominent wave types. For example, in the later arrivals of P and S, the two prominent body waves, will have P-Coda and S-Coda. Analysis of these waves is particularly interesting to study the lateral heterogeneity of the earth. A quality index of the coda waves 'Qc' is supposed to have higher values for the lesser decay. The unconsolidated material highly fractured or otherwise is supposed to cause a greater degree of attenuation of the waves, and that is why it will show lower values of 'Qc'.

Co-seismic subsidence: In tectonically controlled valleys, coinciding with some of the large earthquakes, many rivers might be observed to have course corrections along a particular direction, indicating a zone of subsidence.

Depth index: This is a reach-scale bed–bank elevation difference status of a river compared to the average for the entire length of observation of the river to the difference between the maximum and the minimum values. Expressed by a number without any unit, it helps offer a normalized depth index to present temporal comparability.

$$\text{Depth Index} = \frac{\text{Reach depth} - \text{Average depth}}{\text{Maximum depth} - \text{Minimum depth}}$$

Drill-barrel-sell policy: This is an expression to describe highly irresponsible profit-maximizing oil industrial policy which promotes seeing oil fields as localized pools. As a result, system science approach-based research needed to develop the regional perspective to connect different elements of the tectono-sedimentary history of the evolution of a basin is neglected.

En echelon: Derived from French, *en* which stands for 'upon' and *echelon* for 'stepwise arrangement'. It describes often parallel or subparallel, steplike structural features in outcropping rocks such as reverse faults due to thrusting.

Fast Fourier Transform (FFT): Fourier transform (FT) converts a signal in the time or space domain to the frequency domain. Discretization helps express a continuous signal into a series of numbers. Discrete

FT decomposes a sequence of numbers (or values) into components of different frequencies. FFT is an algorithm to compute discrete FT at a much faster rate by replacing an operation of the type 'N^2' to an operation '$N \log N$' type, where N is the data size. Applications of the FFT cover diverse fields, like engineering, music, science, and mathematics. FFT can also be used to better understand some of the geomorphological problems like identifying controlling factors behind bank-line migration of large rivers by relating different scales of controls in terms of frequencies and measuring planform bank migration for uniform reaches during specific time intervals.

Flood characterization: Flood characterization essentially means reach-scale spatio-temporal variability study in the bed–bank relationship of a river, which due to changing hydro-sedimentary budgeting, makes different reaches of a valley susceptible to inundation due to the 'excess water' over the 'bank-full condition'.

Flood disaster incubation: Flood incubation is the growth potentiality of flood for a given reach of the river valley, which is usually caused by the raising of embankments, which arrests subsequently the formation of riverbank and accelerates the formation of the river bed, thereby reducing the effective accommodation space and also subsequently altering the local base level.

Flood triggering: When due to raising of embankments, bed–bank relationship changes so much so that the normal rise in the river water level which never caused flood earlier start causing the possibility of flash floods mostly due to the breaches in the embankments or due to malfunctioning of other anthropogenic structures like dam failures, and suddenness in the occurrence of such floods is usually attributed as flood triggering.

Freeze–thaw cycle: A process of physical weathering which usually takes place in the high mountains, where during daytime, water enters into the cracks or joints of the rock masses and, during night-time due to fall in temperature, freezes to become ice. Since a given volume of ice is more than the original volume of water, cracks keep on expanding with every cycle and breaks eventually.

Geosynclinal theory: The geosynclinal theory proposes essentially a system that is the aggregate of alternating rises (elementary geanticlines) and sags (elementary geosynclines) within which the geosynclinal process is realized. By the geosynclinal process, what is meant is the formation of the granitic rocks.

Incised valley deposits: When the sea level falls considerably and shore-line regresses to expose the shelf, rivers run extra miles to join the sea, and that additional path is travelled by incising the older marine sediments of the shelf and depositing coarser channel sands. Afterwards, when the sea level rises again, the coastline transgresses towards the continent, and the channel sands get covered with finer marine sediments. This causes an excellent reservoir condition for hydrocarbons which, after being cooked inside the source rocks, could start migrating, make entry into the incised valleys, and get entrapped. This makes the identification of incised valley deposits in the ancient sediment units at depths by using geophysical methods a challenging task.

Indentation model: To explain the aftermath of the collision of the India plate and the Eurasia plate, the indentation model suggests that the Indian lithosphere acted as a rigid indenter, resulting in inhomogeneous thickening in Tibet and areas to the north. It essentially goes against the process of subduction and crustal shortening of the India plate and is currently treated as an obsolete model.

Inselbergs: Coming from the German route (*Insel* for 'island' and *Berg* for 'mountain'), it essentially refers to isolated mountains standing abruptly above the well-developed plains.

Interior sag: Basins located on the interior part of the continental crust in the areas of plate divergence show a simple sagging characteristic, which has sediment deposition thicker and deeper at the central part. No major faults or structural movements are observed.

Inter-seismic uplift: If earthquakes are due to sudden stress release, then it can be assumed that between two earthquakes, stress accumulates, and if the layer lying above the zone of accumulation of stress is sufficiently ductile, it can show upliftment. Since the rate of upliftment is very slow, it was difficult to measure earlier. However, recent evidence shows that in some of the convergent basin margins where active thrusting is present, inter-seismic uplift has been measured.

Mantle delamination and rollback: To explain plate convergence, this model suggests a process of slab subduction. Since the crustal materials are much lighter than mantle materials, the former tends to buoy. Accordingly, it suggests mantle delamination along the plane of subduction and subsequent 'roll back'. There are other alternative models, and it is very difficult to confirm with the available geophysical methods which one fits best for a given situation.

Margin sag–interior sag: These are currently interior sag basins having a high degree of asymmetry (by which these are sometimes separated from the conventional interior sag basins) and a previous cycle of margin sags. Earlier, these basins used to be located near the margin of the plate boundary. Afterwards, due to the formation of a fold belt (orogeny) in the seaward side, it was converted into an interior basin.

Margin sag: This is a type of basin which is located on the outer edges of continental crust blocks in areas of divergence. Most of the margin sag basins studied so far are having at least two basin-forming tectonic origins and are poly-history basins. Margin sag basins are broadly divided into four types: normal clastic, carbonate bank, major delta, and salt tectonics.

Meander cut-offs: See 'channel cut-offs'.

Method of least squares: Between two sets of correlatable variables, if we know which one is the independent variable, then the respective changes can be plotted, which will give us a distribution pattern. Next, we can have a straight line through this maze of points to assign a simple 'characteristic' to this distribution. There are many possibilities of slope variations as well as the magnitudes of intercepts. The distance by which any arbitrary point lies away from the line is the amount of 'misfit'. If we square all these misfits and sum up, the situation for which we get the least value is taken as the best-fit solution for the problem. This is known as the method of least squares. There can be ordinary least squares (linear) or nonlinear least squares.

Mountain-fed rivers: Rivers based on their origin and the sediment load are classified by different authors for different places. One classification (Sinha and Friend, 1994) meant principally for the Indo-Gangetic plains proposed four systems: (i) mountain-fed, (ii) foothill-fed, (iii) plain-fed, and (iv) mixed-fed rivers. The mountain-fed rivers are characterized by very high discharge and low suspended load concentration, and the plains-fed rivers have low discharge and low sediment load concentration. The foothill-fed rivers have moderate values of discharge and suspended load concentration.

Orographic precipitation: When moist air is forced upward due to an upland, it expands due to pressure depletion and experiences adiabatic cooling. When the dew point is reached, the condensation of water vapour causes the formation of clouds. When the water droplets become large enough, there is precipitation. Usually, it is observed that the windward side of the high mountain belts has a considerable amount of orographic precipitation, promoting rich vegetation, whereas on the leeward side of the mountains, there exist desert-like conditions.

Oxbow lakes: See 'channel cut-offs'.

Periglacial solifluction: By the impacts due to periglacial environments, usually those landforms situated at the edges of glacial areas are identified, which are significantly modified due to the freeze–thaw cycle. The movements of masses down the slope in such regions (mass wasting), the periglacial solifluction, can be slow as well as rapid and is principally restricted to cold climates.

Polynomial fit: In empirical studies, if the observer suspects possibilities of certain kinds of relations between two parameters, then by dividing the study area into several smaller reaches, data can be generated by meticulous measurements of those parameters and subsequently plotted by putting the independent variable in the x-axis and the dependent variable on the y-axis. Then, polynomial equations of the type $Y = A + BX + CX^2 + \ldots$ can be fitted through the maze of measured values. Despite many limitations, it helps frame tentative 'laws' which can be refined afterwards.

Pop-up theory: This is a theory of orogeny applicable for a class of hills or mountains, identified sometimes as inselbergs, which were found earlier incompatible with some of the major structural elements in the neighbourhood. For example, the explanation of genesis for the Shillong massif given by Bilham and England (2001) is based on uneven sediment loading across a zone of weakness, which ultimately triggers a block upliftment referred to as 'pop-up'.

Proto-Brahmaputra: There are indications from the subsurface studies of geophysical logs and seismic sections in the lower part of the Brahmaputra River [e.g., studies by Ashraf Uddin and Neil Lundberg (1999)] that during Miocene time, there used to be a Brahmaputra-like river flowing through the eastern part of the present-day Shillong massif or perhaps as the different Yarlung Tsangpo–Irrawaddy system (Robinson et al., 2014), which resembled the modern Brahmaputra and is referred as the proto-Brahmaputra.

Reflection coefficient: Imaging the subsurface stratigraphy by seismic reflection methods is principally mapping of boundaries between two rock units depending on the reflectivity or reflection coefficient of the boundary, which is expressed as the ratio of the difference in the acoustic impedance (expressed by the product of the seismic velocity through the medium and its density) of the layer lying above from that of the below divided by the addition of their acoustic impedance values. Generally, with depth, both density and velocity keep on increasing. As a result, the reflection coefficient value is positive. However, for certain situations like high- to medium-porosity gas sands, the reflection coefficient can be negative. This causes phase reversal of the reflected waves. Moreover, this can cause the appearance of a 'bright spot' (or, 'dim spot' depending upon the convention followed like either crest or trough of the reflected seismic waves blackened) over the gas-bearing zones.

Relict island: When the river bank line keeps on migrating, if certain stable landforms are bypassed, and thereby causing the formation of an island since the landform is older in age than the new location of the river; it is called the relict island. Particularly for big braided rivers, some of the recently formed braid bars can have a comparable size to that of the relict islands, which may create certain confusion. During bank-full conditions, usually, the braid bars are completely submerged, but the relict islands do not. While making estimates for the aggradation over time by monitoring the planform changes in the ratio of braid bars and the channel areas, the areas of relict islands should preferably be excluded from the analysis.

Residence time: Rivers keep on changing courses. How long a river will stay at a place depends on many factors and is known as the residence time of the river. In seismic sections running across the river valleys, when sediments of fluvial origin are marked while proceeding along the lateral direction, measurement of the variation in thickness gives us an understanding of the variability of the residence time of a much older river. It becomes an interesting field of research to explore factors responsible for the variability in residence time during a given time interval.

River-linking project: Projects related to deviating channels from the water-surplus provinces to water-scarce provinces are known as the river-linking projects. So far, most of these projects seem to be driven by too simplistic assumptions and popular demands and after execution found to trigger up different types of ecological crises, which were neither anticipated nor incorporated within the cost–benefit analysis based on which the projects were given administrative clearance.

Sheet erosion: Instead of moving through particular channels, when water moves more or less uniformly with a similar thickness over a surface, it is called sheet flow, and the even erosion of the substrate due to the sheet flow is called sheet erosion. Transport by the sheet flow can take place very frequently over a small distance only, and that is why, sheet erosion is treated as a low-magnitude process in contrast to that with sheet floods, which are usually high magnitude, less frequent, and turbulent (sheet flow can be both laminar or turbulent).

Sinuosity versus slope: If a river is divided into several smaller reaches (selection of reach size can be uniform or non-uniform, to be selected by the investigator based on certain practical logic) and then slope and sinuosity values are computed and, subsequently, slope (in the x-axis) versus sinuosity (in the y-axis) values are plotted, a typical channel signature can be generated from the best-fit curve. Moreover, for different time intervals, how the morphological characteristics were changing can also be demonstrated and studied further. During a particular time, valleys having some meandering channels flowing side by side can also be subjected to an inter-river quantitative comparison.

Sinuosity: Sinuosity (usually expressed by a number called sinuosity index) is the ratio of the actual river length divided by the straight valley length. Thus, it is a measure of how curvaceous a channel is. The value cannot be less than 1. Generally, if the value is equal to or less than 1.3, the river is called straight. Meandering rivers have a very high sinuosity index.

Slope failure: Slope failure is referred to the sudden collapse of a landmass due to weakened self-retainability of the earth under the influence of continuous rainfall or triggering an earthquake. Large-scale constructions in the higher mountains for hydroelectricity generation by the construction of large dams can cause slope failure. The accelerated rate of deforestation in the hills and mountains can cause slope failure. Also, global warming and deglaciation can cause slope failure, causing sometimes high numbers of fatalities.

Solifluction: Solifluction is a collective name for very slow mass movements down a slope related to freeze–thaw processes. Solifluction is not restricted only to cold climates; it can happen in any humid climate.

Steno's principles: Modern understanding of the development of sedimentary basins and the stratigraphy started with three key observations of Nicholas Steno (1638–1686), which are known as Steno's principles. Firstly, beds of sediments deposited in water form as horizontal layers due to gravitational settling (*law of original horizontality*). Secondly, in undisturbed strata, the oldest layer lies at the bottom, and the youngest layer lies at the top (*law of superposition*), and thirdly, horizontal strata extend laterally until they thin out to zero thickness at the edge of the basin of deposition (*law of lateral continuity*). Starting with these simple premises, be it in the outcropping rocks or the subsurface from the seismic sections, if the evidence suggests non-conformity of these principles, the implication is that the things changed after the deposition. It brings great tasks to the researchers to explain the mechanism of these changes.

Stream power: The power of a stream to incise the valleys as well as erode the banks are determined by many factors, of which discharge volume and the slope of the relief through which it flows can be treated as the principal. Besides these, sediment load, continuously varying ratio of the suspended and the bed loads, and joining of tributaries at different locations can influence the stream power for a given reach.

Synorogenic flysch facies: Sediments that are usually produced directly with the erosion of uprising and developing mountains are referred to as 'flysch', the composition of which might be argillaceous rocks (finer clays or silts or the mix of both), poorly sorted angular fragments implying short-distance transport from the site of generation (breccias), lime-mixed clays, and conglomerates. Earlier, some of the geologists associated the word 'flysch' principally with only the Alpine orogeny. However, later on, the use was generalized.

Tectono-geomorphic zonation: This zonation scheme is done by the integration of morphometric elements and tectonic forcing retrieved from stratigraphic budgeting measured mostly from geophysical data which essentially helps to identify the probable rates of fluvial aggradation, residence times of rivers, shifting base levels in terms of zone of depression, slopes and uplifts which may or may not have direct corroboration with the analogous visible relief of a basin.

Thin-skinned thrusting: If ridges, as well as mountain belts caused by thrusting, do not involve the basement, it is called thin-skinned thrusting, and by the same logic, basement involved thrusting is called thick-skinned thrusting.

Toe-cutting: Riverbank stratigraphy might be constituted of unconsolidated materials in the lower part, which can continuously be cut by the river. As a result, a stable portion of the bank can collapse all of a sudden. Under those circumstances, bank erosion is said to be caused due to 'toe-cutting'.

Uphole survey: During conventional seismic reflection-based studies of the sedimentary basins, to facilitate the deep transmission of energy, explosives are blasted at depths below the unconsolidated shallow weathered zone. The uphole survey is conducted before the regular experimental work to map the base of the weathered zone so that the required depth of the shot holes can be fixed up in advance. Besides knowing the thickness of the weathered zone below a given point on the surface, seismic wave velocities through the weathered and sub-weathered zones are also computed during the uphole surveys, which are needed for the static correction. The lateral variation in seismic velocities for the weathered and sub-weathered zones can be utilized further to understand Quaternary basin evolution studies, which can also help connect structural elements related to the deeper subsurface.

Wandering of thalweg: For the multi-channel big braided rivers carrying huge sediment load, the location of the deepest channel is observed to change within short time intervals. Coleman (1969), while studying the channel processes and sedimentation of the Brahmaputra River, observed this and mentioned it as the 'wandering of thalweg'.

Width index: This is a reach-scale measurement of channel belt widths compared to the average channel belt width for the entire length of observation of the river concerning the difference between the maximum and the minimum values. Expressed by a number without any unit, it helps to offer a normalized width index to facilitate temporal comparability of the river dynamics.

$$\text{Width Index} = \frac{\text{Reach width} - \text{Average width}}{\text{Maximum width} - \text{Minimum width}}$$

References

Acharya, S. K., Sastry, M. V. A., 1976. Stratigraphy of Eastern Himalaya. *Himalayan Geology Seminar* 41(1), 49–66.

Ahmed, A. E., Murthy, R. V. S., Bharktya, D. K. 1993. Depositional environment, structural style and hydrocarbon habitat in Upper Assam Basin. In: Biswas, S. K., Dave, A., Garg, P., Pandey, J., Maithani, A., Thomas, N. J. (Eds.), *Proceedings of the Second Seminar on Petroliferous Basins of India*, Vol. 1, pp. 437–458. Dehradun: Indian Petroleum Publishers.

Allen, J. R. L., 1965. A review of the origin and characteristics of recent alluvial sediments. *Sedimentology* 5, 89–191.

Allen, P. A., Densmore, A. L., 2000. Sediment flux from an uplifting fault block, *Basin Research* 12, 367–380.

Ambraseys, N., 2000. Reappraisal of North Indian earthquakes at the turn of the 20th century. *Current Science* 79(9), 1237–1250.

An Yin, W., Harrison, T. M., 2000. Geologic evolution of the Himalayan–Tibetan orogen. In: Jeanloz, R., Albee, A.L., Burke, K.C. (Eds.), *Annual Reviews of Earth and Planetary Science*, Vol. 28, pp. 211–280. https://doi.org/10.1146/annurev.earth.28.1.211

Anderson, R. S., Densmore, A. L., 1997. Tectonic geomorphology of the Ash Hill fault, Panamint Valley, California. *Basin Research* 9, 53–63.

Aslan, A., Blum, M. C., 1999. Contrasting styles of Holocene avulsion, Texas Gulf Coastal Plain, USA. In: Smith, N.D., Rogers, J. (Eds.), *Fluvial Sedimentology VI: Special Publication, International Association of Sedimentology*, Vol. 28, pp. 193–209. Oxford: Blackwell.

Bachrach, R., Nur, A., 1998. High-resolution shallow-seismic experiments in the sand, Part I: Water table, fluid flow and saturation. *Geophysics* 63 (4), 1225–1233.

Bally, A. W., 1997. Hydrocarbon potential of Assam-Arakan and in India's Tertiary folded belt and foreland basins – a ranking of exploration opportunities, Unpublished report Oil & Natural Gas Corporation Limited, India, pp. 1–109.

Barman, S., Bhattacharjya, R. K., 2015. Change in snow cover area of Brahmaputra river basin and its sensitivity to temperature. Environmental Systems Research 4, 1–10.

Barton, D. C., 1929. The seismic method of mapping geological structure. *Transactions of the American Institute of Mining, Metallurgical and Petroleum Engineers, Incorporated (AIME)* 81, 572–624.

Beaumont, C., Fullsack, P., Hamilton, J., 1992. Erosional control of active compressional orogens. In: McClay, K. R. (Ed.), *Thrust Tectonics*, pp. 1–18, London: Chapman & Hall.

Beck, R. H., Lehner, P., 1974. Oceans, new frontier in exploration. *AAPG Bulletin* 58, 376–395.

Beghoul, N., Barazangi, M., Isacks, B. L., 1993. Lithospheric structure of Tibet and Western North America: Mechanisms of uplift and a comparative study. *Journal of Geophysical Research* 98, 1997–2016.

Bhandari, L. I., Fuloria, R. C., Sastri, V. V., 1973. Stratigraphy of Assam valley. *AAPG Bulletin* 57(4), 643–654.

Bhimasankaram, V. L. S., 1977. *Exploration Geophysics-An Outline.* Hyderabad: Association of Exploration Geophysicists (AEG).

Bilham, R., England, P., 2001. Plateau 'pop-up' in the great 1897 Assam earthquake. *Nature* 410, 806–809.

Bird, P., 1978. Initiation of intracontinental subduction in the Himalaya. *Journal of Geophysical Research* 83, 4975–4987.

Bond, J., 1899. Annual report of triangulation 1897–1898. In: Strahan, C. (Ed.), *Part II*, pp. xii–xiii. Calcutta: Survey of India Department.

Bora, A. K., 2004. Fluvial Geomorphology. In: Singh, V. P., Sharma, N., Ojha, C.S.P. (Eds.) *The Brahmaputra Basin Water Resources.* pp. 88–122. Kluwer Academic Publishers, Dordrecht, The Netherlands.

Bristow, C. S., 1987. Brahmaputra River: Channel migration and deposition. In: Ethridge, F.G., Flores, R.M., Harvey, M.D. (Eds.), *Recent Developments in Fluvial Sedimentology*, pp. 63–74. Tulsa, OK: Society of Economic Paleontologists & Mineralogists. Special Publication No. 39.

Bristow, C. S., 1993. Sedimentary structures exposed in bar tops in the Brahmaputra River, Bangladesh. *Geological Society of London Special Publications* 75, 277–289.

Bull, W. B., 2007. *Tectonic Geomorphology of Mountains: A New Approach to Paleoseismology.* Oxford: Blackwell Publishing.

Burbank, D. W., Anderson, R. S., 2012. *Tectonic Geomorphology*, 2nd edition. Wiley-Blackwell.

Burbank, D. W., Blythe, A. E., Putkonen, J., Pratt-Sitaula, B., Gabet, E., Oskin, M., Barros, A., Ojha, T. P., 2003. Decoupling of erosion and precipitation in the Himalayas. *Nature* 426, 652–655.

Burke, K. B. S., 1973. Seismic techniques in exploration of quaternary deposits. *Geoexploration* 11, 207–231.

Claerbout, J. F., 1971. Toward a unified theory of reflector mapping. *Geophysics*, 36, 467–481.

Coakley, B. J., Watts, A. B., 1991. Tectonic controls on the development of unconformities: The North Slope, Alaska. *Tectonics* 10(1), 101–130.

Coleman, J. M., 1969. Brahmaputra River channel processes and sedimentation. *Sedimentary Geology* 3, 129–239.

Cowan, D. R., Cowan, S., 1991. Analytical techniques in the interpretation of regional aeromagnetic data. *Exploration Geophysics* 22, 81–84.

Crompton, S. L., Allen, P. A., 1995. Recognition of forebulge unconformities associated with early stage foreland basin development: Example from the North Alpine foreland basin. *Bulletin of American Association of Petroleum Geologists* 79, 1495–1514.

Crowell, J.C., 1974. Origin of late Cenozoic basins in southern California. In: *Tectonics and Sedimentation: SEPM Special Publication*, Vol. 22, pp. 190–204, Tulsa, OK: Society of Economic Paleontologists and Mineralogists.

Curray, J.R., 1994. Sediment volume and mass beneath the Bay of Bengal. *Earth and Planetary Science Letters* 125, 371–383.

Dadson, S. J., Hovius, N., Chen, H., Dade, W. B., Hsieh, M.L., Willett, S.D., Hu, J. C., Horng, M.J., Chen, M.C., Stark, C.P., Lague, D., Lin, J.C., 2003. Links between erosion, runoff variability and seismicity in the Taiwan orogen. *Nature* 426, 648–651.

Dadson, S. J., Hovius, N., Chen, H., Dade, W. B., Lin, J. C., Hsu, M. L., Lin, C. W., Horng, M.J., Chen, T.C., Milliman, J., Stark, C. P., 2004. Earthquake-triggered increase in sediment delivery from an active mountain belt. *Geology* 32, 733–736.

Darby, S.E., Dunn, F.E., Nicholls, R.J., Rahman, M., Riddy, L., 2015. A first look at the influence of anthropogenic climate change on the future delivery of fluvial sediment to the Ganges-Brahmaputra-Meghna delta. *Environmental Science: Processes & Impacts* 17 (9), 1587–1600.

Das Gupta, A. B., Biswas, A. K., 2000. *Geology of Assam*, p. 169. Bangalore: Geological Society of India.

Das Gupta, S., Nandy, D.R., 1982. Seismicity and tectonics of Meghalaya Plateau, NE India. In: *Seventh Symposium on Earthquake Engineering*, pp. 19–24, University of Roorkee.

Dasgupta, S., Narula, P. L., Acharyya, S. K., Banerjee, J., 2000. *Seismotectonic Atlas of India and its Environs*, pp. 1–40. Calcutta: Geological Survey of India.

Datta, B., Singh, V. P., 2004. Hydrology. In: Singh, V. P., Sharma, N., Ojha, C.S.P. (Eds.) *The Brahmaputra Basin Water Resources*. pp. 139–195. Kluwer Academic Publishers, Dordrecht, The Netherlands.

DeCelles, P. G., Gehrels, G. E., Quade, J., Ojha, T. P., 1998. Eocene-early Miocene foreland basin development and the history of Himalayan thrusting, western and central Nepal, *Tectonics* 17, 741–765.

DeCelles, P. G., Giles, K. A., 1996. Foreland basin systems. *Basin Research* 8, 105–123.

DeCelles, P. G., Lawton, T. F., Mitra, G., 1995. Thrust timing, growth of structural culmination, and synorogenic sedimentation in the type Sevier orogenic belt, western United States, *Geology* 23, 699–702.

Dewey, J. F., Shackleton, R. M., Chang, C., Sun, Y., 1988. The tectonic evolution of the Tibetan plateau. *Philosophical Transactions of the Royal Society, London, A* 327, 379–413.

Dickinson, W.R., 1974b, Plate tectonics and sedimentation, in Dickinson, W.R., (Ed.), *Tectonics and Sedimentation: Society of Economic Paleontologists and Mineralogists (SEPM)* Special Publication 22, p. 1–27, https://doi.org/10.2110/pec.74.22.0001.

Dunn, J. A., Auden, J. B., Ghosh, A. M., Wadia, D. N., 1939. The Bihar-Nepal earthquake of 1934. *Geological Survey of India* 73, 1–139.

England, P. C., Houseman, G., 1988. The mechanics of the Tibetan plateau. *Philosophical Transactions of the Royal Society, London, A* 326, 301–319.

England, P., Molnar, P., 1990. Surface uplift, uplift of rocks, and exhumation of rock. *Geology* 18, 1173–1177.

Evans, P., 1932. Explanatory notes to accompany a table showing the Tertiary succession in Assam. *TRANSACTIONS, The Mining, Geological and Metallurgical Institute of India*, 27, 168–248.

Evans, P., 1935. General report of Geological Survey of India. *Records of the Geological Survey of India* 69, 83.

Evans, P., 1964. The tectonic framework of Assam. *Journal of Geological Society of India* 5, 80–96.

Evans, P., 1965. Structure of NEFA area of Assam. Min. Geol. Me HI. Inst. Ind. Dr Wadia Commemorative Volume, p. 693.

Evans, P., Mathur, L. P., 1964. 'Oil in India'. International Geological Cong. 22 Session, India, 1964.

Flemings, P. B., Jordan, T. E., 1990. Stratigraphic modelling of foreland basins: Interpreting thrust deformation and lithospheric rheology. *Geology* 18, 430–434.

Freund, R., 1965. A model of the structural development of Israel and adjacent areas since upper Cretaceous times. *Geological Magazine* 102, 189–205.

Friend, P. F., Sinha, R., 1993. Braiding and meandering parameters. In: Best, J. L., Bristow, C. S. (Eds.), 1993, *Braided Rivers*, pp. 105–111. Geological Society, Special Publication 75.

Gahalaut, V. K., Chander, R., 1992. A rupture model for the great earthquake of 1897, northeast India, and its geodynamic and seismic hazard implication. *Tectonophysics*, 204(1–2), 163–174. Available from: https://doi.org/10.1016/0040-1951(92)90277-D

Gansser, A., 1964. *Geology of the Himalayas*. Wiley-Interscience, New York.

Gansser, A., 1980. The significance of the Himalayan Suture Zone. *Tectonophysics* 62, 37–52.

Gardner, L. W., 1939. An aerial plan of mapping a subsurface structure by refraction shooting. *Geophysics* 4, 247–259.

Garzanti, E., Vezzoli, G., Ando, S., France-Lanord, C., Singh, S.K., Foster, G., 2004. Sand petrology and focused erosion in collision orogens: The Brahmaputra case. *Earth and Planetary Science Letters* 220, 157–174.

Ghosh, A. M. N., 1935. The Mishmi Hills. *Records Geological Survey of India* 69(1), 84–87.

Ghosh, A. M. N., 1956. Recent advances in geology and structure of Eastern Himalaya. In: *Proceedings of the Indian Science Congress*, Vol. 44, pp. 85–99.

Godwin Austen, M. H., 1869. Notes from Assaloo, North Cachar, on the Great Earthquake of 10 January 1869. In: *Proceedings of the Asiatic Society of Bengal*. pp. 91–99.

Gogoi, C., 2013. Sedimentology, Geochronology and Neotectonics of the alluvium in parts of the Brahmaputra plains in North East India (Unpublished PhD Thesis, Dibrugarh University, India).

Gogoi, P. R., Chetia, M., Borgohain, S., Bora, M., Kumar, A., Lahiri, S. K., 2022. Morphotectonic forcing and anthropogenic impact behind a recently emerged relict island of the Brahmaputra River. *Earth Surface Processes and Landforms*, 1–21. Available from: https://doi.org/10.1002/esp.5443.

Goodbred Jr., S. L., Kuehl, S. A., 1998. Floodplain processes in the Bengal Basin and the storage of Ganges–Brahmaputra river sediment: An accretion study using 137Cs and 210Pb geochronology. *Sedimentary Geology* 121, 239–258.

Goodbred Jr., S. L., Kuehl, S. A., 2000. The significance of large sediment supply, active tectonism, and eustasy on margin sequence development: Late Quaternary stratigraphy and evolution of the Ganges–Brahmaputra delta. *Sedimentary Geology* 133, 227–248.

Goodbred Jr., S. L., Kuehl, S. A., Steckler, M. S., Sarker, M. H., 2003. Controls on facies distribution and stratigraphic preservation in the Ganges–Brahmaputra delta sequence. *Sedimentary Geology* 155, 301–316.

Gordon, I., Heller, P. L., 1993. Evaluating major controls on basinal stratigraphy, Pine valley, Nevada: Implications for syntectonic deposition. *Geological Society of America Bulletin*. 105, 47–55.

Goswami, D. C., 1985. Brahmaputra River, Assam, India: Physiography, basin denudation and channel aggradation. *Water Resources Research* 21, 959–978.

Goswami, D. C., 1998. Fluvial regime and flood hydrology of the Brahmaputra River, Assam. *Memoirs-Geological Society of India* 41, 53–75.

Goswami, P. C., Goswami, P., 2007. Structural style & timing of structural deformation in Naga Foothills Region—a discussion. Association of Petroleum Geologists, India, Bulletin, pp. 153–160.

Grall, C., Steckler, M. S., Pickering, J. L., Goodbred, S., Sincavage, R., Paola, C., Akhter, S. H., Spies, V., 2018. A base-level stratigraphic approach to determining Holocene subsidence of the Ganges-Meghna-Brahmaputra Delta plain. *Earth and Planetary Science Letters* 499, 23–36.

Green, C. M., Fairhead, J. D., Maus, S., 1998. Satellite-derived gravity: Where are we and what's next. *The Leading Edge* 17, 77–79.

Green, R., 1974. The seismic refraction method- a review. *Geoexploration* 12, 259–284.

Gregory, D. I., Schumm, S. A., 1987. The effect of active tectonics on alluvial river morphology. In: Richards, K.S. (Ed.) *River Channels: Environment and Process*, pp. 41–68. London: Blackwells.

Grelle, G., Guadagno, F. M., 2009. Seismic refraction methodology for groundwater level determination: "Water seismic index". *Journal of Applied Geophysics* 68, 301–320.

Guiseppe, A. C., Heller, P. L., 1998. Long-term river response to regional doming in the Price River formation, central Utah. *Geology* 26, 239–242.

Gupta, H. K., Singh, H. N., 1982. Is the Shillong region, Northeast India undergoing a dilatancy stage precursory to a large earthquake? *Tectonophysics* 85, 31–33.

Gupta, S., 1997. Himalayan drainage patterns and the origin of fluvial megafans in the Ganges foreland basin. *Geology* 25, 11–14.

Gutenberg, B., Richter, C. F., 1954. *Seismicity of the Earth and Associated Phenomenon*. Princeton, NJ: Princeton University Press.

Handique, G. K., Sethi, A. K., Sarma, S. C., 1989. *Review of Tertiary Stratigraphy of Parts of Upper Assam Valley*, pp. 23–36. Geological Survey of India, Special Publication No. 23.

Hasserlström, B., 1969. Water prospecting and rock investigation by the seismic refraction method. *Geoexploration* 7, 113–132.

Hayden, H. H., 1910. Some coalfields in the Northeastern Assam. *Rec. Geol. Surv. Ind.* 40(4), 283–319.

Hazarika, D., Baruah, S., Gogoi, N. K., 2009. Attenuation of coda waves in the northeastern region of India. *Journal of Seismology*. 13, 141–160.

Heiland, C. A., 1929. Modern instruments and methods of seismic prospecting. *A.I.M.E. Geophys. Prosp.* 81, 625–653.

Holbrook, J., Schumm, S. A., 1999. Geomorphic and sedimentary response of rivers to tectonic deformation: A brief review and critique of a tool for recognizing subtle epeirogenic deformation in modern and ancient settings. *Tectonophysics* 305, 287–306.

Holt, W. E., Stern, T. A., 1994. Subduction, platform subsidence, and foreland thrust loading: The late Tertiary development of Taranaki basin, New Zealand. *Tectonics* 13, 1068–1092.

Hovius, N., 1998. Controls on sediment supply by larger rivers. In: Shanley, K.W., McCabe, P.J. (Eds.), *Relative Role of Eustasy, Climate, and Tectonism in Continental Rocks*, pp. 3–16, Society for Sedimentary Geology (SEPM), Tulsa, Oklahoma, Special Publication 59. https://doi.org/10.2110/pec.98.59.0002

Hunter, W. W., 1879. *A Statistical Account of Assam*. London: Trubner & Co.

Jain, V., Sinha, R., 2004. Fluvial dynamics of an anabranching river system in Himalayan foreland basin, Baghmati River, north Bihar plains, India. *Geomorphology* 60, 147–170.

Johnson, M. R. W., 2002. Shortening budgets and the role of continental subduction during the India-Asia collision. *Earth Science Reviews* 59, 101–123.

Jong, S. D., Auping, W. L., Oosterveld, W. T., Usanov, A., Abdalla, M., Bovenkamp, A. V. D., and Frattina, C. F. D., 2017. *The Geopolitical Impact of Climate Mitigation Policies: How Hydrocarbon Exporting Rentier States and Developing Nations can Prepare for a Sustainable Future.* The Hague Centre for Strategic Studies, Lange Voorhout 16/2514 EE/The Hague/The Netherlands.

Karunakaran, C., Ranga Rao, A., 1979. Status of exploration for hydrocarbons in the Himalayan region: Contribution to stratigraphy and structure. *Geological Survey of India Miscellaneous Publication* 41, 1–66.

Kayal, J. R., Arefiev, S. S., Saurabh, B., Hazarika, D., Gogoi, N., Gautam, J. L., Santanu, B., Dorbath, C., Tatevossain, R., 2012. Large and great earthquakes in the Shillong plateau–Assam valley area of northeast India region: Pop-up and transverse tectonics. *Tectonophysics*, 532–535, 186–192. Available from: https://doi.org/10.1016/j.tecto. 2012.02.007

Keller, E. A., Pinter, N., 2002. *Active Tectonics-Earth Quakes, Uplift, and Landscape* (2nd edition). London: Prentice-Hall.

Kent, W. N., Dasgupta, U., 2004. Structural evolution in response to fold and thrust belt tectonics in northern Assam - A key to hydrocarbon exploration in the Jaipur anticline area. *Marine and Petroleum Geology* 21, 785–803.

Kent, W. N., Hickman, R. G., Dasgupta, U., 2002. Application of a ramp/flat-fault model to interpretation of the Naga thrust and possible implications for petroleum exploration along the Naga thrust front. *American Association of Petroleum Geologists Bulletin* 86(12), 2023–2045.

Khan, P. K., 2005. Variation in dip-angle of the Indian plate subducting beneath the Burma plate and its tectonic implications. *Journal of Geosciences* 9(3), 227–234.

Khattri, K. N., 1987. Great earthquake, seismicity gaps and potential for earthquake disaster along Himalayan plate boundary. *Tectonophysics* 138, 79–92.

Kingston, D. R., Dishroon, C. P., Williams, P. A., 1983. Global basin classification system. *AAPG Bulletin* 67(12), 2175–2193.

Klemme, H. D., 1975. Giant oil fields related to their geologic setting- a possible guide to exploration. *Bulletin of Canadian Petroleum Geology* 23, 30–66.

Koefoed, O., 1965, Direct methods of interpreting resistivity observations. *Geophysical Prospecting* 13, 568–591.

Kotoky, P., Bezbaruah, D., Baruah, J., Sarma, J. N., 2005. Nature of bank erosion along the Brahmaputra river channel, Assam, India. *Current Science* 88(4), 634–640.

Kotoky, P., Bezbaruah, D., Sarma, J. N., 2003. Erosion activity on Majuli-the largest river island in the world. *Current Science*, 84(7), 929–932.

Kraus, M. J., Wells, T. M., 1999. Recognizing avulsion deposits in the ancient stratigraphical record. In: Smith, N.D., Rogers, J. (Eds.), *Fluvial Sedimentology VI*, pp. 251–268, International Association of Sedimentologists, [TS] Ghent University, Belgium Special Publication No. 28.

Kunte, S. V., Rao, S. V., 1989. Tectonic model for hydrocarbon exploration in the Naga foothills, ONGC Bulletin 26, 1–16.

La Touche, T. H. D., 1883. Notes on a traverse through Khasi Jaintia and North Cachar Hills. *Records Geological Survey of India* 10, 198–203.

La Touche, T. H. D., 1885. Coal and limestone in Doigrung river near Golaghar, Assam. *Records Geological Survey of India* 17, 31–32.

La Touche, T. H. D., 1886. Geology of upper Dihing basin in the Singpho Hills. *Records Geological Survey of India* 19, 111–115

La Touche, T. H. D (Ed.), 1910. *The Journals of Major James Rennell, 1910*. Calcutta: The Asiatic Society.

Lahiri, S. K., 2013. Laws of erosion in Majuli: A statistical approach on GIS-based data. *South East Asian Journal of Sedimentary Basin Research* 1(1), 80–89.

Lahiri, S. K., 2018. Riverbank erosion in Rohmoria: Impact, conflict and peoples' struggle. In: Joy, K. J., Das, P. J., Chakroborty, G., Mahanta, C., Paranjape, S., Vispute, S. (Eds.), *Water Conflicts in Northeast India*, pp. 165–177, Routledge: Taylor & Francis Group.

Lahiri, S. K., Borgohain, J., 2011. Rohmoria's challenge: Natural disasters, popular protest and state apathy. *Economic and Political Weekly* 46(2), 31–35.

Lahiri, S. K., Sarma, A., Wasson, R. J., 2021. Bed-bank relationship and flood characterisation in the Upper reach of the Brahmaputra Valley, Assam. In: Bhuiyan, C. et al. (eds.), *Water Security and Sustainability, Lecture Notes in Civil Engineering 115*, pp. 191–206, Singapore: Springer Nature. Available from: https://doi.org/10.1007/978-981-15-9805-0_16

Lahiri, S. K., Sinha, R., 2012. Tectonic controls on the morphodynamics of the Brahmaputra River system in the upper Assam valley, India. *Geomorphology* 169–170, 74–85.

Lahiri, S. K., Sinha, R., 2014. Morphotectonic evolution of the Majuli Island in the Brahmaputra valley of Assam, India inferred from geomorphic and geophysical analysis. *Geomorphology* 227, 101–111.

Lahiri, S. K., Sinha, R., 2015. Application of Fast Fourier Transform (FFT) in fluvial dynamics - a case study from the upper Brahmaputra valley in Assam. *Current Science* 108(1), 90–95.

Lane, E. W., 1957. A study of the shape of channels formed by natural streams flowing in erodible material. Missouri River Division Sediment Series No. 9. U.S. Army Engineer Division, Missouri River, Corps of Engineers, Omaha, Nebraska.

Latrubesse, E., 2008. Patterns of anabranching channels: The ultimate end-member adjustment of mega rivers. *Geomorphology* 101, 130–145.

Latrubesse, E. M., Stevaux, J. C., Sinha, R., 2005. Tropical rivers. *Geomorphology* 70, 187–206.

Lattman, L. H., 1959. Geomorphology: A new tool for finding oil. *Oil Gas Journal* 57, 231–236.

LeDain, A. Y., Tapponnier, P., Molnar, P., 1984. Active faulting and tectonics of Burma and surrounding regions, *Journal of Geophysical Research*, 89, 453–472.

Leeder, M. R., 1993. Tectonic controls upon drainage development, river channel migration and alluvial architecture: Implications for hydrocarbon reservoir development and characterization. Geological Society London Special Publications 73(1): 7–22. DOI: 10.1144/GSL.SP.1993.073.01.02.

Leopold, L. B., Wolman, M. G., 1957. River Channel Patterns: Braided, Meandering and Straight. Geological Survey Professional Paper 282-B, United States Government Printing Office, Washington, 39–85.

M'Cosh, J., 1837. Topography of Assam. Printed by order of Government: G. H. Huttmann, Bengal Military Orphan Press, Calcutta. Second Indian Reprint 2000, LOGOS Press, Darya Ganj, New Delhi.

Mahanta, C., et al. 2014. *Physical Assessment of the Brahmaputra River*. Dhaka, Bangladesh: International Union for Conservation of Nature.

Mallet, F. R., 1875. Note on coals recently found near Moflong, Khasia Hills. *Records Geological Survey of India* 18(3), 65–92.

Mallet, F. R., 1876. On the coalfield of Naga Hills, bordering the Sibsagar and Lakhimpur districts. *Geological Survey of India Memoirs* 12(2), 1–95.

Mallet, F. R., 1882. On Iridesmine from Noa-Dihing river, Assam. *Rec. GSI* 1, 53–55.

Mallick, R. K., 1992. Exploration for Palaeocene-Lower Eocene Hydrocarbon prospects in the eastern parts of upper Assam Basin. *Indian Journal of Petroleum Geology* 1(1), 117–129.

Mallick, R. K., Raju, S.V., 1995. Application of Wireline Logs in characterization and evaluation of generation potential of Paleocene-Lower Eocene source rocks in parts of Upper Assam Basin, India, The Log Analyst 36(3). [Paper Number: SPWLA-1995-v36n3a3]

Mallick, R. K., Raju, S. V., Gogoi, K. D., 1997. The Langpar-Lakadong petroleum system: Upper Assam Basin, India. *Indian Journal of Petroleum Geology* 6(2), 1–18.

Mallick, R., Hubbard, J. A., Lindsey, E. O., Bradley, K. E., Moore, J. D. P., Ahsan, A., Alam, A. K. M. K., Hill, E. M., 2020. Subduction initiation and rise of the Shillong plateau. *Earth and Planetary Science Letters* 543, 116351. Available from: https://doi.org/10.1016/j.epsl.2020.116351

Mathur, L. P., Evans, P., 1964. Oil in India. I.G.C. 22nd Session, India.

Matte, P., Mattauer, M., Olivet, J. M., Griot, D. A., 1997. Continental subductions beneath Tibet and the Himalayan orogeny: A review. *Terra Nova* 9(5/6), 264–270.

McLaren, J. M., 1904. Geology of upper Assam. *Rec. GSI* 31(4), 179–204.

Medlicott, H. B., 1865. The coal of Assam, results of a brief visit to the coalfields that province in 1865; with Geological note on Assam and the Hills to the south of it. *Memoir of Geological Survey of India* 4(3), 388–442.

Meyer, B., Tapponnier, P., Bourjot, L., Metivier, Y., Gaudemer, G., Peltzer, G., Shunmin, G., Zhitai, C., 1998. Crustal thickening in Gansu –Qinghai, lithospheric mantle subduction, and oblique, strike-slip controlled growth of the Tibetan plateau. *Geophysical Journal International* 135, 1–48.

Miall, A. D., 1981. Alluvial sedimentary basins: Tectonic setting and basin architecture. In: Miall, A. D. (Ed.) *Sedimentation and Tectonics in Alluvial Basins*, pp. 1–33. Geological Association of Canada, Toronto Special Paper 23.

Miall, A. D., 1985. Architectural-element analysis: A new method of facies analysis applied to fluvial deposits. *Earth Science Review* 22, 261–308.

Miall, A. D., 1991. Hierarchies of architectural units in clastic rocks, and their relationship to sedimentation rate. In: Miall, A.D., Tyler, N. (Eds.) *The Three-Dimensional Facies Architecture of Terrigenous Clastic Sediments, and Its Implications for Hydrocarbon Discovery and Recovery*, Vol. 3, pp. 6–12. Society of Economic Paleontologists and Mineralogists (Society for Sedimentary Geologists), Tulsa, Oklahoma, USA.

Mitrovica, J. X., Beaumont, C., Jarvis, G. J., 1989. Tilting of continental interiors by the dynamical effects of subduction. *Tectonics* 8, 1079–1094.

Molnar, P., Tapponnier, P., 1975. Cenozoic tectonics of Asia: Effects of continental collision. *Science* 189, 419–426.

Morino, M., Kamal, A. S. M. M., Akhter, S. H., Rahman, M. Z., Ali, R. M. E., Talukder, A., Khan, M. M. H., Matsuo, J., Kaneko, F., 2014. A paleo- seismological study of the Dauki fault at Jaflong, Sylhet, Bangladesh: Historical seismic events and an attempted rupture segmentation model. *Journal of Asian Earth Sciences*, 91, 218–226. Available from: https://doi.org/10.1016/j.jseaes.2014.06.002

Morozova, G. S., Smith, N. D., 1999. Holocene avulsion history of the lower Saskatchewan fluvial system, Cumberland Marshes, Saskatchewan-Manitoba, Canada. In: Smith, N. D., Rogers, J. (Eds.), *Fluvial Sedimentology VI*, pp. 231–249. https://doi.org/10.1002/9781444304213.ch18

Mukhopadhyay, D., 2015. Geology of the Shillong plateau and the Brahmaputra valley. *Current Science*, 108(3), 317–319.

Mukhopadhyay, M., 1984. Seismotectonics of transverse lineaments in the eastern Himalaya and its foredeep, *Tectonophysics* 109, 227–240.

Mukhopadhyay, M., Dasgupta, S., 1988. Deep structure and tectonics of the Burmese arc: Constraints from earthquake and gravity data. *Tectonophysics* 149, 299–322.

Murthy, K. N., 1983. Geology and hydrocarbon prospects of assam shelf- recent advances and present status. *Petroleum Asia Journal*, 6(4), 1–14.

Murthy, M. V. N., Talukdar, S. C., Bhattacharya, A. C., Chakrabarti, C., 1969. The Dauki Fault of Assam. *ONGC Bulletin*, 6(2), 57–64.

Muscat, M., 1933. The theory of refraction shooting. *Physics* 4, 14–38.

Nandy, D. R., 1980. Tectonic patterns in Northeastern India. *Indian Journal of Earth Sciences* 7, 103–107.

Nandy, D. R., 1981. Discussion on Tectonic pattern in Northeastern India, reply. *Indian Journal of Earth Sciences* 8, 82–86.

Nandy, D. R., 1983. The Eastern Himalaya and the Indo-Buraman orogen in relation to the Indian plate movement. *Miscellaneous Publication - Geological Survey of India* 43, 153–159.

Nandy, D. R., 2001. *Geodynamics of Northeastern India and Adjoining Region*, p. 209, Kolkata, India: ABC Publications.

Nandy, D. R., 2017. *Geodynamics of Northeastern India and Adjoining Region (Revised Edition)*. Guwahati: Scientific Book Centre.

Nandy, D. R., Dasgupta, S., 1991. Seismotectonic domains of Northeastern India and adjacent areas. *Physics and Chemistry of the Earth* 18, 371–384.

Ni, J., Barazangi, M., 1984. Seismotectonics of the Himalayan collision zone: Geometry of the underthrusting Indian plate beneath the Himalaya, *Journal of Geophysical Research: Solid Earth* 89, 1147–1163.

Oldham, R. D., 1899. Report on the Great Earthquake of 12 June 1897. *Memoir of Geological Survey of India* 29, 1–379.

Pakiser, L. C., Black, R. A., 1957. Exploring for ancient channels with the refraction seismograph. *Geophysics* 22, 32–47.

Park, Y. H., Doh, S. J., Yun, S. T., 2007. Geoelectric resistivity sounding of riverside alluvial aquifer in an agricultural area at Buyeo, Geum River watershed, Korea: An application to groundwater contamination study. *Environmental Geology* 53(4), 849–859.

Pascoe, E. H., 1912. Coal in the Namchik Valley, Upper Assam. *Geological Survey of India (Records)* 41(3), 214–216.

Pascoe, E. H., 1914. The Petroleum occurrences of Assam and Bengal. *Memoir of Geological Survey of India* 40(2), 270–329.

Payton, C. E. (Ed.), 1977. *Seismic Stratigraphy – Applications to Hydrocarbon Exploration (AAPG Memoir 26)*. Tulsa, OK: Association of Petroleum Geologists.

Pickering, J. L., Diamond, M. S., Goodbred, S. L., Grall, C., Martin, J. M., Palamenghi, L., Paola, C., Schwenk, T., Sincavage, R. S., Spieß, V., 2019. Impact of glacial-lake paleofoods on valley development since glacial termination II: A conundrum of hydrology and scale for the lowstand Brahmaputra-Jamuna paleo valley system. *GSA Bulletin* 131(1–2), 58–70.

Platt, J., England, P. C., 1993. Convective removal lithosphere beneath Mountain belts: Thermal and mechanical consequences. *American Journal of Science* 293, 307–336.

Poddar, M.C., 1953. A short note on the Assam earthquake of Aug. 15, 1950. In: Ramachandra Rao, M.B. (Ed.), A compilation of papers on the Assam Earthquake of August 15, 1950. Publication No. 1 Central Board of Geophysics, Calcutta, Gov. of India, 38-42.

Posamentier, H. W., Allen, G. P., 1993. Siliciclastic sequence stratigraphic patterns in foreland ramp-type basins. *Geology* 21, 455–458.

Powell, C. M., 1986. Continental underplating model for the rise of the Tibetan plateau. *Earth and Planetary Science Letters* 81, 79–94.

Powell, C., Conaghan, P., 1973. Plate tectonics and the Himalaya. *Earth and Planetary Science Letters* 20, 1–12.

Prasad, B. N., Mani, K. S., 1983. Distribution of seismic velocities as related to basin configuration in Upper Assam Valley. *Journal of Association of Exploration Geophysicists* III(4), 25–33.

Purkait, B., 2004. Hydrogeology. In: Singh, V. P., Sharma, N., Ojha, C.S.P. (Eds.) *The Brahmaputra Basin Water Resources*. pp. 113–138. Kluwer Academic Publishers, Dordrecht, The Netherlands.

Rajendran, C. P., Rajendran, K., Duarah, B. P., Baruah, S., Earnest, A., 2004. Interpreting the style of faulting and paleoseismicity associated with the 1897 Shillong Northeast India, earthquake: Implications for regional tectonism. *Tectonics*, 23(4), TC4009. Available from: https://doi.org/10.1029/2003TC001605

Rajendran, C. P., Rajendran, K., Duarah, B. P., Baruah, S., Earnest, A., 2006. Reply to comment by R. Bilham on "Interpreting the style of faulting and paleoseismicity associated with the 1897 Shillong Northeast India, earthquake: Implications for regional tectonism". *Tectonics*, 25, TC2002. Available from: https://doi.org/10.1029/2005TC001902

Ranga Rao, A., 1983. Geology and hydrocarbon potentials of a part of Assam-Arakan basin and its adjoining region. In: *Symposium on Petroliferous basins of India*, pp. 127–158.

Ranga Rao, A., Samanta, M. K., 1987. Structural style of the Naga overthrust belt and its implications on exploration. *Bulletin of Oil & Natural Gas Commission* 24(1), 69–109.

Rao, M. P., Cook, E. R., Cook, B. I. et al. 2020. Seven centuries of reconstructed Brahmaputra River discharge demonstrate underestimated high discharge and flood hazard frequency. *Nature Communications* 11, 6017. https://doi.org/10.1038/s41467-020-19795-6

Rayleigh, L., 1885. On waves propagated along a plane surface of an elastic solid. *Proceedings of the London mathematical Society* 17, 4–11.

Raymo, M. E., Ruddiman, W.F., 1992. Tectonic forcing of late Cenozoic climate. *Nature* 359, 117–122.

Reitz, M. D., Pickering, J. L., Goodbred, S. L., Paola, C., Steckler, M. S., Seeber, L., Akhter, S. H., 2015. Effects of tectonic deformation and sea level on river path selection: Theory and application to the Ganges-Brahmaputra-Meghna River Delta. *Journal of Geophysical Research: Earth Surface* 120(4), 671–689.

Rennell, J., 1765. *A General Map of the River Baramputrey, From Its Confluence with the Ifsamuty near Dacca towards Assam*. London: India Office Library and Records.

Richards, K., Chandra, S., Friend, P., 1993. Avulsive channel systems: Characteristics and examples. In: Best, J. L., Bristow, J. L. (Eds.), *Braided Rivers: Special Publication*, Vol. 75, pp. 195–203, London: Geological Society of London.

Richardson, W. R., Thorne, C. R., 2001. Multiple thread flow and channel bifurcation in a braided river: Brahmaputra-Jamuna River, Bangladesh. *Geomorphology* 38, 185–196.

Robert, A., 2003. *River Processes – An Introduction to Fluvial Dynamics*. London: Arnold, A Member of the Hodder Headline Group.

Robinson, R. A. J., Brezina, C. A., Parrish, R. R., Horstwood, M. S. A., Oo, N. W., Bird, M. I., Thein, M., Walters, A. S., Oliver, G. J. H., Zaw, K., 2014. Large rivers and orogens: The evolution of the Yarlung Tsangpo-Irrawaddy system and the eastern Himalayan syntaxis. *Gondwana Research*, 26, 112–121.

Robinson, W., 1841. A Descriptive Account of Assam: With a sketch of the local geography, and a concise history of the tea plant of Assam: To which is added, a short account of the neighbouring tribes, exhibiting their history, manners and customs.

Ruddiman, W. F., 2008. *Earth's Climate-Past and Future* (2nd edition). NewYork: W.H. Freeman and Company.

Sale, H. M., Evans, P., 1940. The geology of British oilfields: 1, The Geology of Assam-Arakan oil Region (India and Burma). *Geological Magazine* 77, 337–363.

Sangree, J. M., Widmier, J. M., 1979. Interpretation of depositional facies from seismic data. *Geophysics* 44, 131–160.

Sarma, J. N., 2005. Fluvial process and morphology of the Brahmaputra River in Assam, India. *Geomorphology* 70, 226–256.

Sarma, J. N., 2008. Bank erosion of the Brahmaputra River around Rohmoria in Dibrugarh District, Assam. In: Chandan Kumar Sarma (Ed.), *North East India History Association Souvenir, 29th Annual Session*, pp. 318–323.

Sarma, J. N., Phukan, M. K., 2004. Origin and some geomorphological changes of the river island Majuli of the Brahmaputra in Assam, India. *Geomorphology* 60, 1–19.

Sarma, J. N., Phukan, M. K., 2006. Bank erosion and bankline migration of the Brahmaputra River in Assam during the twentieth century. *Journal of Geological Society of India* 68, 1023–1036.

Schumm, S. A., 1981. Evolution and response of the fluvial system: Sedimentologic implications. *Society of Economic Paleontologists and Mineralogists Special Publications* 31, 19–29.

Schumm, S. A., 1986. *Alluvial River Response to Active Tectonics, Studies in Geophysics, Active Tectonics*, pp. 80–94. Washington, DC: National Academy Press.

Schumm, S. A., 1993. River response to base level change: Implications for sequence stratigraphy, *Journal of Geology* 101, 279–294.

Schumm, S. A., Dumont, J. F., Holbrook, J. M., 2000. *Active Tectonics and Alluvial Rivers*. Cambridge: Cambridge University Press.

Seeber, L., Armbruster, J. G., Quittmeyer, R. C., 1981. Seismicity and continental sub-duction in the Himalayan arc. In: Gupta, H. K., Delany, F. M. (Eds.), *Zagros–Hindukush–Himalaya: Geodynamic Evolution*. Washington, DC: American Geophysical Union, pp. 215–242.

Selley, R. C., 2000. *Applied Sedimentology* (2nd edition). San Diego: Academic Press.

Sheriff, R. E., 1973. Factors affecting amplitudes - A review of physical principles in lithology and direct detection of hydrocarbons using geophysical methods. Geophysical Prospecting 25, 125–138.

Sheriff, R. E., 1982. Seismic stratigraphy, First Indian Reprint, EBP Educational Reprint, Dehradun.

Sheriff, R. E., 2002. *Encyclopedic Dictionary of Applied Geophysics* (4th edition). SEG Publications.

Sheriff, R., Geldart, L. P., 1983. *Exploration Seismology*, Vol. 2. Cambridge: Cambridge University Press.

Sheriff, R. E., Geldart, L. P., 1995. *Exploration Seismology* (2nd edition). Cambridge: Cambridge University Press.

Shrestha, A. B., Agrawal, N. K., Alfthnan, B., Bajracharya, S. R., Maréchal, J., van Oort, B. (Eds.), 2015. The Himalayan Climate and Water Atlas: Impacts of Climate Change on Water Resources in Five of Asia's Major River Basins. Published jointly by the International Centre for Integrated Mountain Development (ICIMOD), Lalitpur, Nepal; Global Resource Information Database (GRID)-Arendal, Norway; and the Centre for International Climate and Environmental Research-Oslo (CICERO), Norway.

Simpson, R. R., 1896. The geology of Mikir Hills. *Memoir of Geological Survey of India* 28(1), 71–95.

Simpson, R. R., 1906a. Jaipur and Nazira Coalfields, upper Assam. *Geological Survey of India* 34(4), 199–238.

Simpson, R. R., 1906b. Notes on upper Assam coalfields. *Geological Survey of India (Records)* 34(4), 199–241.

Singh, K., Chaudhury, A., Bhattacharyya, N., 1996. *Seven decades of Geophysics in Northeast India*, published by Association of Exploration Geophysicists, Hyderabad, India. pp. 1–87.

Singh, S. K., 2006. Spatial variability in erosion in the Brahmaputra basin: Causes and impacts. *Current Science* 90, 1272–1276.

Singh, S. K., Kumar, A., Lanord, C. F., 2006. Sr and 87Sr/86Sr in waters and sediments of the Brahmaputra river system: Silicate weathering, CO_2 consumption and Sr flux. *Chemical Geology* 234, 308–320.

Sinha, R., 1996. Channel avulsion and floodplain structure in the Gandak-Kosi inter-fan, north Bihar plains, India. *Zeitschrift für Geomorphologie, N.F. Supplementband* 103, 249–268.

Sinha, N., Kumar, V., and Kalita, P. C., 1997. Hydrocarbon plays in the frontal part of Naga thrust and fold Belt, Dimapur Low, Dhansiri Valley, Assam. In: *Proceedings Second International Petroleum Conference & Exhibition, PETRO TECH-97: New Delhi*, Vol.1, pp. 365–372.

Sinha, R., Friend, P.F., 1994. River systems and their sediment influx, Indo-Gangetic plains, Northern Bihar, India. *Sedimentology* 41, 825–845.

Sukhija, B. S., Rao, M. N., Reddy, D. V., Nagabshanam, P., Hussain, S., Chadha, R. K., Gupta, H. K., 1999. Timing and return of major paleoseismic events in the Shillong Plateau, India. *Tectonophysics* 308, 53–65.

Tandon, A. N., 1954. Study of the great Assam earthquake of August 1950 and its aftershocks. *Indian Journal of Meteorology and Geophysics* 5, 95–137.

Tandon, S.K., Sinha, R., 2007. Geology of large river systems, In: Gupta, A. (Ed.), *Large Rivers: Geomorphology and Management*, John Wiley & Sons, Ltd., West Sussex, England. pp. 7–28.

Telford, W. M., Geldart, L. P., Sheriff, R. F., 1990. *Applied Geophysics* (2nd edition), Cambridge: Cambridge University Press.

Thakur, V. C., Jain, A. K., 1975. Some observation on deformation, metamorphism and tectonic significance of the rocks of some parts of Mishmi hills, Lohit district (NEFA), Arunachal Pradesh, *Himalayan Geology* 5, 339–364.

Thorne, C. R., Russell, A. P. G., Alam, M. K., 1993. Planform pattern and channel evolution of the Brahmaputra River, Bangladesh. *Geological Society, London, Special Publications* 75, 257–276.

Tucker, G. E., Slingerland, R. L., 1996. Predicting sediment flux from the fold and thrust belts. *Basin Research* 8, 329–349.

Tucker, G. E., Slingerland, R. L., 1997. Drainage basin responses to climate change. *Water Resources Research* 33, 2031–2047.

Uddin, A., Lundberg, N., 1999. A paleo-Brahmaputra? Subsurface lithofacies analysis of Miocene deltaic sediments in the Himalayan–Bengal system, Bangladesh. *Sedimentary Geology* 123, 239–254.

Valdiya, K. S., 1999. Why does the river Brahmaputra remain untamed? *Current Science* 76(10), 1301–1305.

Valdiya, K. S., 2002. Emergence and evolution of Himalaya: Reconstructing history in the light of recent studies. *Progress in Physical Geography* 26(3), 360–399.

Venkatachary, K. V., Bandyopadhyay, K., Bhanumurthy, V., Rao, G. S., Sudhakar, S. Pal, D. K., Das, R. K., Sarma, U., Manikiam, B., MeenaRani, H. C., Srivastava, S. K., 2001. Defining a space-based disaster management system for floods: A case study for damage assessment due to 1998 Brahmaputra floods. *Current Science* 80(3), 369–377.

Verma, R. K., Mukhopadhyay, M., 1976. Tectonic significance of anomaly-elevation relationships in north-eastern India, *Tectonophysics* 34, 117–133.

Verma, R. K., Mukhopadhyay, M., 1977. An analysis of the gravity field in northeastern India, *Tectonophysics* 42, 283–317.

Verma, R. K., Mukhopadhyay, M., Ahluwalia, M. S., 1976a. Seismicity, gravity, and tectonics of northeast India and northern Burma. *Bulletin of the Seismological Society of America* 66, 1683–1694.

Verma, R. K., Mukhopadhyaya, M., Ahluwalia, M. S., 1976b. Earthquake mechanisms and tectonic features of northern Burma. *Tectonophysics*, 32, 387–399.

Verma, R. K., Mukhopadhyay, M., Bhuin, N. C., 1978. Seismicity, gravity and tectonics in the Andaman Sea. *Journal of Physics of the Earth* 26(suppl.), 5233–5248.

Verma, R. K., Roonwal, G. S., Gupta, Y., 1993. Statistical analysis of seismicity of NE India and northern Burma during the period 1979–1990. *Journal of Himalayan Geology*, 4(1), 71–79.

Wadia, D. N., 1968. The Himalayan Mountains: Its origin and Geographical Relations, Mountains and Rivers of India. In: *21st International Geographical Congress India*, p. 42.

Waschbusch, P. J., Royden, L. H., 1992. Spatial and temporal evolution of foredeep basins: Lateral strength variations and inelastic yielding in continental lithosphere. *Basin Research* 4, 179–196.

Wasson, R. J., 2003. A sediment budget for the Ganga-Brahmaputra catchment. *Current Science* 84(8), 1041–1047.

Wasson, R., Acharjee, S., Rakshit, R., 2022. Towards the identification of sediment sources and processes of sediment production, in the Yarling-Tsangpo-Brahmaputra River catchment for reduction of fluvial sediment loads. *Earth Science Reviews* 226, 1–15. https://doi.org/10.1016/j.earscirev.2022.103932

Westerway, R., 1995. Crustal volume balance during the India – Eurasian collision and altitude of the Tibetan plateau: A working hypothesis. *Journal of Geophysical Research* 100(B8), 15173–15192.

Whitehead, P. G., Barbour, E., Futter, M. N., Sarkar, S., Rodda, H., Caesar, J., Butterfield, D., Jin, L., Sinha, R., Nicholls, R., Salehin, M., 2015. Impacts of climate change and socio-economic scenarios on flow and water quality of the Ganges, Brahmaputra and Meghna (GBM) river systems low flow and flood statistics. *Environmental Science: Processes & Impacts* 17(6), 1057–1069.

Wilcox, R., 1830. *Map of the Brahmaputra and Ichamati Rivers, Reduced and Drawn by M.H. Dias.* London: India Office Library and Records.

Wilcox, R., 1832. IV memoir of a survey of Assam and neighbouring countries executed in 1825-6-7-8. *Asiatic Researches* 17, 370–386.

Willett, S. D., Beaumont, C., 1994. Subduction of Asian lithospheric mantle beneath Tibet inferred from models of continental collision. *Nature* 369, 642–645.

Wobus, C., Heimsath, A., Whipple, K., Hodges, K., 2005, Active out-of-sequence thrust faulting in the central Nepalese Himalaya. *Nature* 434, 1008–1011.

Woodward, D., Menges, C. M., 1991. Application of uphole data from petroleum seismic surveys to groundwater investigations, Abu Dhabi (United Arab Emirates). *Geoexploration* 27, 193–212.

Yadav, G. S., Dasgupta, A. S., Sinha, R., Lal, T., Srivastava, K. M., Singh, S. K., 2010. Shallow sub-surface stratigraphy of interfluves inferred from vertical electric soundings in western Ganga plains, India. *Quaternary International* 227, 104–115.

Zelt, A. C., Azaria, A., Levander, A., 2006. 3D seismic refraction traveltime tomography at a groundwater contamination site. *Geophysics* 58(9), 1314–1323.

Zhou, W. L., Morgan, W. J., 1995. Uplift of the Tibetan plateau. *Tectonics* 4, 359–369.

Index

For Product Safety Concerns and Information please contact our EU
representative GPSR@taylorandfrancis.com
Taylor & Francis Verlag GmbH, Kaufingerstraße 24, 80331 München, Germany

www.ingramcontent.com/pod-product-compliance
Lightning Source LLC
Chambersburg PA
CBHW060327220326
41598CB00023B/2628